Ecosystems and Human Activity

Second Edition

Judith Woodfield

Contents

1 Ecosystems and communities

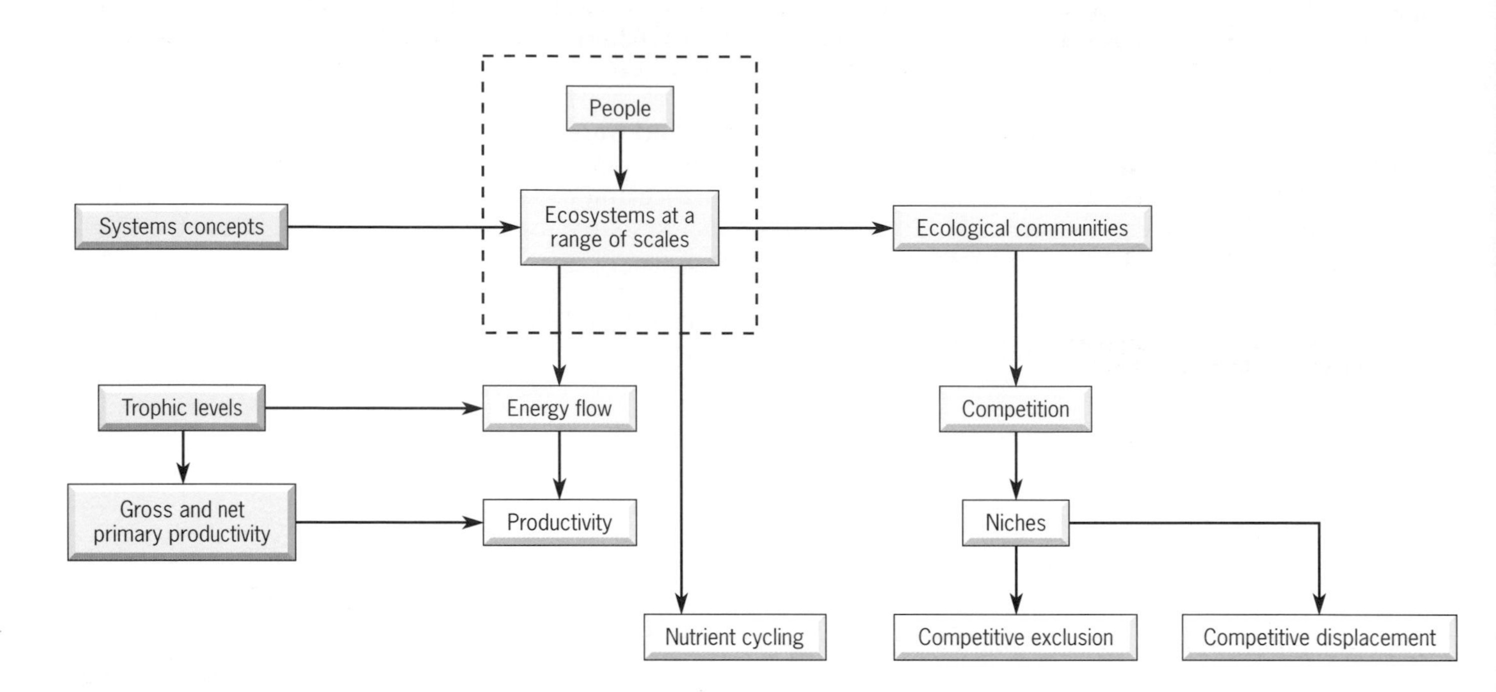

Figure 1.1 Human impact on the remote and inhospitable continent of Antarctica

Figure 1.2

1.1 People and ecosystems

There are very few places on earth which are not influenced by the culture and economy of people (Figs 1.1, 1.2). At low population densities and with limited technology, people have a more direct relationship with an area's 'Innate environmental quality' than at high population densities and with developed technology. When populations grow and environmental management becomes more complex, soon there is a need for policies to direct that management.

To understand how the environment works requires not only knowledge about **ecosystem** structure, dynamics and distribution but also a detailed understanding of people's decision-making processes and the effects of their decisions upon the ecosystem.

> People do not lie outside the global ecosystem, but are an intrinsic part of it. How we answer a particular environmental challenge depends largely on our own animal physiology, though our response is modified to some degree by our culture and the economic system. Hence we cannot directly equate the utility of an area with its innate environmental qualities.
>
> P Haggett, Professor of Urban and Regional Geography, Bristol University

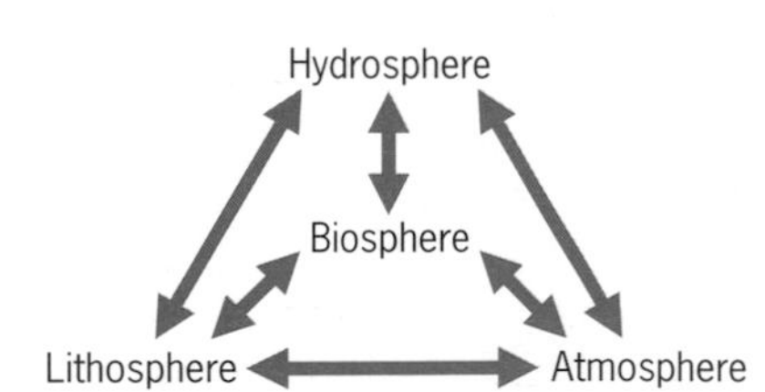

Figure 1.3 The relationship of the biosphere to the hydrosphere, atmosphere and lithosphere

1.2 Understanding ecosystems

The living world or **biosphere** reaches to the bottom of the oceans and stops in the lower layers of the atmosphere at the highest point where birds, insects and microorganisms are found.

The biosphere spans three **abiotic** environments – the atmosphere, **hydrosphere** and **lithosphere** – and relies on an exchange of materials from all three (Fig. 1.3).

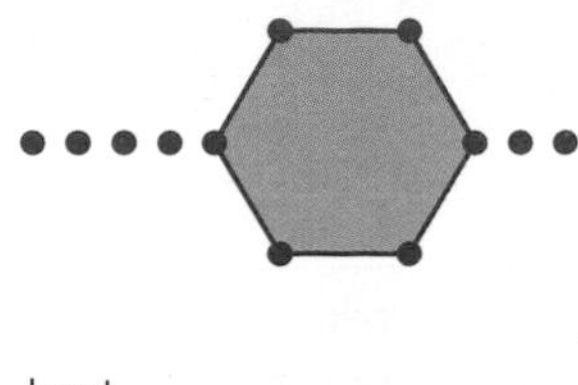

Systems

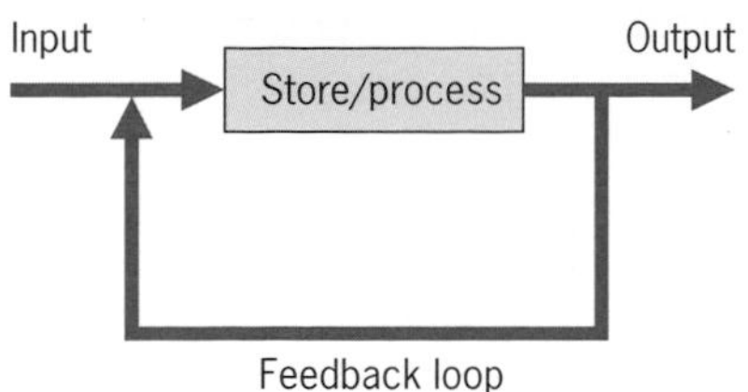

Figure 1.4 A simplified view of how a system works

The biosphere is the global ecosystem. A system is a set of parts that are linked by some kind of relationship. There are **inputs** and **outputs** and areas for **storage** or processing. Relationships where the outputs can affect the inputs are called **feedback** loops (Fig. 1.4).

In ecosystems we study living things functioning with their environment rather than just looking at individual types of animals and plants (Fig. 1.5). We also look at the processes involved in maintaining these relationships.

Figure 1.5 A generalised ecosystem showing inputs, outputs and stores

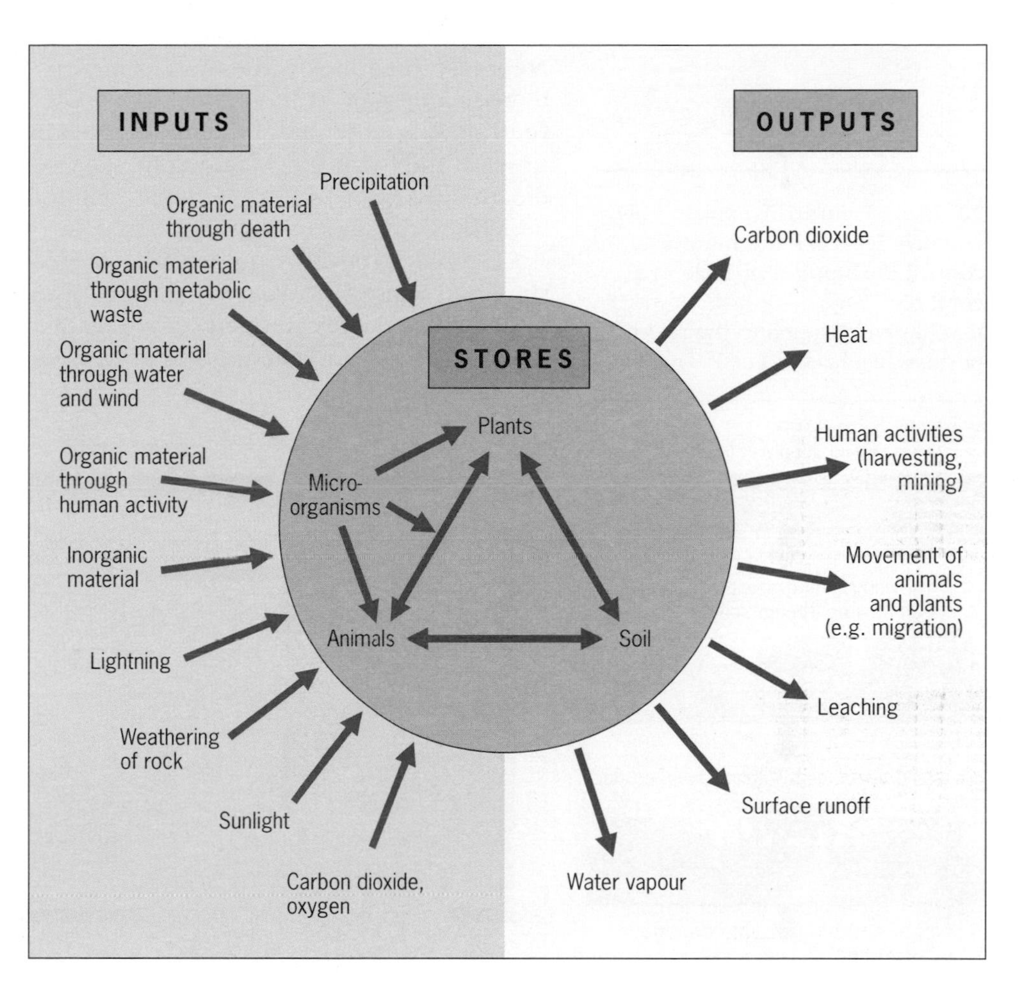

Figure 1.6

Systems at different scales

All natural systems are open systems where inputs and outputs are freely exchanged with other systems. Ecologists have identified a number of ecosystems at different scales which have a distinctive pattern of relationships. Many of these are described within this book. By studying ecosystems at a variety of scales, ecologists can focus on international, regional or local issues. On a global scale, ecologists have mapped out world biomes which include tropical rainforests (Chapter 2), temperate deciduous forests (Chapter 4), boreal coniferous forests (Chapter 5) and temperate grasslands (Chapter 8). At a regional scale, human and physical variations have produced ecosystems such as peat bogs (Chapter 9), sand dunes (Chapter 6), lowland wet grasslands (Chapter 9), moorland and heathland (Chapter 7) and estuaries (Chapter 11). Each of the regional ecosystems is made up of smaller local ecosystems such as hedges and watercourses (Chapter 10).

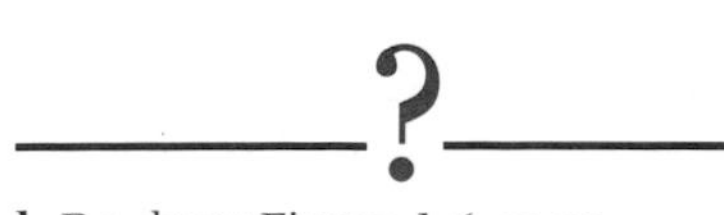

1 Re-draw Figure 1.6 as an ecosystem. Identify and label the inputs and outputs.

However, none of these ecosystems can be viewed separately, because inputs, outputs and transfers operate among them. For example, a pond will often have marginal vegetation that contains 'dry land' **species** such as trees and long grass; a woodland edge may contain components of the grassland that surrounds it. These 'blurred' edges are called **ecotones** and are generally rich in species because they contain a mixture of **habitats** characteristic of the ecosystems on either side of them.

How ecosystems work: feedback

When an ecosystem is in balance or harmony, it is in steady state or **dynamic equilibrium**. This is achieved by self-regulation. Certain parts of the system change when other parts change. This feeds back to affect the first change, although there is inevitably some delay between one event and the next. **Negative feedback** is the more common. It decreases the amount of change by reducing some of the inputs. This leads to stability (Fig. 1.7). **Positive feedback** is more rare. It increases the amount of change. This leads to an upset of the balance or steady state. For example, if a lake is polluted and fish die, the dead fish will cause further pollution and the death of more fish, because decay uses up oxygen.

Examples of positive and negative feedback in a coral reef ecosystem are shown in Figure 1.8. Many of the creatures which produce calcium carbonate, of which coral reefs are made, are sensitive to water depth. Increased depth means reduced sunlight and less growth of coral and calcium-producing algae (Chapter 14).

?

2a Using Figure 1.8, explain how negative feedback mechanisms control the amount of algae in a coral reef.

b What will happen to the reef if positive feedback occurs?

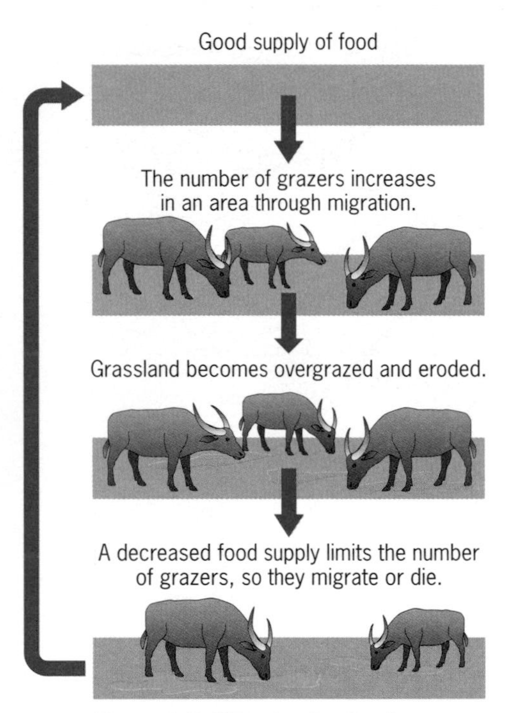

Figure 1.7 Negative feedback

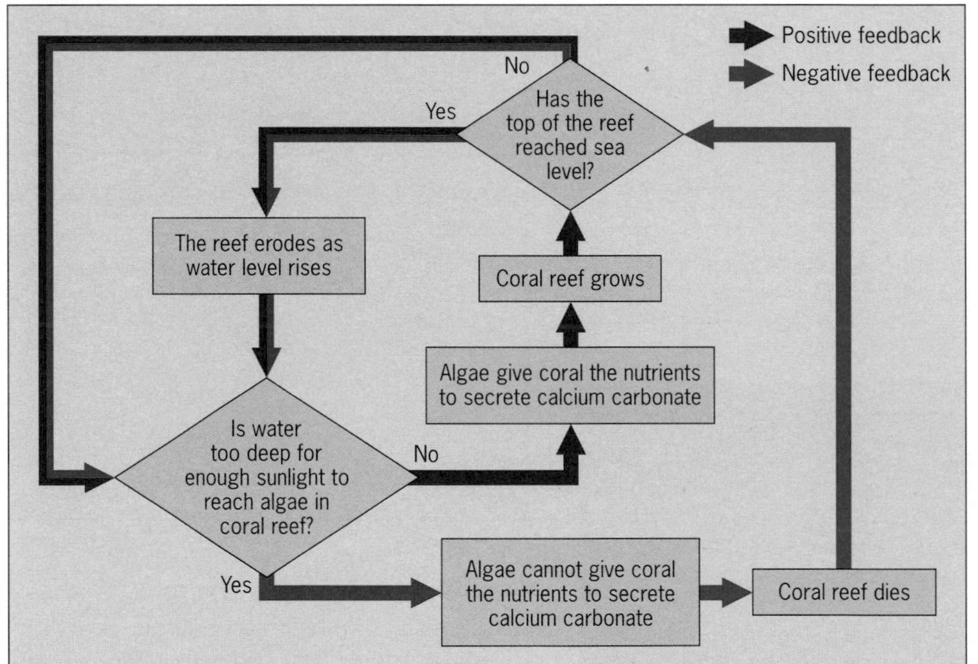

Figure 1.8 Feedbacks in a coral reef ecosystem

1.3 Energy flow

Energy cannot be created or destroyed; it can only be transferred from one state to another. All energy from the sun, fixed by the plants during **photosynthesis**, must do one of three things:

- be stored as chemical energy in plants or animal materials (living or dead),
- be passed through the ecosystem along **food chains** and **webs**, or
- escape from the system as outputs of material or heat energy.

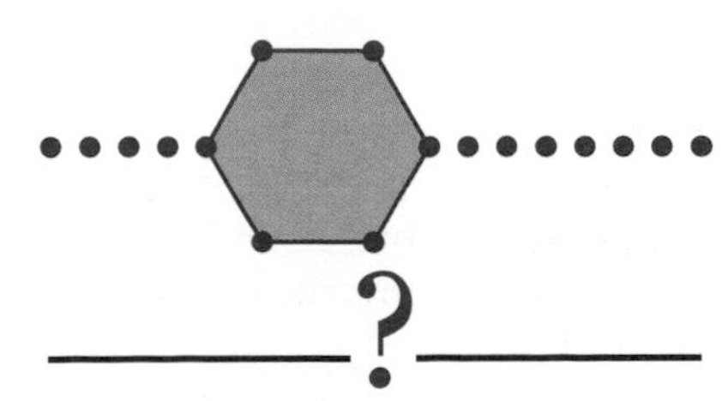

?

3 Using Figure 1.10, explain why the biomass of living materials per unit area (Fig. 1.9) decreases with each higher **trophic level** in the food chain.

4 Draw food chains to show the energy flow (starting with energy from sunlight) for the following meals:

- Fish and chips
- Roast beef and carrots.

Trophic levels

The energy stored in plant and animal material is called the **biomass** and this can be measured as dry weight, ash weight (the weight after burning) or in calories (Fig. 1.9).

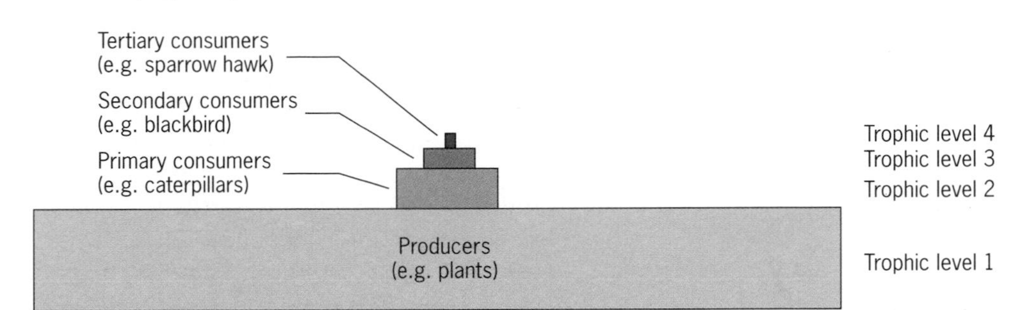

Figure 1.9 The relationship of biomass to trophic level

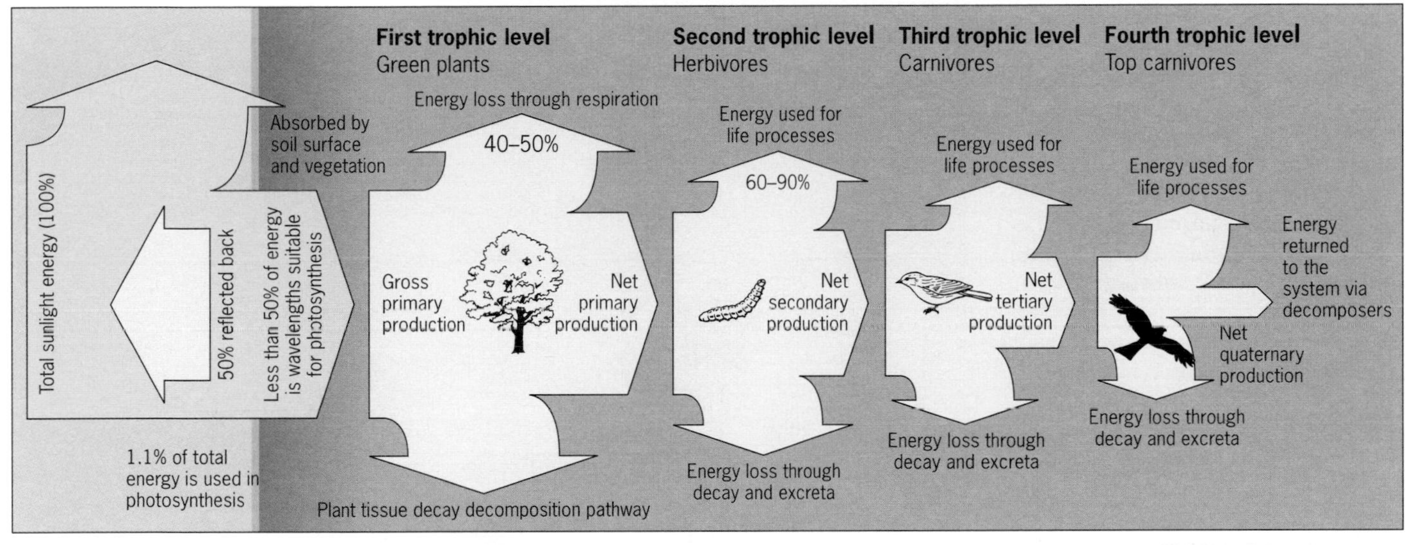

Figure 1.10 Energy flow within a general ecosystem (after Chapman, 1993)

?

5a Tabulate the creatures in Figure 1.11 into trophic levels shown on Figure 1.9.
b Suggest which trophic level **decomposers** represent.
c What happens to the size of individuals as you move up the food chain? State reasons for these size differences.
d What is the average number of links which make up the food web shown in Figure 1.11?
e Explain why links in food chains are seldom longer than five.

6 The sundew (Fig. 9.10) traps insects but is classified as a **producer**. Suggest reasons for this.

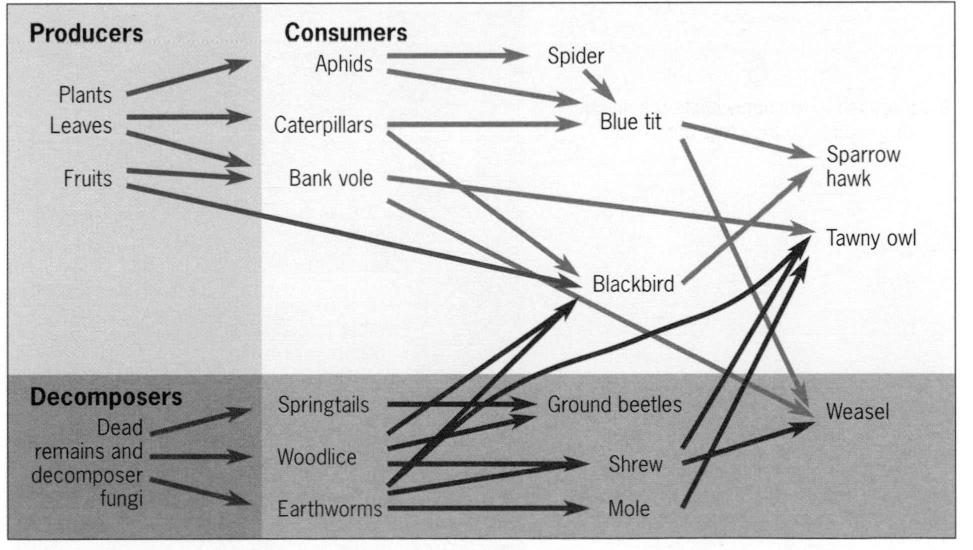

Figure 1.11 A simplified food web for a woodland ecosystem (after Abbott, 1990)

1.4 Productivity

Once we have put organisms into trophic levels, the rate at which each level can build up energy can be worked out. This helps us to discover the potential of an ecosystem for food production.

Gross and net primary productivity

Gross **primary productivity** (GPP) in tonnes per hectare per year is a measure of all the photosynthesis that occurs in an ecosystem. Net primary productivity (NPP) is the energy fixed in photosynthesis (GPP) minus energy lost by respiration.

?

7a Study Figure 1.12. How much energy is stored within the above-ground plant tissues each year?
b How much energy is transferred to the root system each year?
c The sum of respiration and energy fixed in photosynthesis shown in Figure 1.12 represents the net primary productivity of the system. What is the net primary productivity of the sheep-grazed ecosystem?
d How much energy at any given time is stored within living plants above ground?
e Give an explanation as to why this figure is so low compared to the net primary productivity of the sheep-grazed system.
f Explain where all the energy has gone.

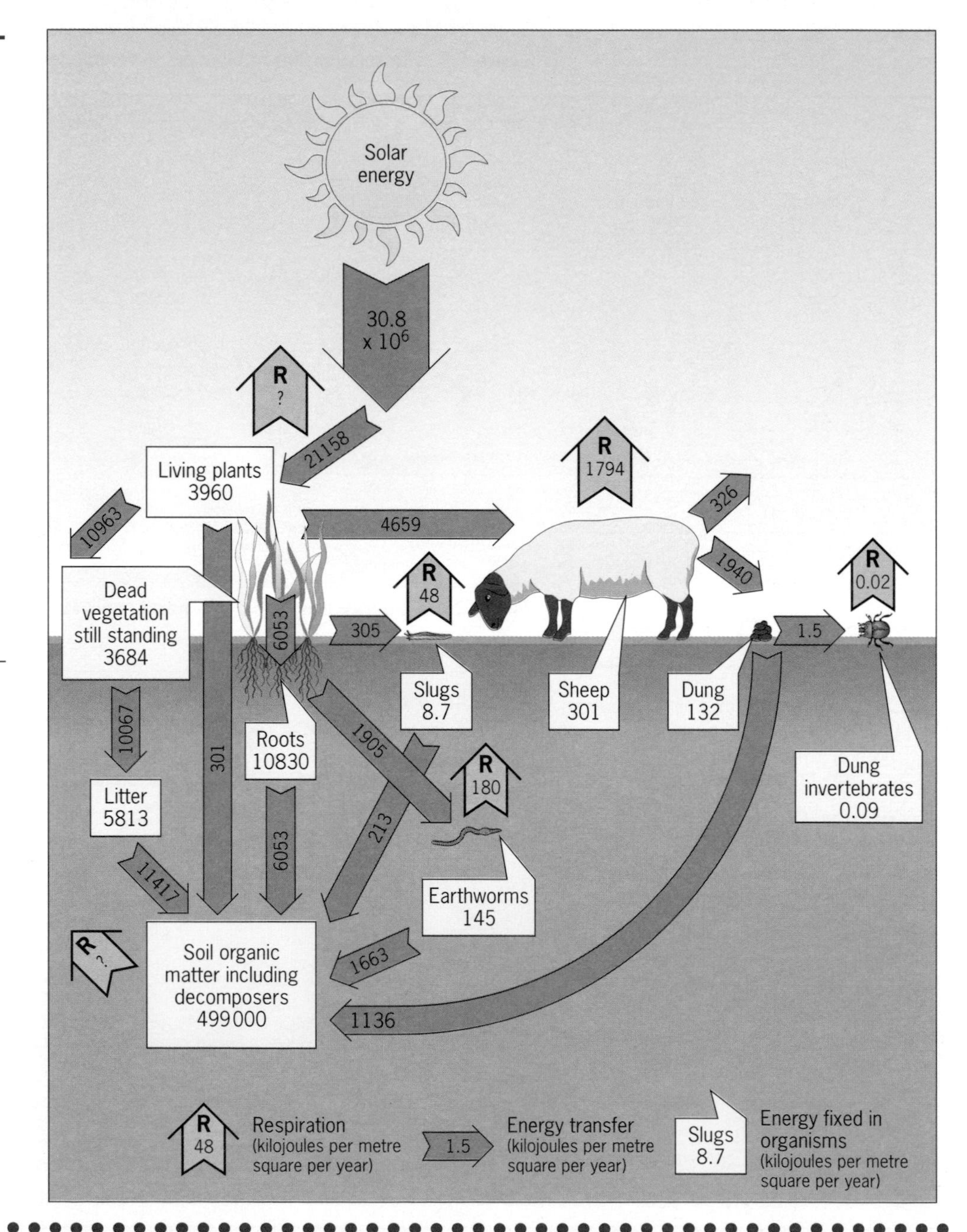

Figure 1.12 Energy flow in a sheep-grazed grassland ecosystem (after Cox and Moore, 1980)

Comparing the productivity of ecosystems

Once the rate of primary productivity in an ecosystem is established, we can compare it with other ecosystems, allowing us to give a figure for the potential of different ecosystems for food production (Table 1.1, Fig. 1.13).

Table 1.1 Net primary production and plant biomass for different types of ecosystems (*Source:* C. Krebs, 1985)

Ecosystem	Area (million km²)	Net primary productivity per unit area (g/m²/year)		World net primary production (billion tonnes/year)	Biomass or standing crop (kg/m²)		World biomass (billion tonnes)
		Normal range	Mean		Normal range	Mean	
1 Tropical rain forest	17.0	1000 – 3500	2200	37.4	6 – 80	45	765
2 Tropical seasonal forest	7.5	1000 – 2500	1600	12.0	6 – 60	35	260
3 Temperate evergreen forest	5.0	600 – 2500	1300	6.5	6 – 200	35	175
4 Temperate deciduous forest	7.0	600 – 2500	1200	8.4	6 – 60	30	210
5 Boreal forest	12.0	400 – 2000	800	9.6	6 – 40	20	240
6 Woodland and shrubland	8.5	250 – 1200	700	6.0	2 – 20	6	50
7 Savanna	15.0	200 – 2000	900	13.5	0.2 – 15	4	60
8 Temperate grassland	9.0	200 – 1500	600	5.4	0.2 – 5	1.6	14
9 Tundra and alpine	8.0	10 – 400	140	1.1	0.1 – 3	0.6	5
10 Desert and semidesert scrub	18.0	10 – 250	90	1.6	0.1 – 4	0.7	13
11 Extreme desert, rock, sand and ice	24.0	0 – 10	3	0.07	0 – 0.2	0.02	0.5
12 Cultivated land	14.0	100 – 3500	650	9.1	0.4 – 12	1	14
13 Swamp and marsh	2.0	800 – 3500	2000	4.0	3 – 50	15	30
14 Lake and stream	2.0	100 – 1500	250	0.5	0 – 0.1	0.02	0.05
Total continental	*149.0*		*773*	*115.2*		*12.3*	*1837*
15 Open ocean	332.0	2 – 400	125	41.5	0 – 0.005	0.003	1.0
16 Upwelling zones	0.4	400 – 1000	500	0.2	0.005 – 0.1	0.02	0.00
17 Continental shelf	26.6	200 – 600	360	9.6	0.001 – 0.04	0.01	0.27
18 Algal beds and reefs	0.6	500 – 4000	2500	1.6	0.04 – 4	2	1.2
19 Estuaries	1.4	200 – 3500	1500	2.1	0.01	1	1.4
Total marine	*361.0*		*152*	*55.0*		*0.01*	*3.9*
Full total	*510.0*		*333*	*170.2*		*12.31*	*1841*

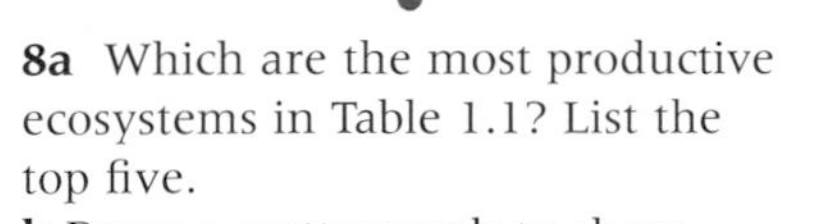

8a Which are the most productive ecosystems in Table 1.1? List the top five.
b Draw a scattergraph to show biomass and net primary productivity per unit area.
c Describe the relationship between these two variables.

Figure 1.13 Ecosystems with very different levels of productivity: extreme desert (Antarctica) and the adjacent continental shelf

?

9 Compare Figure 1.10 with Figure 1.14:
a What is the difference between the transfer of energy and that of nutrients?
b Which additional substances are needed for the nutrient system to work?

1.5 Nutrient cycling

Energy flows through an ecosystem, but nutrients cycle within it. Nutrient cycling can also be affected by materials moving between ecosystems, for example, soil blown across fields or birds migrating across large distances. Humans can disrupt this cycling by removing nutrients.

Nutrients are the chemical elements and compounds needed for organisms to grow and function. In the living world, energy flow and nutrient cycling are interdependent (Fig. 1.14). The rate of nutrient cycling may limit the rate at which energy can be trapped. For example, plants cannot make new cells to trap energy if essential nutrients are absent.

The amount of energy flow may also limit the rate of nutrient cycling that takes place. In living plants, the rate of nutrient uptake will depend on available energy. If a plant is not capturing enough energy for life processes, then its death will result in a breakdown of its organic matter.

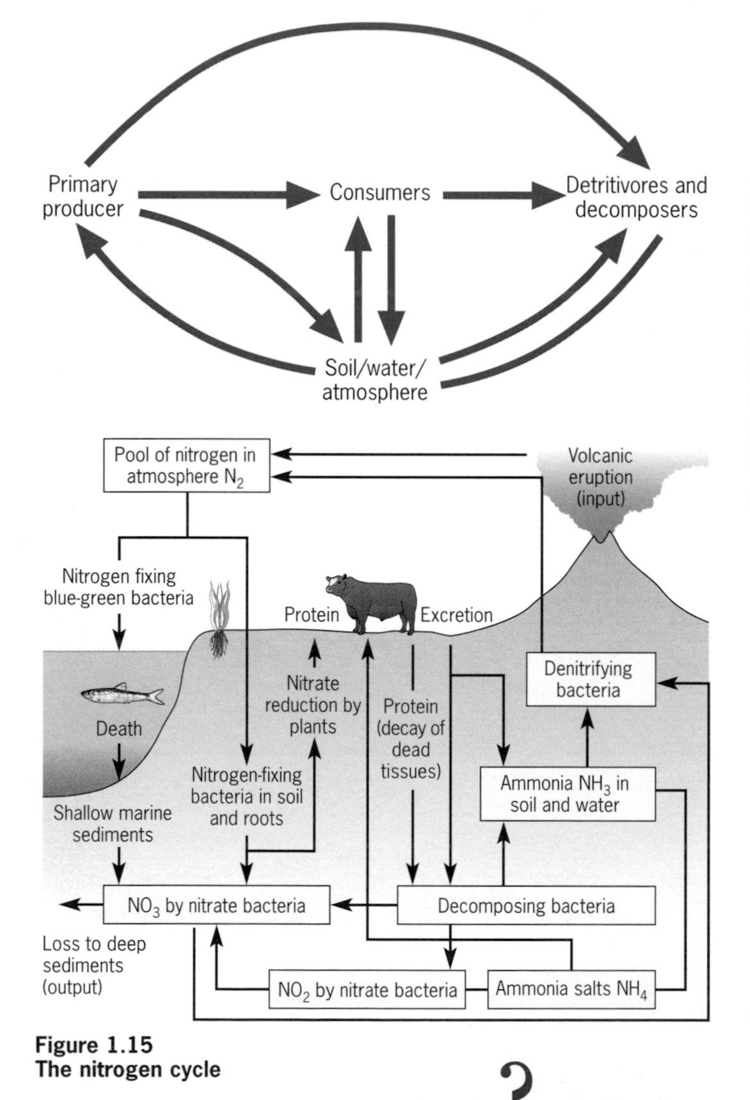

Figure 1.14 A simplified view of nutrient cycling within an ecosystem

Figure 1.15 The nitrogen cycle

1.6 Ecological communities

What is a community?

An 'ecological community' is a term used by ecologists to describe the living components of an ecosystem. It is a highly structured collection of plant and animal populations occurring in the same place at the same time. For simplicity, each community can be recognised as a separate unit but, in reality, few plant and animal populations can survive independently of those in adjacent communities. Ecologists usually use the most abundant or dominant species in a community when they name it, for example 'heathland' communities, or 'grassland' communities.

Community regulation: competition

Competition is one of the most important interactions within communities, and its study is essential to understanding community structure. All organisms require energy, nutrients, space and water for survival. The ability of a species to survive, reproduce and extend its range into new environments depends upon its ability to find food and suitable abiotic inputs – carbon dioxide or oxygen, water and shelter – perhaps by competing with and displacing other species.

In the wild, the effects of competition *between* species (interspecific competition) are difficult to observe because these effects can occur over a long period of time. Any changes that might be caused by competition can be difficult to separate from the effects of other events, such as predation and changes in climate.

Intraspecific competition, between individuals *within* a species, causes enhanced survival of the species. Those fittest to survive do so, and then breed, producing individuals which are better adapted to their habitat. Intraspecific competition is much fiercer than interspecific, because the needs of competing individuals match exactly.

?

10 Figure 1.16 aims to show the difference between 'individuals', 'populations', 'communities' and 'ecosystems'. What do you consider to be the differences between these four terms?

Figure 1.16 The difference among individuals, populations, communities and ecosystems (after Shreeve, 1983)

1.7 Niches

Think of the habitat as the address and the **niche** as the profession of a living organism. A niche is formed when competition forces differences in lifestyles and feeding methods among species. Where there are enough resources, this can lead to a greater variety of species within a habitat; in other words, a greater diversity (Chapter 2). Some animals and plants **co-evolve**, developing physiological relationships for survival which allow both species to benefit. This is known as **symbiosis** (Chapters 2, 7, 12 and 15). Other organisms develop **adaptations** which form mutual-benefit or commensal associations (Chapter 2).

Species diversity is beneficial to the community, because the more species that exist within the community, each with its own special adaptation for its role in the community (its niche), then the less likelihood there will be of any single species becoming so numerous as to displace others. Ecologists suggest that the more complex a food web is, the more stable it is (see Chapters 2 and 12).

Community co-evolution gives rise to **resource partitioning**. For example, diurnal predatory birds such as kestrels and sparrow-hawks are replaced at night by nocturnal predators with a similar niche, such as tawny and long-eared owls. This form of resource partitioning is known as temporal separation.

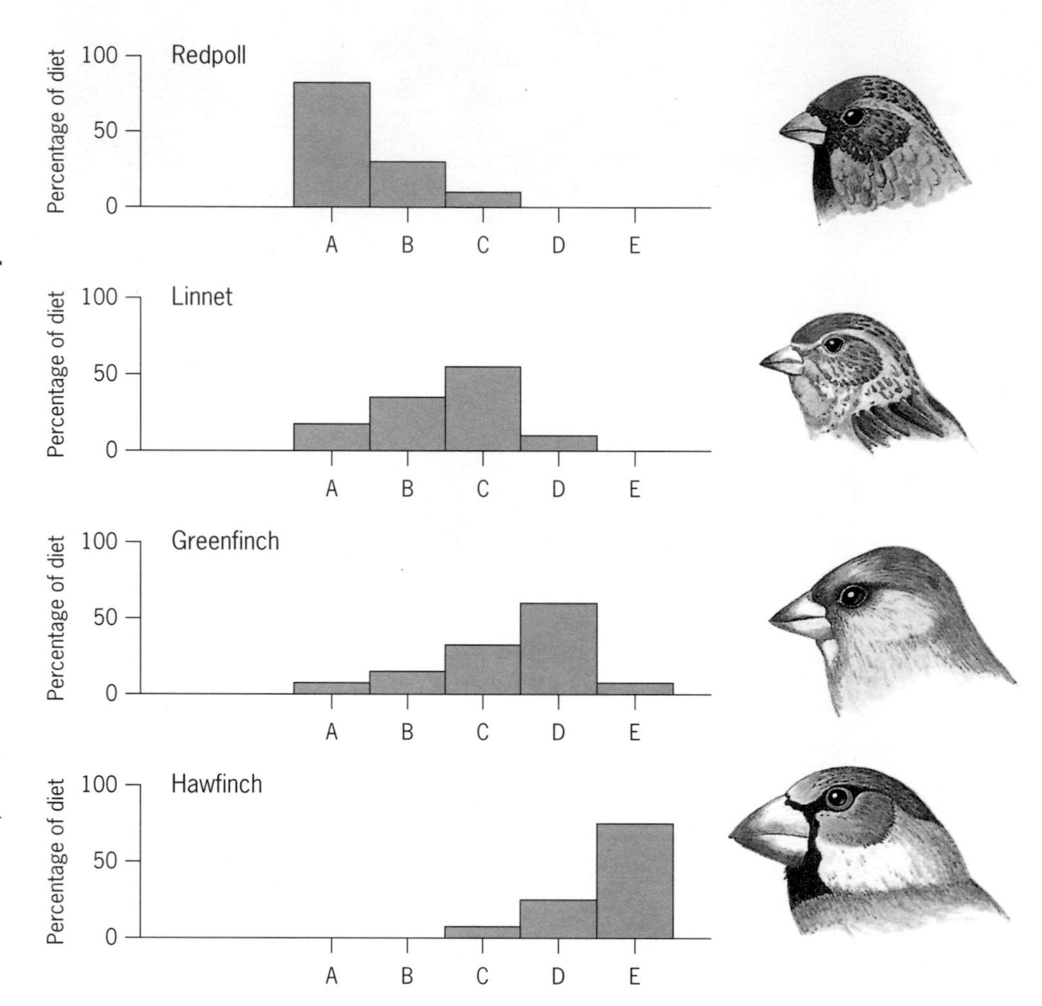

Figure 1.17 Variation in seed size eaten by and bill size among British finches (after RSPB, 1980)

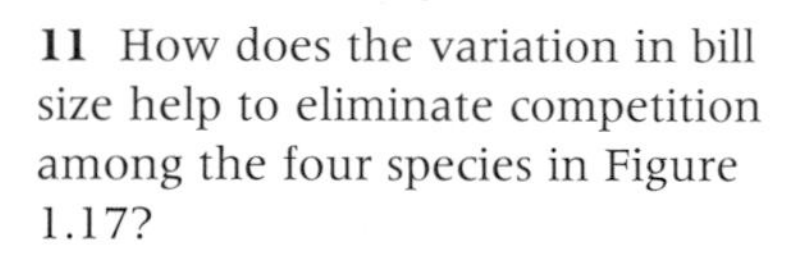

11 How does the variation in bill size help to eliminate competition among the four species in Figure 1.17?

12a Draw a graph of the figures in Table 1.2 to relate the size of insect taken to the bill size.
b List the differences between the species.
c To what extent do prey items correlate with differences in bill size?

13 Using Figure 1.18, explain what this indicates about the feeding sites for each of these three birds.

Table 1.2 Prey characteristics for three species of tit (*Source:* RSPB, 1990)

Size range of prey (mm)	Percentage of prey of each size		
	Blue tit	Marsh tit	Great tit
>0 – 2	59	22	27
>3 – 4	29	52	20
>5 – 6	3	16	22
>6	10	11	32
Length of bill (mm)	9.3	10.4	13.1

Figure 1.18 Variation in feeding sites of three tit species in Marley during the winter (after Lach, 1971)

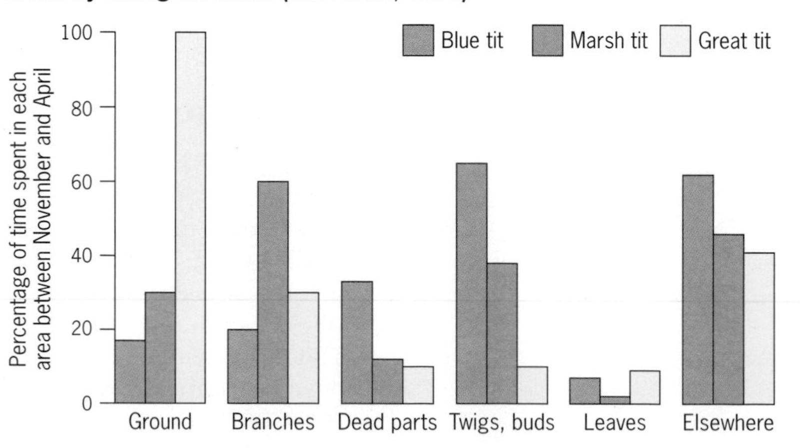

1.8 Competitive exclusion

Ecologists have discovered that species cannot co-exist if their needs are too similar. When this is observed in a community, species often occupy a narrower niche than the potential niche available to them. For example, in the past the chaffinch, a bird species widely distributed all over Europe and North Africa, invaded the islands on Grand Canary and Tenerife. It evolved into a new species, the blue chaffinch. A second wave of the parent species reached the islands and now competes with the blue chaffinches for food and nesting sites. As a result, resource partitioning has occurred. The blue chaffinch breeds in coniferous forests, while the descendants of the second wave breed in deciduous forests. The presence of the original colonist restricts the other species to the habitats in which it is the more effective competitor. This process is known as competitive exclusion and is confirmed by the fact that, on Palma, where the blue chaffinch is absent, the new invader occupies the same range of habitats as it does elsewhere in the rest of its range – deciduous and coniferous forests, parks and gardens.

1.9 Competitive displacement

Competitive displacement occurs when a new species invades a community and eliminates an established one. An example of this can be seen in the fossil record. When North and South America were joined by the Isthmus of Panama in the Pliocene period (2 million years ago), South and North American species crossed the land bridge. North American species proved to be the more successful competitors and virtually eliminated South American fauna. Another example is how the introduction of the non-native grey squirrel into Great Britain has virtually eliminated the population of native red squirrels.

Summary

- An ecosystem is a system of organisms functioning together with their non-living environment. Although ecosystems appear to be classified in discrete units, they are always open rather than closed.
- As open systems, ecosystems receive and lose inputs and outputs of nutrients, energy, water and gas.
- Energy flows through a system. Originating from the sun, it is trapped by producers, then transferred to consumers, but a certain proportion is lost at each trophic level. Nutrients are cycled within the biosphere.
- Organisms form links between each other within the ecosystem. These combine to form webs. A pyramid of biomass of organisms exists, with a decrease in numbers of individuals at each higher trophic level.
- Ecosystems are self-regulating, but mechanisms of positive and negative feedback create or prevent changes, Negative feedback can restore ecosystems to stability, whereas positive feedback can lead to instability.
- The ability of a species to survive within a community depends upon finding food, often by competing with and sometimes displacing other species. Competition forces differences in lifestyles and feeding methods among species. This leads to species diversity. Within a species, competition can lead to the survival of the fittest.
- Each species has its own special adaptations for its role in the community. This role is known as its niche and prevents displacement by other species.

2 Tropical rainforests: ecology

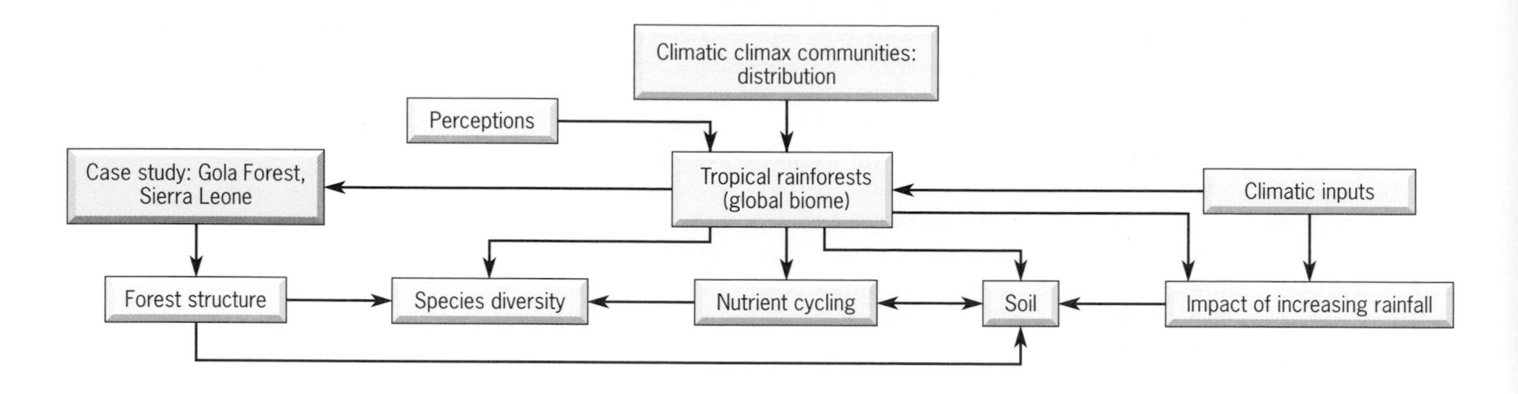

2.1 Community distribution

Climatic climax communities such as tropical rainforests that occur in similar climate zones (Fig. 2.1) tend to resemble each other in their various structures and the habits and life-histories of the organisms they contain. This is due to **convergence**, the process whereby unrelated **species** have evolved to resemble each other because they experience similar environmental

Figure 2.1 Five major types of climates (after Cox and Moore, 1980)

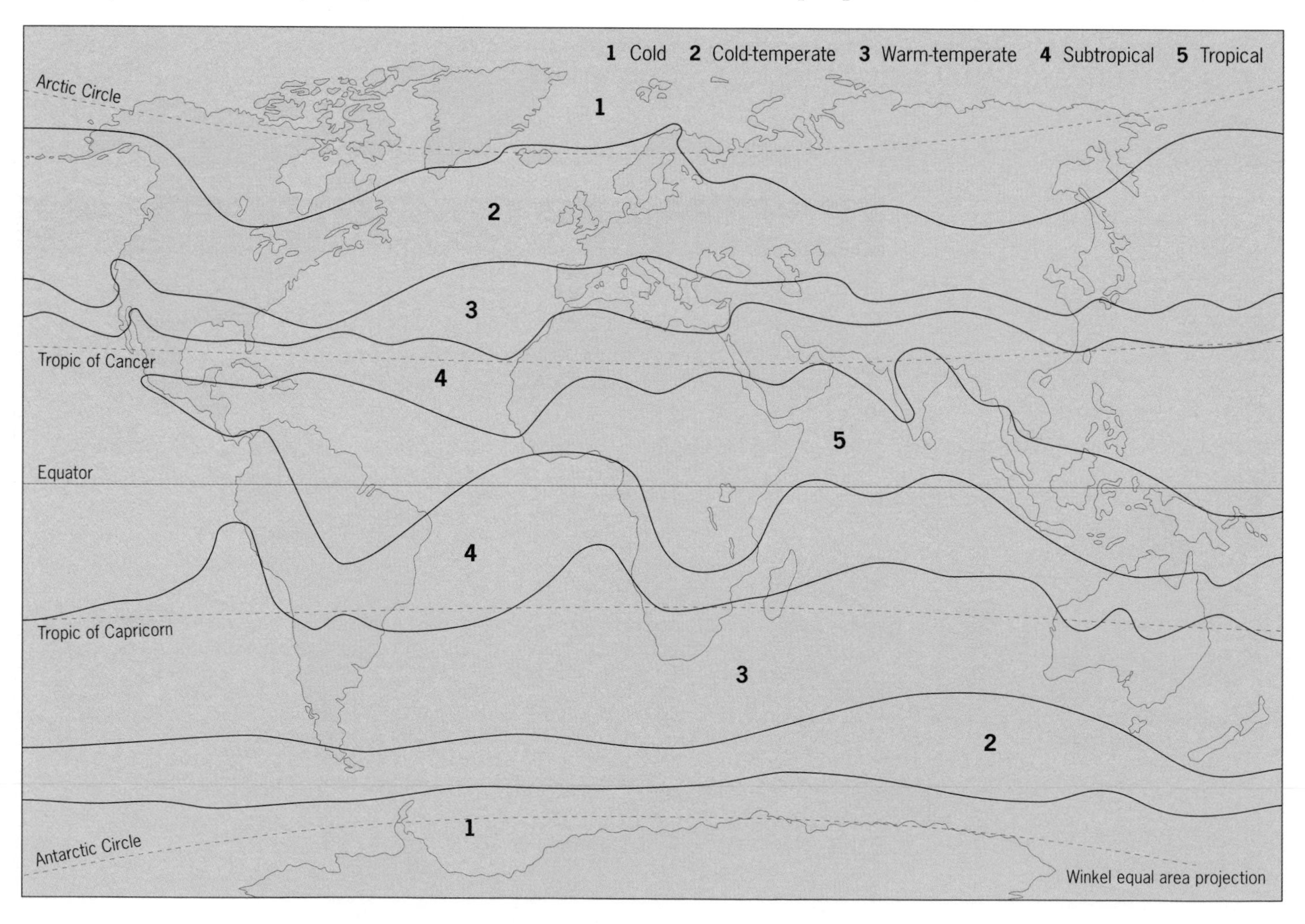

?

1a Trace Figure 2.1 and overlay your map on Figure 2.2. List any similarities and differences.
b To what extent do terrestrial biomes reflect patterns of climatic zones?
c How do the **primary productivity** and **biomass** appear to vary with latitude for forest biomes (see Table 1.1)?
d How do primary productivity and biomass vary with latitude for grassland biomes?

2 Using Figure 2.3 explain how latitude could partly account for the productivity of biomes.

variables. The same is true of communities such as temperate **deciduous** forests (Chapter 4), **boreal coniferous** forests (Chapter 5), temperate grasslands (Chapter 8) and coral reefs (Chapter 14). **Ecologists** have called these convergent communities **biomes** or biome types (Fig. 2.2). Opinions vary as to the number of biomes in the world, largely because it is often difficult to work out whether a vegetation type represents a true climatic community or a **seral stage** (see Chapter 6).

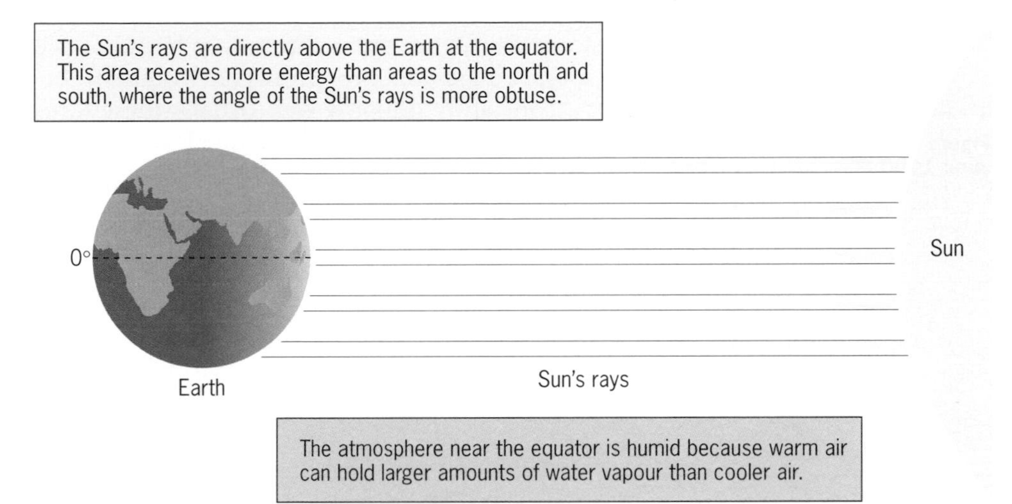

The Sun's rays are directly above the Earth at the equator. This area receives more energy than areas to the north and south, where the angle of the Sun's rays is more obtuse.
0°
Sun
Earth
Sun's rays
The atmosphere near the equator is humid because warm air can hold larger amounts of water vapour than cooler air.

Figure 2.3 The relationship of climate and latitude

Figure 2.2 Map of world vegetation zones (after Cox and Moore, 1980)

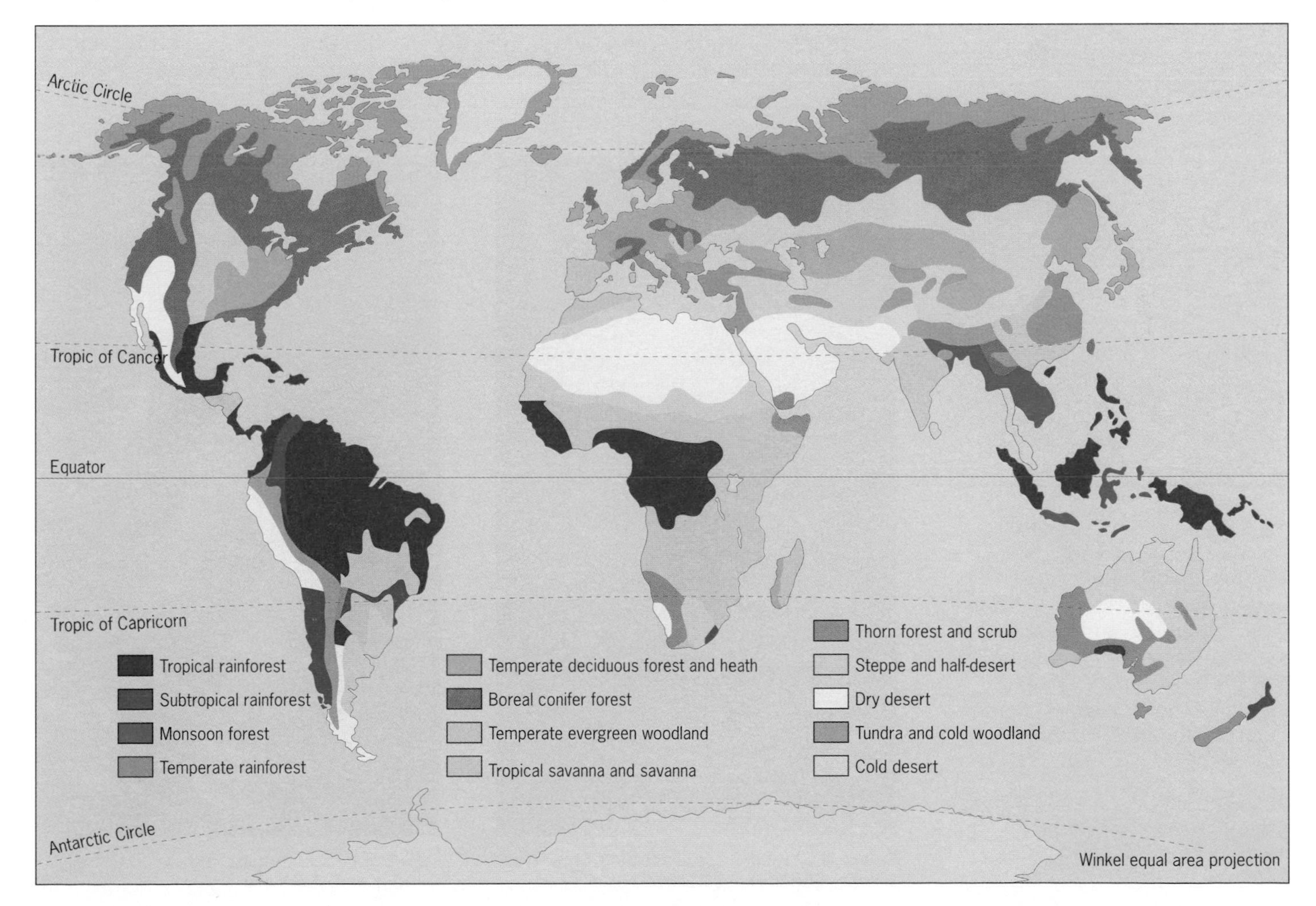

2.2 The tropical rainforest ecosystem

From the window of the light aircraft, a carpet of peppercorn-green forest stretched towards the end of the earth. Here and there I could glimpse a ribbon of silver snaking amid the trees, or occasionally a vast expanse of water, mirroring our progress in its depths. This I thought, is rainforest – for now, at least, paradise on Earth.

Figure 2.4 A quote by Damien Lewis, June 1990

Perceptions of the rainforest

The tropical rainforest (Fig. 2.4) is one **ecosystem** which has had a lot of media coverage since about 1980. However, few people really understand how this complex ecosystem works. People have many different perceptions of the rainforest.

Temperatures

Some people seem to think it is permanently hot inside a rainforest, but much higher temperatures are found outside it. High humidity and low wind speed are common in this environment, but not every tropical moist forest is constantly dripping wet. Usually these forests are well drained, and are often dry underfoot. Only sometimes are they swampy.

Climate

The belief that the climate does not vary at all is also wrong. Within the forest there are daily temperature changes. There are also micro-climatic differences between the upper canopy and the forest floor. Moreover, at higher altitudes and latitudes there are noticeable seasons (Fig. 2.3).

The sun's rays can reach through some parts of the forest more easily than others, so there is a variety of light patterns. The forest is not dark everywhere.

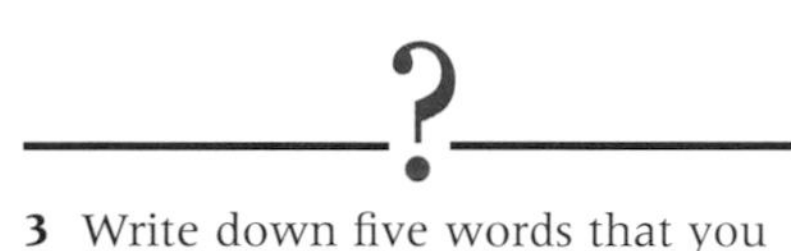

3 Write down five words that you think describe the physical characteristics of tropical rainforests. Check your list as you work through Section 2.2.

Vegetation

The perception of the forest as a tangled mass of vegetation, requiring an army with machetes to get through it, is usually wrong. Many early travellers viewed the forest superficially from the road or riverside. The edge of the forest in Figure 2.5 is certainly more dense than the interior in Figure 2.6 because more sunlight is available. It is often possible to walk freely on the floor of a rainforest, since it is open, apart from fallen logs.

Figure 2.5 Rainforest vegetation seen from a path, Queensland, Australia

Figure 2.6 Rainforest interior, Queensland, Australia

?

4a Explain what seasonality means.

b Using an atlas find climate graphs for areas close to 0° north, 5° north and 10° north. Do they show any evidence of seasonality?

c Find climate graphs for areas close to 0° north which are at different altitudes. Do they show any evidence of seasonality?

5 Using Figures 2.5 and 2.6, list the differences between the two views of tropical rainforest.

Plant growth

The belief that plants grow all the year round is also largely wrong. Plants which grow continuously throughout the year can be found in tropical forests, but they are mostly limited to bamboos and tree ferns. Few trees grow continuously beyond the seedling stage. Growth spurts tend to occur during a few periods of the year. The reasons for this are not clearly understood.

2.3 How climate defines the rainforest

Tropical rainforests are found throughout the world (Fig. 2.7). Most of the differences between tropical and temperate forests are linked to climate. Tropical rainforests are climatic climax communities (see Chapter 6) and therefore reflect the nature of the **inputs** associated with a tropical environment.

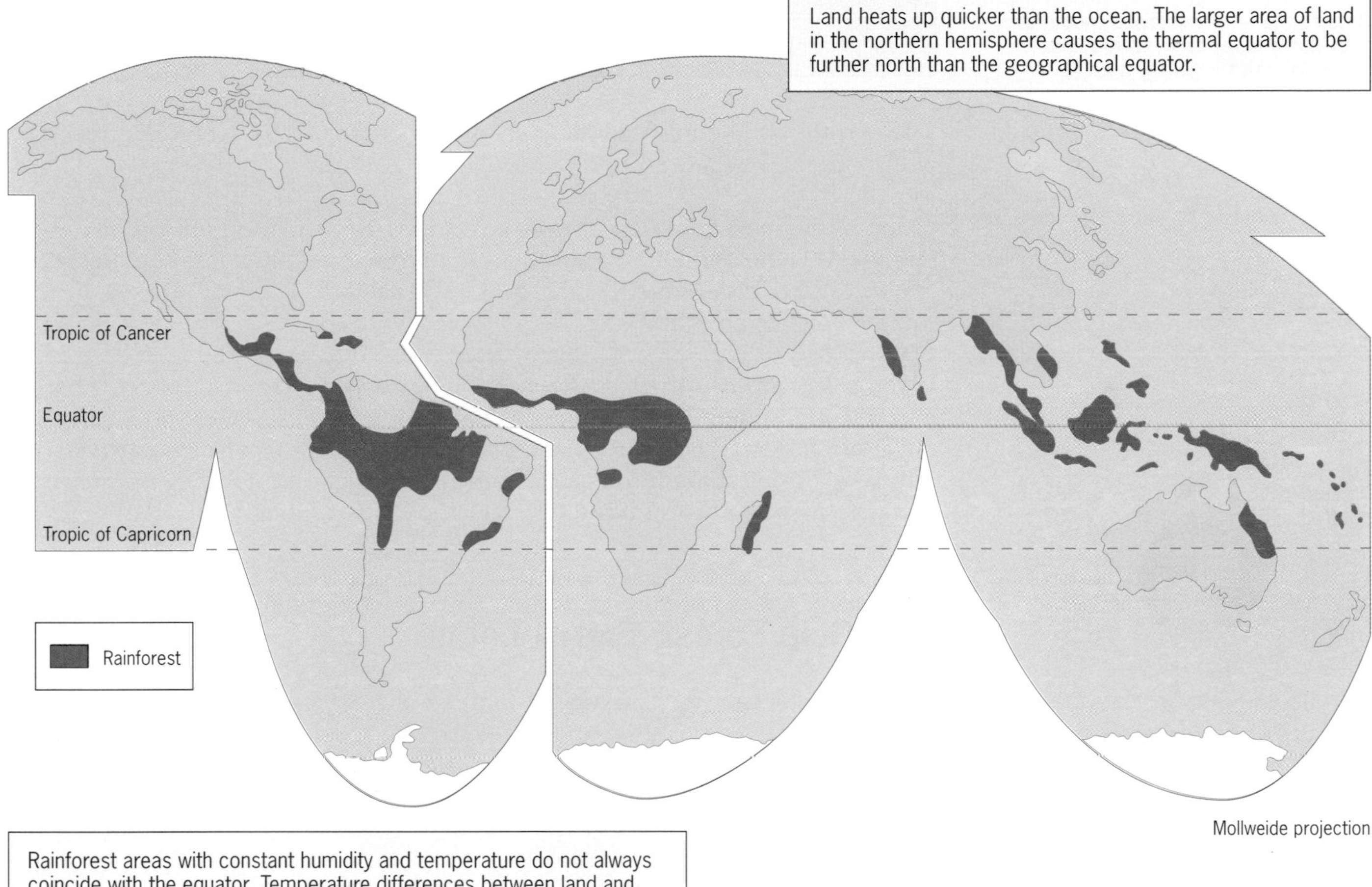

Land heats up quicker than the ocean. The larger area of land in the northern hemisphere causes the thermal equator to be further north than the geographical equator.
Tropic of Cancer
Equator
Tropic of Capricorn
Rainforest
Mollweide projection
Rainforest areas with constant humidity and temperature do not always coincide with the equator. Temperature differences between land and sea affect the circulation of air masses and ocean currents. These cause tropical air to move north or south, thus extending its climatic influence.

Figure 2.7 World distribution of tropical forest land that could be theoretically covered by rainforests (after Longman and Jenik, 1987)

Several characteristics of a tropical rainforest help to distinguish it from other forests on earth:

1 A climatologist named Köppen defined a tropical lowland climate as that where monthly mean temperatures are above 18°C and daily differences are greater than seasonal ones.

2 Distribution of rainfall further subdivides forest communities into: (a) those that receive continuous precipitation, where there are no arid seasons (moist tropical rainforest) and (b) subhumid evergreen forest with a few dry months (dry tropical rainforest).

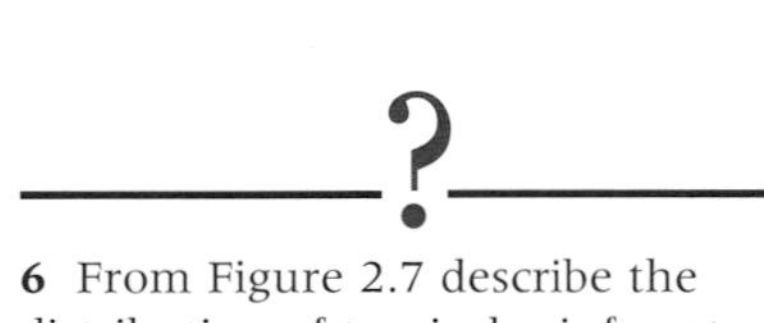

6 From Figure 2.7 describe the distribution of tropical rainforests.

?

7 From Figure 2.8 explain how the forest affects the rainfall.

8 Use Figure 2.3 to explain why tropical rainforests have high energy inputs.

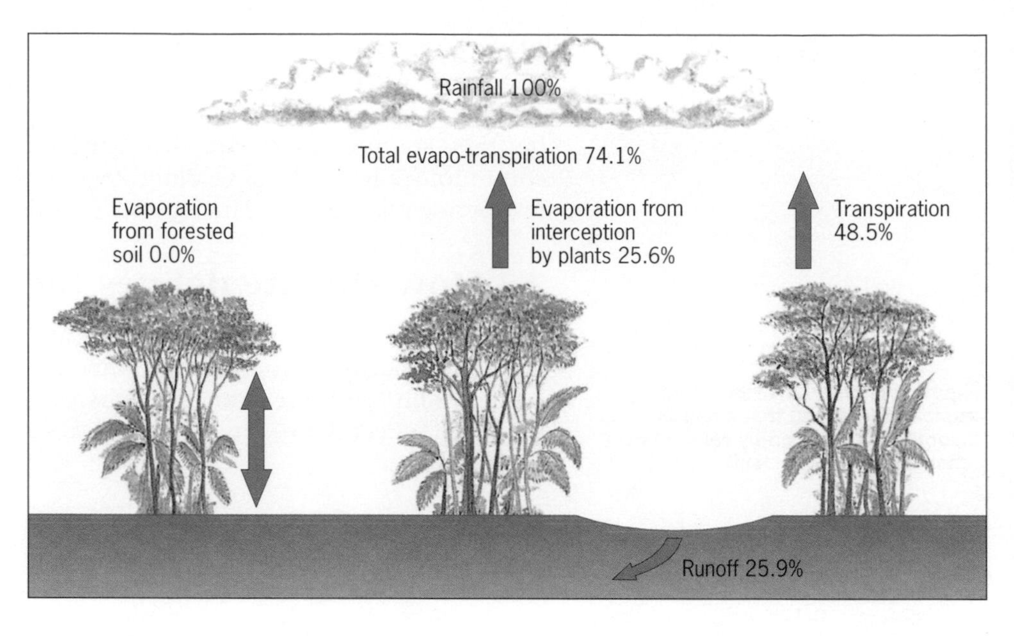

Figure 2.8 A generalised model of the water balance in an Amazon rainforest (after Salati et al., 1984)

3 True tropical rainforests are defined as containing mostly evergreen trees which are not resistant to cold or drought.

4 One of the most important factors governing the formation of tropical rainforests is the rainfall. Annual totals are more than 2000mm. The presence of a forest area also influences the amount of rainfall.

Productivity

When solar **radiation** strikes the forest, long-wave radiation is reflected back into the atmosphere as it hits the vegetation or ground surface. Short-wave radiation, however, is absorbed and used in the process of **photosynthesis** (Fig 1.10). Tropical rainforests have some of the highest total of net primary productivity in the world (see Table 1.1).

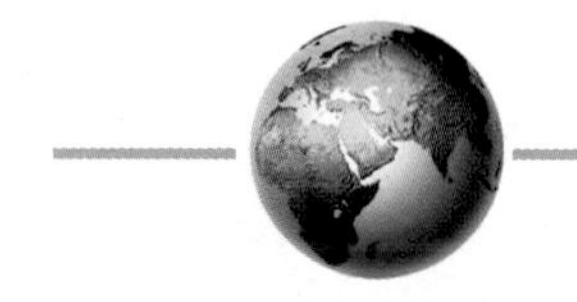

The Gola Forest, Sierra Leone

The Gola Forest is the only area of lowland rainforest left in Sierra Leone. It covers 748 square kilometres and is divided into three forest reserves: Gola North, East and West (Fig. 2.9). Records of climatic variables have been kept at Kenema Farm north-west of the reserves 7°53′ N, 11°11′ W (Fig. 2.10). The Gola rainforest is 7°N of the Equator and displays some seasonality.

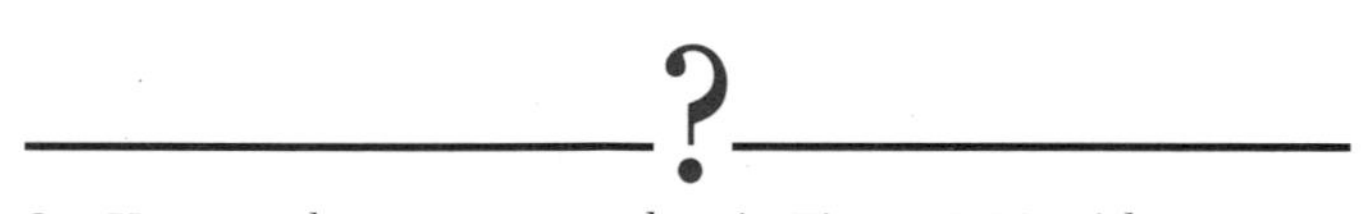

9a Use an atlas to compare data in Figure 2.10 with a temperate region such as the British Isles.
b What major differences are there between rainfall and temperature characteristics?

10a Calculate an index of plant productivity for the Gola Forest using Figure 2.10.
b Compare this with the figures for Paris.

You can calculate an index of net primary productivity using the formula:

$$I = \frac{T_m P}{120 T_r}$$

Where:
I = (estimated) index of plant productivity
T_m = average temperature of the warmest month (°C)
T_r = Annual temperature range between warmest and coldest month (°C)
P = annual precipitation (cm)

Example:
Paris (temperate zone)

$$I = \frac{35^{\circ}\text{C} \times 57.4 \text{ cm}}{120 \times 32^{\circ}\text{C}} = \frac{2009}{3840}$$

$I = 0.52$

Figure 2.9 The Gola Forest, Sierra Leone

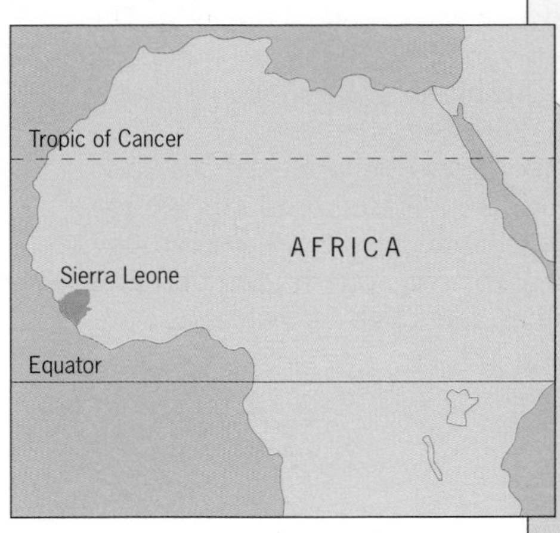

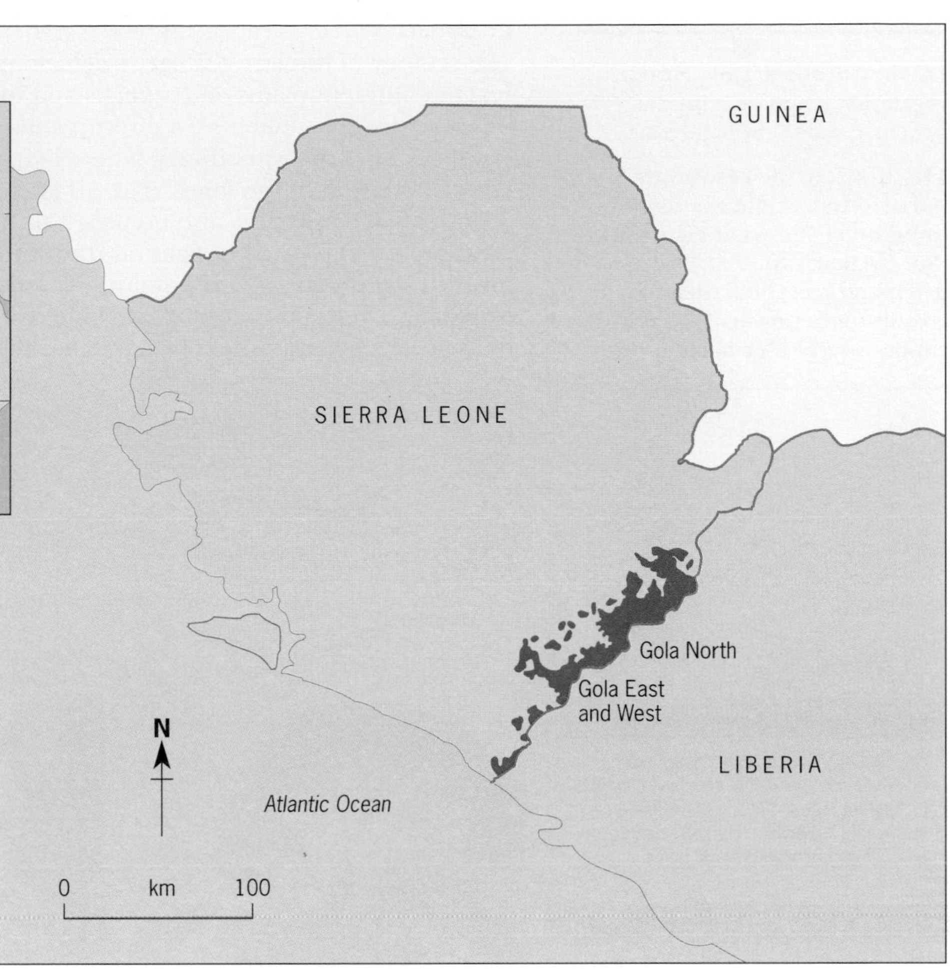

Figure 2.10 Climate details for Kenema Farm, Gola Forest

Altitude 152.4 metres
Minimum monthly temperature range: 15–20.5°C
Maximum monthly temperature range: 31.5–37°C
Mean annual rainfall: 2770 mm

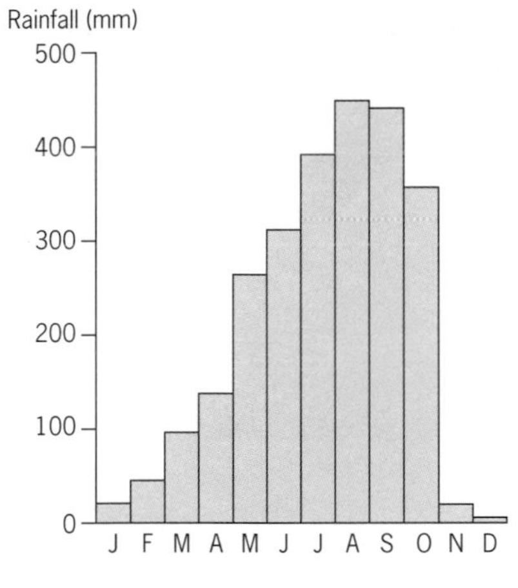

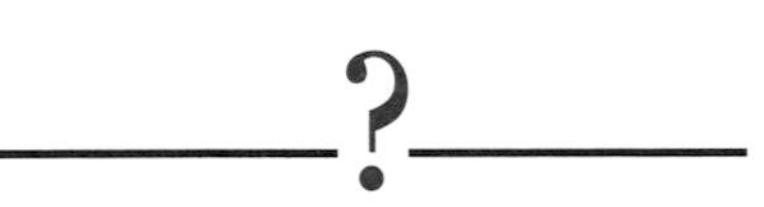

11 Using Table 2.1:

a Draw a cross-section of the slope and position the trees along it.

b Draw on trees to scale, adding buttress roots if appropriate (see Figure 2.19).

c Label the trees using the following classification to help:

Upper tree layer (emergents)	36.5m and over
Middle layer	21.3 – 36.49m
Lower layer	9.1 – 21.29m
Shrublayer	1.8 – 9.09m

Table 2.1 Index to trees recorded on a transect of Gola Forest West (*Source:* D. Small, 1952)

Height above sea level (metres)	Distance along transect (metres)	Tree number and name	Total height (metres)	Bole height (metres)	Girth at 1.29m (metres)	Crown diameter (metres)	Buttress height and circumference
21.3	2.68	1. Korlei	29.8	16.4	1.3	14.9	1.7/0.3
	4.69	2. Ndorkei	25.6	11.6	0.6	10.6	–/1.4
	8.04	3. Yawi	40.2	21.9	2.8	13.7	2.4/14.6
	7.37	4. Korlei	10.6	9.1	0.3	2.4	–/0.4
	8.71	5. Unknown	7.6	4.6	0.03	3.0	–/0.3
	10.05	6. Mampa	15.2	12.2	0.3	2.4	–/0.5
	12.73	7. Buni	12.2	6.7	0.6	3.6	–/0.7
19.8	13.40	8. Dani-Kpavi	22.8	15.2	1.2	19.8	1.2/5.6
	14.74	9. Unknown	9.1	4.6	0.6	6.1	–/1.1
	15.41	10. Unknown	6.1	1.8	0.33	5.5	0.3/1.3
	16.08	11. Mampa	16.8	14.9	0.6	2.4	–/0.7
	16.75	12. Yawi	36.0	10.7	2.4	15.2	2.9/10.2
	18.76	13. Pah	7.6	5.5	0.27	3.6	–/0.3
	19.43	14. Dengbehawi	14.6	11.9	0.5	10	–/0.6
	20.1	15. Marnpa	12.2	10.7	0.5	2.4	–/0.6
	20.77	16. Tel i	8.5	6.1	0.4	1.5	–/0.5
	24.77	17. Dema-Kpavi	10.7	7.6	0.5	4.6	2.4/1.3
	25.46	18. Korlei	10.7	9.1	0.4	2.4	–/0.8
	28.81	19. Mampa	18.3	12.1	0.3	6.1	0.6/2.2
18.3	36.18	20. Ngonyai	9.1	4.6	0.03	1.5	–/0.4

?

12 How does the Gola Forest vegetation compare with the structure shown in Figure 2.11?

13a How are the **resources partitioned** within the rainforest community shown in Figure 2.11? (See Section 1.8).
b Why do you think there is a greater species diversity at the euphotic level (See Section 1.8)?

Forest structure

The ecology of the forest is partly determined by how much of the sun's energy reaches different plants and animals and by the waveband of the light. The forest structure is therefore a direct result of variations in energy inputs. This produces **stratum specificity** where living things are found at a particular layer or stratum of the forest (Fig. 2.11).

Horizontal differences also occur within the forest when a tree falls and creates a sudden patch of light on the forest floor. As the forest **succeeds** towards its climatic climax community (see Chapter 6), extra light causes lower-level vegetation to increase in density. This vegetation then becomes less dense again when the canopy is established and less light is available.

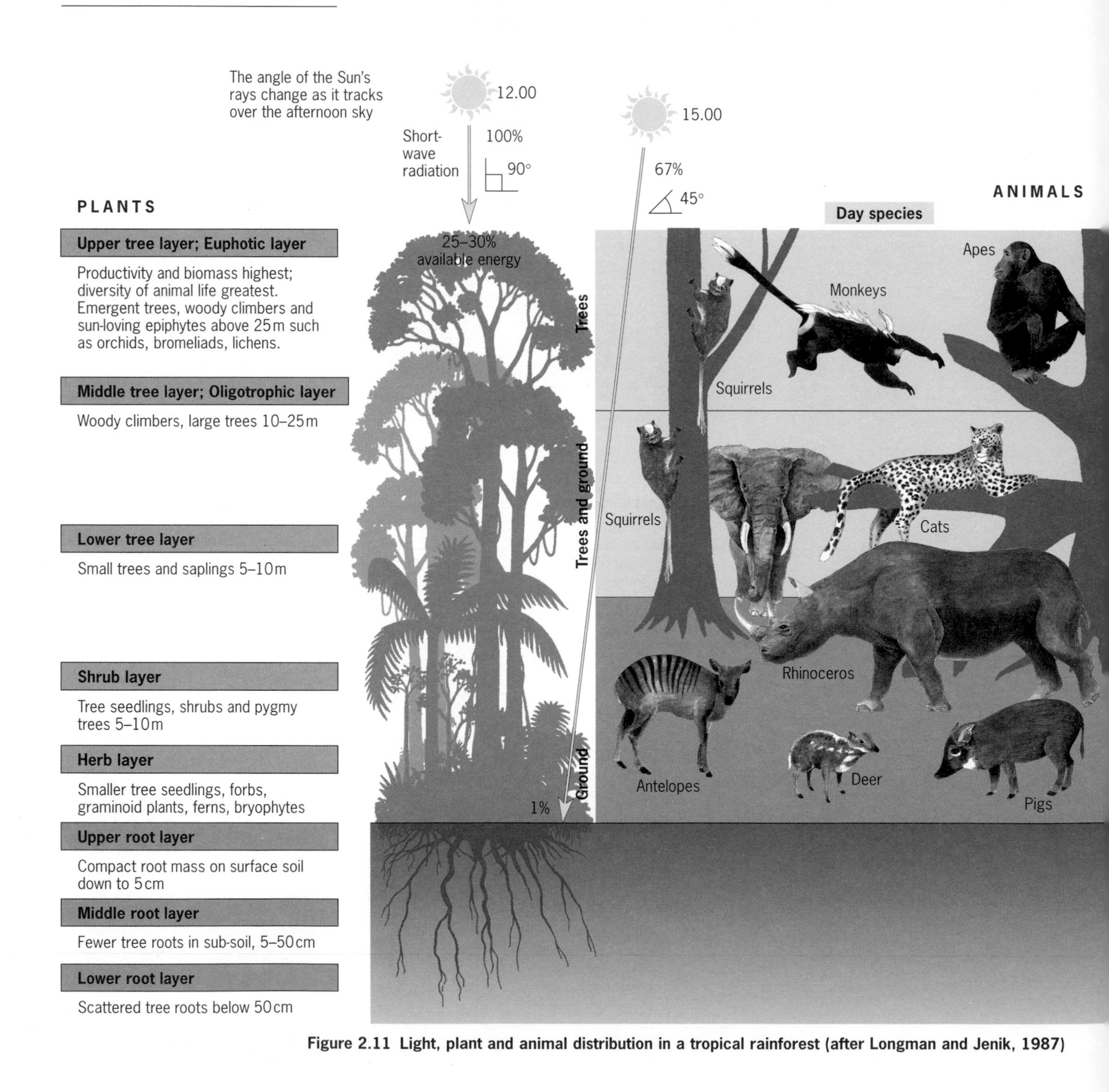

Figure 2.11 Light, plant and animal distribution in a tropical rainforest (after Longman and Jenik, 1987)

?

14 Study Figure 2.12 and, using your knowledge of tropical rainforests, explain the evidence that leads you to believe that this area is a tropical rainforest environment.

If extra solar energy is combined with moisture and nutrients there is, in theory, the potential for a larger biomass and a species-rich community. This can only happen if key species do not dominate an ecosystem. Trees are the life forms that dominate the climatic climax community of the tropical rainforest, but no single tree species dominates the forest. The reason for this lack of dominance could be explained by the impact of insects on trees. In rainforests, predation by insects has left individual trees widely spaced and prevents one species from dominating throughout. A wide variety of species further reduces damage to the ecosystem as a whole, since insects do not occur as plagues. In plantation areas, trees can be badly affected by insects, since a large area of one species can provide a continuous food supply. There will be a time lag before **negative feedback** can restore the system to a steady state or **dynamic equilibrium** (see Section 1.3).

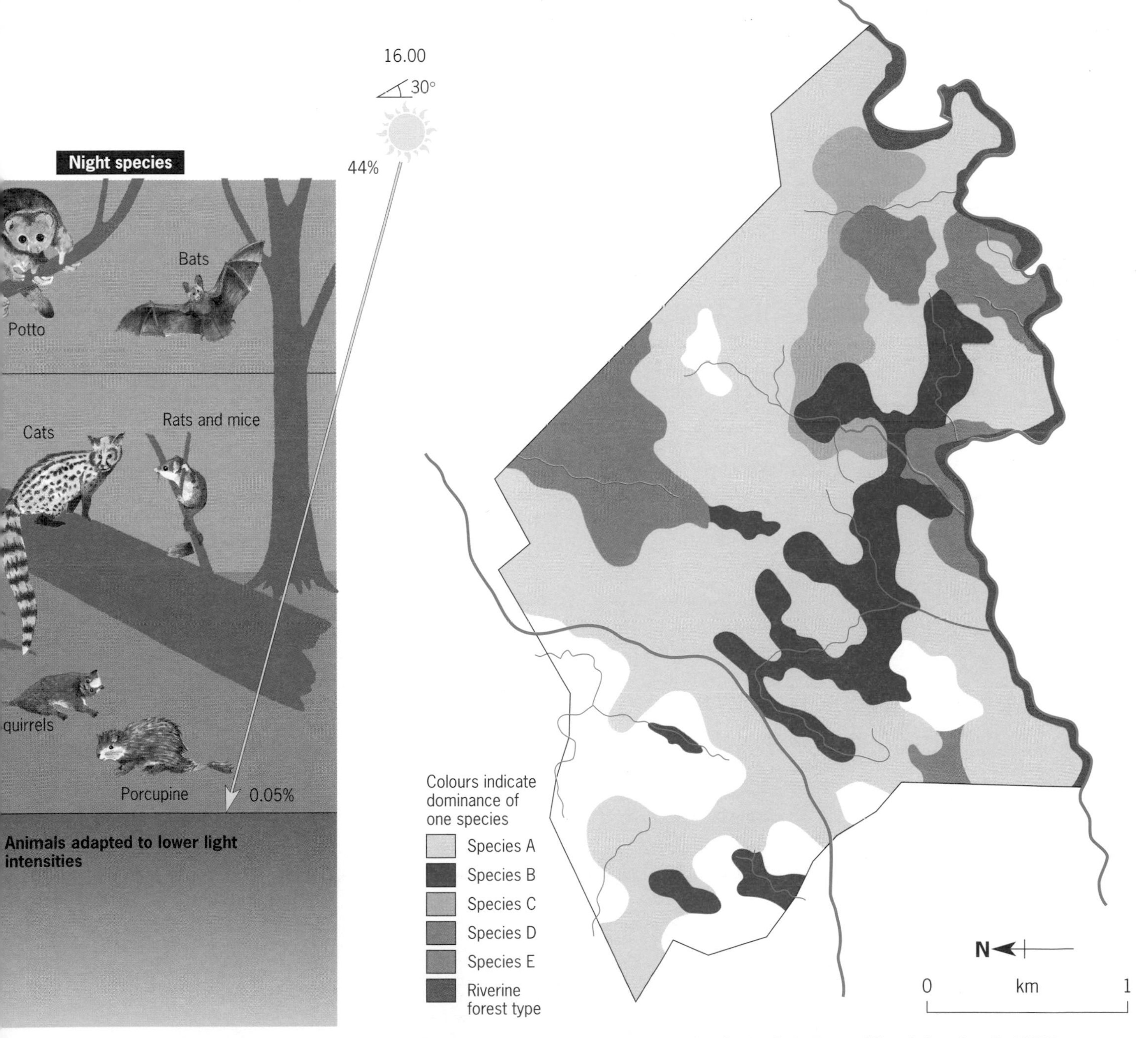

Figure 2.12 The principal tree species in the Gola Forest West (after Small, 1952)

2.4 Species diversity

Many species follow the pattern shown in Figure 2.13, in particular vascular plants and some **invertebrates.** However, other life forms show an increase in diversity at higher latitudes: algae, fungi, lichen and some insects.

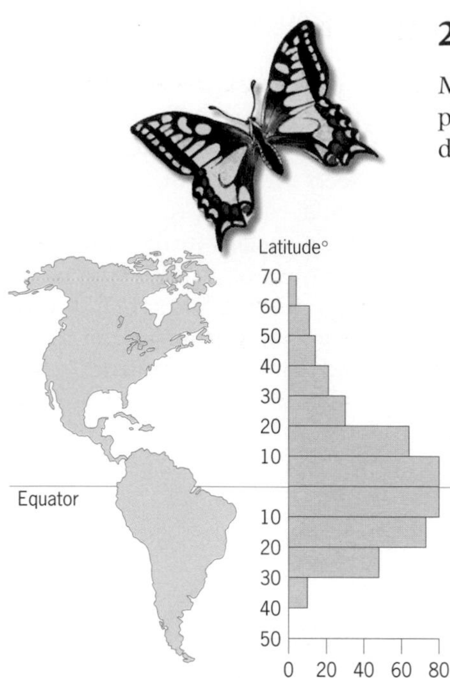

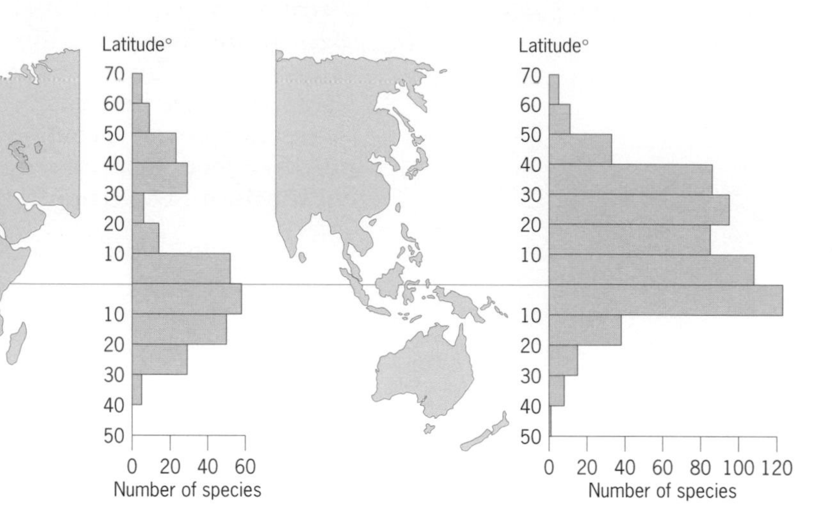

Figure 2.13 Global swallowtail butterfly distribution (after Collins and Morris, 1985)

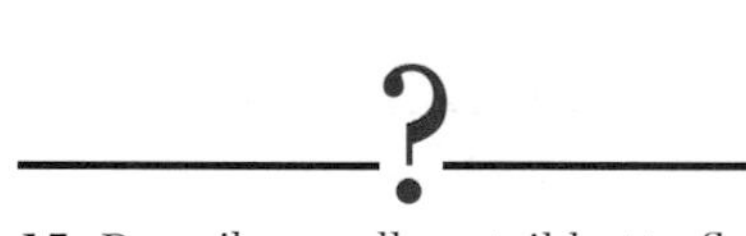

15 Describe swallow-tail butterfly distribution as shown in Figure 2.13.

16 From Figure 2.14 explain why delaying germination through allelopathy would help survival.

Generally there is an increase in species diversity at lower latitudes (near the Equator). Half of the world's **gene pool** is thought to exist in tropical rainforests. This is explained by three factors:

1 The greater rate of mutations per 1000 individuals and evolution of new species. One theory is that levels of ultraviolet-B radiation increase at the Equator and can change DNA. This can lead to more frequent mutations and therefore more rapid evolution of new species.

2 The opportunities for co-existence within the tropical rainforest ecosystem.

3 Many of the rainforest areas in the low latitudes are surrounded by few physical barriers, which allows immigration of species to take place. This is particularly true in the Zaire and Amazon basins. They are very ancient ecosystems, up to 100 million years old in some parts, and do not lose the species they already have, while gaining extra ones.

The huge variety of plants provides many opportunities for specialised animals. Many of these animals have been forced by interspecific **competition** (see Section 1.7) to form very narrow **niches**. To co-exist within the forest, many of them have adaptations to improve their chances of survival (Fig. 2.14).

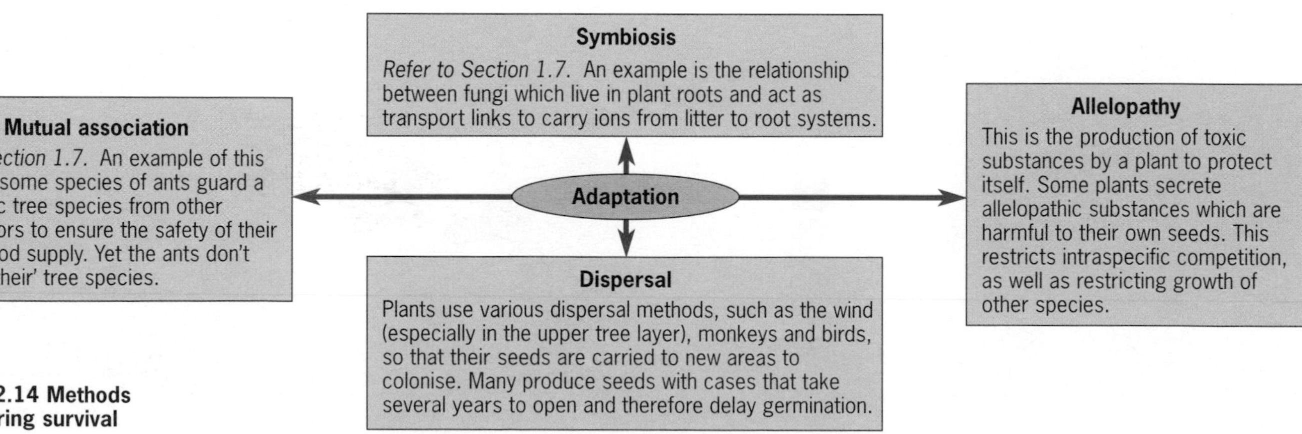

Symbiosis
Refer to Section 1.7. An example is the relationship between fungi which live in plant roots and act as transport links to carry ions from litter to root systems.
Mutual association
See Section 1.7. An example of this is that some species of ants guard a specific tree species from other predators to ensure the safety of their own food supply. Yet the ants don't harm 'their' tree species.
Adaptation
Allelopathy
This is the production of toxic substances by a plant to protect itself. Some plants secrete allelopathic substances which are harmful to their own seeds. This restricts intraspecific competition, as well as restricting growth of other species.
Dispersal
Plants use various dispersal methods, such as the wind (especially in the upper tree layer), monkeys and birds, so that their seeds are carried to new areas to colonise. Many produce seeds with cases that take several years to open and therefore delay germination.

Figure 2.14 Methods of ensuring survival

2.5 The impact of increased rainfall

Precipitation falling in tropical rainforests (Fig. 2.15) can be expressed using this equation:

$$P = I + SF + TF$$

Where:

P = precipitation
I = water intercepted and evaporated directly
SF = water intercepted, reaching ground level by stem flow
TF = water falling through directly to the ground or dripping from foliage.

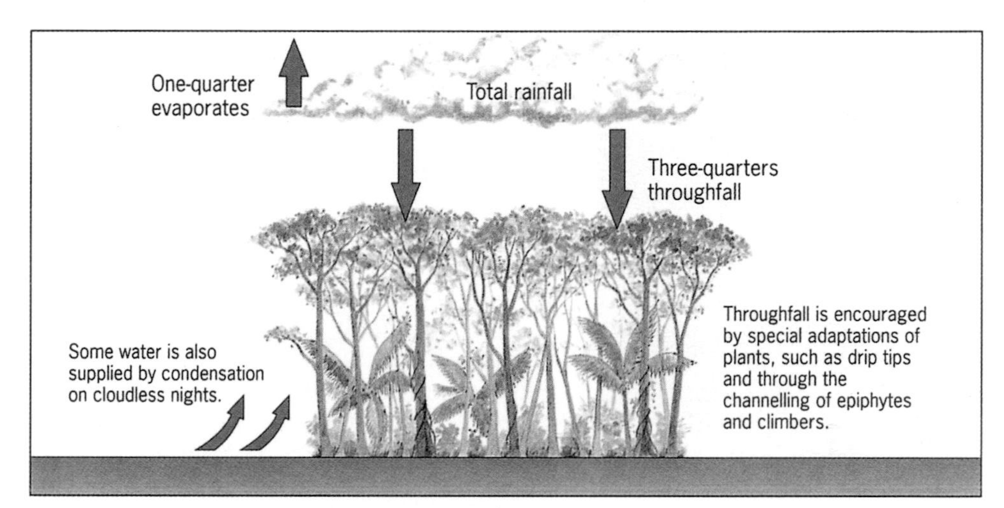

One-quarter evaporates
Total rainfall
Three-quarters throughfall
Some water is also supplied by condensation on cloudless nights.
Throughfall is encouraged by special adaptations of plants, such as drip tips and through the channelling of epiphytes and climbers.

Figure 2.15 Water flow in a tropical rainforest

?

17a Using Figures 2.16 and 2.17, identify parts of plants which are adapted to:

- encourage evaporation and interception,
- encourage throughfall and stem flow.

b Where would each of these appear in the rainforest structure? Give reasons why.

High levels of radiation alone cannot create high productivity levels. For example, hot deserts have some of the highest solar inputs, yet low productivity rates (see Table 1.1).

Water increases productivity rates. The added water input increases the size of rainforest plants. Indeed, cell expansion in plants relies on turgidity. This is when the cells are full of water. Extra water allows stems and leaves to lengthen. Some bamboo species grow by 1m per day. Flushing of new growth begins when water is in sufficient supply and is likely to be linked with variations in the rainfall pattern. Young leaves can cope with lack of water by hanging limply, thereby reducing surface area available for intercepting radiation and **evapotranspiration**. Loss of turgidity also closes stomata, which will further reduce evapotranspiration.

Figure 2.16

Figure 2.17

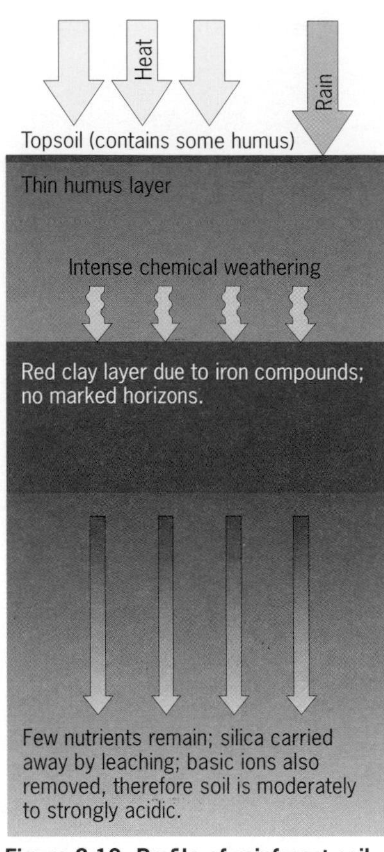

Figure 2.18 Profile of rainforest soil

2.6 Rainforest soil

High temperatures and high rainfall within tropical rainforests lead to rapid weathering of the **ferrisols** which commonly occur beneath them. Minerals are quickly weathered and leached out of the system, leaving few nutrients for the rainforest to use (Fig. 2.18). Those nutrients that remain are taken up quickly by the plants.

Sometimes the iron compounds in some tropical soils become concentrated in a horizon to form a hard layer, known as a **laterite.** Plant roots cannot push through laterite, affecting plant growth. This usually happens during a dry season when evaporation and upward movement of minerals occur.

The root systems are shallow and grow towards the available nutrients in the thin layer of **humus** at the top of the soil profile. Some trees need extra support if their roots are too shallow. These trees grow buttress or aerial roots (Fig. 2.19).

2.7 Nutrient cycling

There are many different nutrient cycles. Some nutrients such as nitrogen, carbon, water and oxygen form gaseous compounds in the atmosphere. Others such as calcium, potassium and sodium (which occur as compounds, not as metals) are largely cycled between organisms and the soil. An element in short supply within an ecosystem may limit the rate of ecosystem productivity. The rate and character of nutrient cycling is therefore of considerable interest to humans, who often wish to crop ecosystems at various **trophic levels**.

The overall rate of growth within a rainforest is very rapid. The most important factors controlling this are energy, water and nutrient inputs. Ten per cent of matter is decomposed above the forest floor and is re-used by epiphytes. (Epiphytes are herbaceous plants that grow on other plants but do not feed on their hosts – for example orchids.) The lack of nutrients within the soil system is compensated for by very rapid recycling of leaf litter. Humidity and high temperatures allow for rapid transfer of nutrients.

Leaf litter in tropical rainforests can decompose within three to four months, whereas at temperate latitudes this can take two or more years. On the forest floor there can be about 38 000 microscopic arthropods per square metre which help to break this material down. Mycorrhizal fungi, living in association with plant roots, can act as transport links to carry ions from litter to root systems. Termites, with symbiotic protozoa in their digestive tract, also rapidly decompose wood. The protozoa help the termites to digest the wood, which allows them to decompose a large tree rapidly.

Figure 2.19 Buttress roots and aerial roots

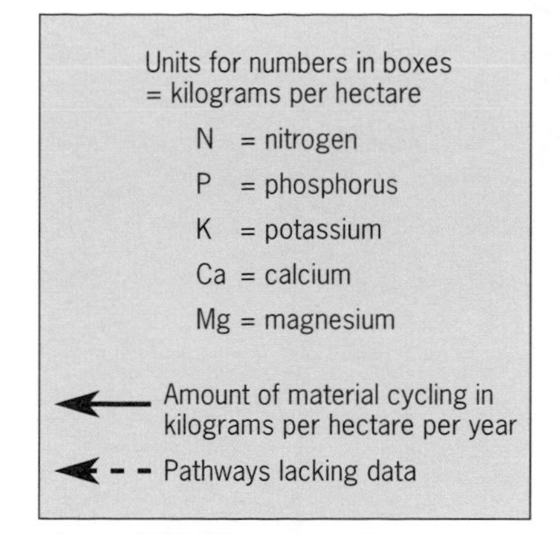

Units for numbers in boxes
= kilograms per hectare
N = nitrogen
P = phosphorus
K = potassium
Ca = calcium
Mg = magnesium
Amount of material cycling in kilograms per hectare per year
Pathways lacking data

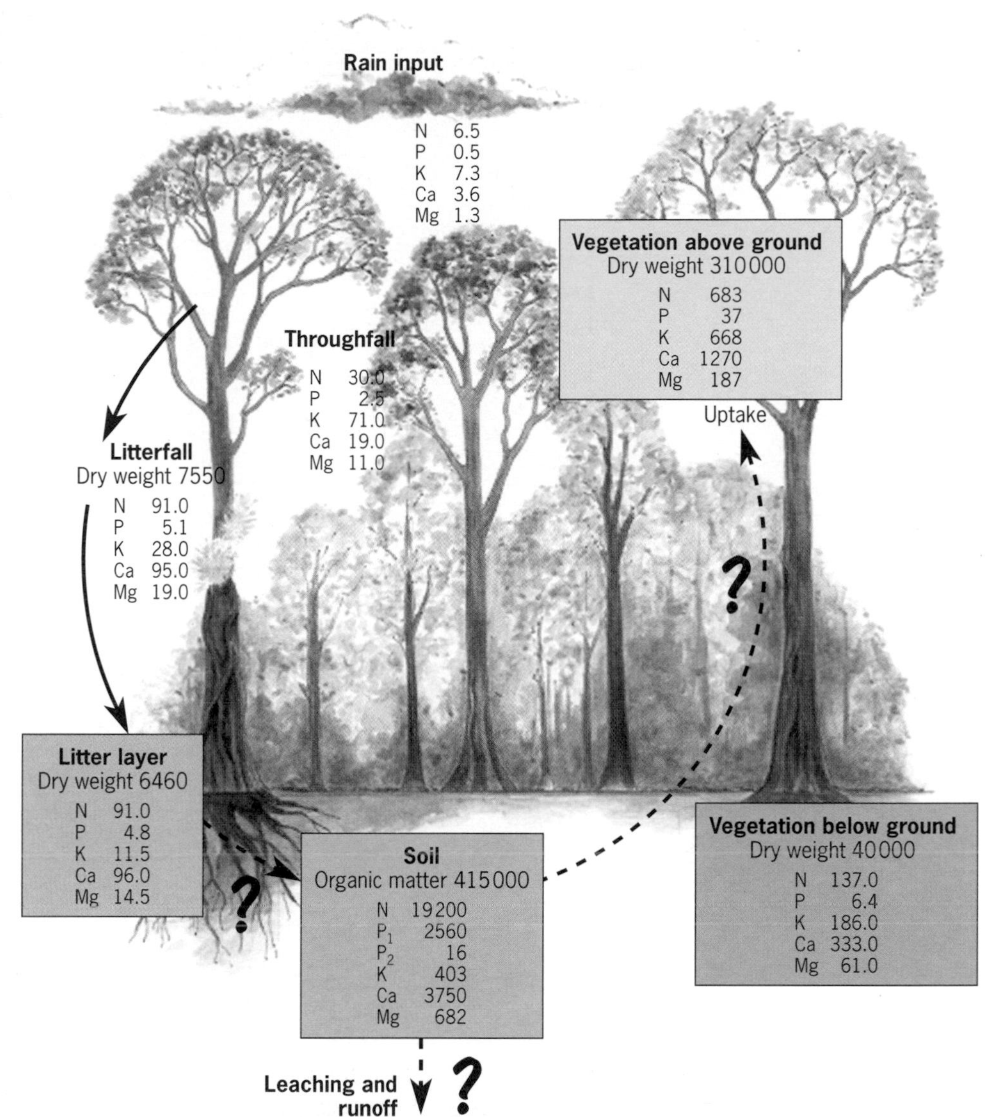

Figure 2.20 Nutrient cycling in a rainforest of New Guinea (after Edwards, 1982)

?

18 Using Figure 2.20:

a Draw proportional rectangles to represent biomass figures in:

- vegetation above ground
- litter fall
- vegetation below ground
- the litter layer
- the soil.

b Now, using proportional arrows, show the cycling of nitrogen and carbon.

c Add together the totals for **biotic** stores of nutrients.

d Add together the totals for **abiotic** stores.

e Where are most nutrients stored?

f How does this relate to your knowledge of rainforest soils and their structure, humus and nutrient content?

19 Essay: Climate determines the characteristics of the rainforest ecosystem. Discuss.

Summary

- There is a strong positive correlation between the net primary productivity of biomes and their biomass. These appear to be related to latitude and solar input.
- Tropical rainforests are climatic climax communities and have evolved due to high inputs of solar radiation and precipitation. Solar energy combined with moisture and nutrient availability provides high net primary productivity, a large biomass and a species-rich community.
- Differential availability of solar radiation determines the structure of the rainforest. This in turn affects the distribution of the fauna.
- Half of the world's gene pool is thought to exist in tropical rainforests due to the extremely long period of stability and continuity in some tropical rainforests. High immigration rates, opportunities for co-existence and a higher rate of mutations and evolution of new species add to this gene pool.
- Most nutrients are stored in the rainforest biomass; rainforest soils are weathered rapidly and contain few nutrients. The shallow roots of rainforest trees reflect the rapid cycling of nutrients, which takes place within the biomass and on the forest floor.

3 Tropical rainforests: management

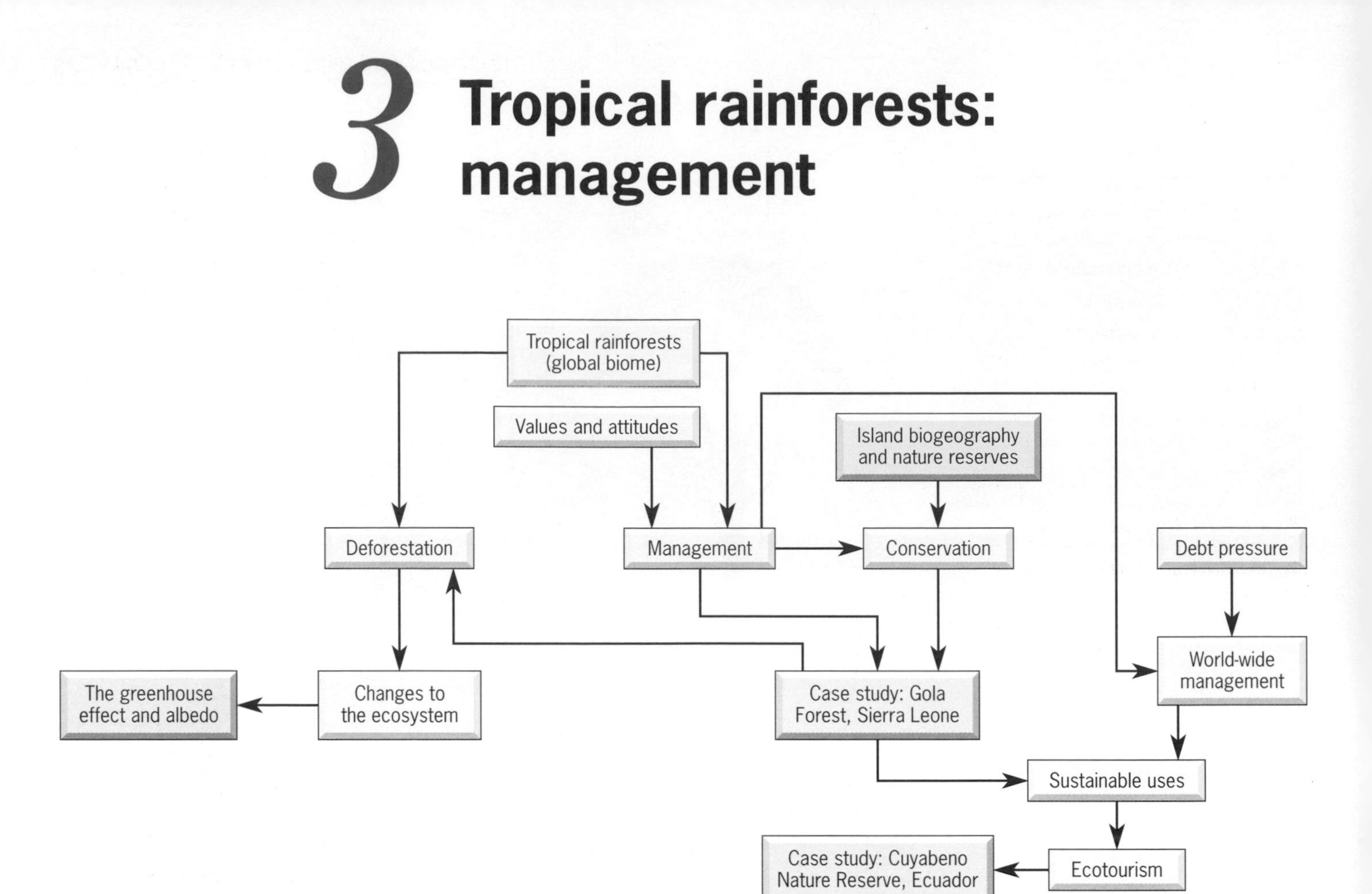

3.1 Deforestation

There are many short-term benefits to clearing tropical rainforests (Fig. 3.1). However, if **deforestation** continues at current rates, then all tropical forests will disappear in less than 200 years. In 1993, 80 per cent of timber came from Asia and Pacific countries and was sold to economically developed countries.

Figure 3.1 Main reasons for clearing rainforest areas

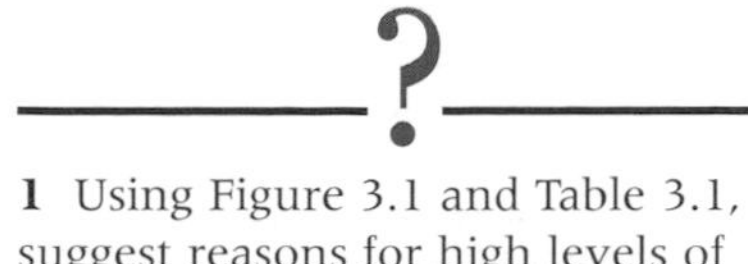

1 Using Figure 3.1 and Table 3.1, suggest reasons for high levels of deforestation in the countries listed.

2a Write down your own attitudes to the clearing of tropical rainforests.

b Compare your attitudes with those in Figure 3.2. Use Figure 3.2. to help you identify your value positions. In which category would you put yourself?

c Suggest possible value positions for countries such as those in Table 3.1.

Table 3.1 Top ten countries with the largest area of tropical forest being destroyed, 1989–90

Country	Area of forest lost each year (km²)	Percentage of total forest area	GNP per person (US$)	Debt per person (US$)
Brazil	34 500	0.7	2 680	772
Indonesia	12 000	1.0	560	355
Bolivia	10 000	2.1	620	586
Mexico	9 000	1.9	2 490	1 096
Venezuela	8 500	1.5	2 560	1 722
Zaire	7 000	0.6	230	256
Peru	5 500	0.9	1 160	912
Myanmar (Burma)	5 000	1.6	No data	112
Sudan	4 500	1.0	No data	515
Malaysia	3 500	2.0	2 340	1 089

GNP per person for the UK in 1990 was US $ 16 070.

ENVIRONMENTALISM

ECOCENTRISM ◄——► TECHNOCENTRISM

Deep ecologists

1 Intrinsic importance of nature for the humanity of man

2 Ecological (and other natural) laws dictate human morality

3 Biorights – the right of endangered species or unique landscapes to remain unmolested

Self-reliance soft technologists

1 Emphasis on smallness of scale and hence community identity in settlement, work and leisure

2 Integration of concepts of work and leisure through a process of personal and communal improvement

3 Importance of participation in community affairs, and of guarantees of the rights of minority interests. Participation seen both as a continuing education and a political function

4 Lack of faith in modern large-scale technology and its associated demands on élitist expertise, central state authority and inherently anti-democratic institutions

5 Implication that materialism for its own sake is wrong and that economic growth can be geared to providing for the basic needs of those below subsistence levels

Environmental managers

1 Belief that economic growth and resource exploitation can continue assuming:
a suitable economic adjustments to taxes, fees, etc.
b improvements in the legal rights to a minimum level of environmental quality
c compensation arrangements satisfactory to those who experience adverse environmental and/or social effects

2 Acceptance of new project appraisal techniques and decision review arrangements to allow for wider discussion or genuine search for consensus among representative groups of interested parties

Cornucopians

1 Belief that man can always find a way out of any difficulties, either political, scientific or technological

2 Acceptance that pro-growth goals define the rationality of project appraisal and policy formulation

3 Optimism about the ability of man to improve the lot of the world's people

4 Faith that scientific and technological expertise provides the basic foundation for advice on matters pertaining to economic growth, public health and safety

5 Suspicion of attempts to widen basis for participation and lengthy discussion in project appraisal and policy review

6 Belief that all impediments can be overcome given a will, ingenuity and sufficient resources arising out of growth

Figure 3.2 The range of ideologies concerning humans and the environment (after O'Riordan, 1981)

3.2 Changes to the ecosystem

Clearing large areas of tropical rainforest produces changes in the physical environment: the soil, **biosphere**, **hydrosphere** and atmosphere (Fig. 3.3). The rainforest is a **climatic climax community**, and therefore any major changes in the **inputs** will have an impact on the whole **ecosystem**. **Negative feedback** mechanisms will act to maintain a **dynamic equilibrium**, but major changes will trigger **positive feedback** mechanisms that will cause the whole system to alter or break down.

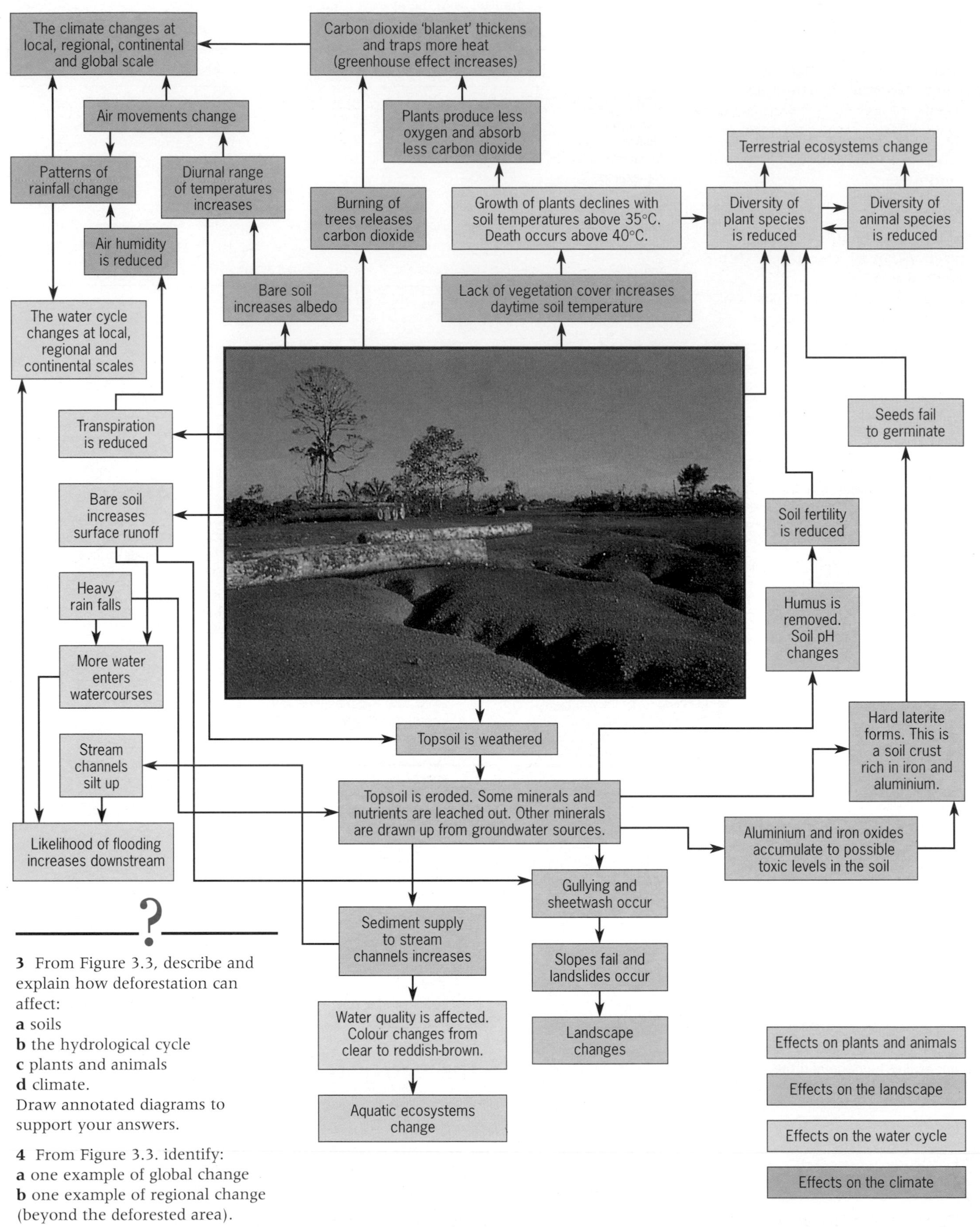

The climate changes at local, regional, continental and global scale
Carbon dioxide 'blanket' thickens and traps more heat (greenhouse effect increases)
Air movements change
Plants produce less oxygen and absorb less carbon dioxide
Terrestrial ecosystems change
Patterns of rainfall change
Diurnal range of temperatures increases
Burning of trees releases carbon dioxide
Growth of plants declines with soil temperatures above 35°C. Death occurs above 40°C.
Diversity of plant species is reduced
Diversity of animal species is reduced
Air humidity is reduced
Bare soil increases albedo
Lack of vegetation cover increases daytime soil temperature
The water cycle changes at local, regional and continental scales
Seeds fail to germinate
Transpiration is reduced
Bare soil increases surface runoff
Soil fertility is reduced
Heavy rain falls
More water enters watercourses
Humus is removed. Soil pH changes
Topsoil is weathered
Hard laterite forms. This is a soil crust rich in iron and aluminium.
Stream channels silt up
Topsoil is eroded. Some minerals and nutrients are leached out. Other minerals are drawn up from groundwater sources.
Likelihood of flooding increases downstream
Aluminium and iron oxides accumulate to possible toxic levels in the soil
Gullying and sheetwash occur
Sediment supply to stream channels increases
Slopes fail and landslides occur
Water quality is affected. Colour changes from clear to reddish-brown.
Landscape changes
Aquatic ecosystems change
Effects on plants and animals
Effects on the landscape
Effects on the water cycle
Effects on the climate

?

3 From Figure 3.3, describe and explain how deforestation can affect:
a soils
b the hydrological cycle
c plants and animals
d climate.
Draw annotated diagrams to support your answers.

4 From Figure 3.3. identify:
a one example of global change
b one example of regional change (beyond the deforested area).

Figure 3.3 Environmental impacts of large-scale deforestation

Table 3.2 Percentage contribution of greenhouse gases to each unit of extra human-induced warming (*Source:* D. Toke, 1988)

Carbon dioxide from fossil fuel burning	40
Carbon dioxide from deforestation	15
Chlorofluorocarbons	20
Methane	15–20
Nitrous oxide	5–10

The greenhouse effect and albedo

Many predictions have been made based on the theory of the **greenhouse effect** and global warming of the atmosphere. Rainforest deforestation does contribute to these phenomena (Fig. 3.3 and Table 3.2).

Some researchers believe that there will be few temperature changes at the Equator, but major increases will occur at the North and South Poles (Fig. 3.4). These predictions depend on various assumptions. For instance, predictions about the size of these climatic changes depend on the feedback mechanisms which some researchers think are part of the greenhouse effect.

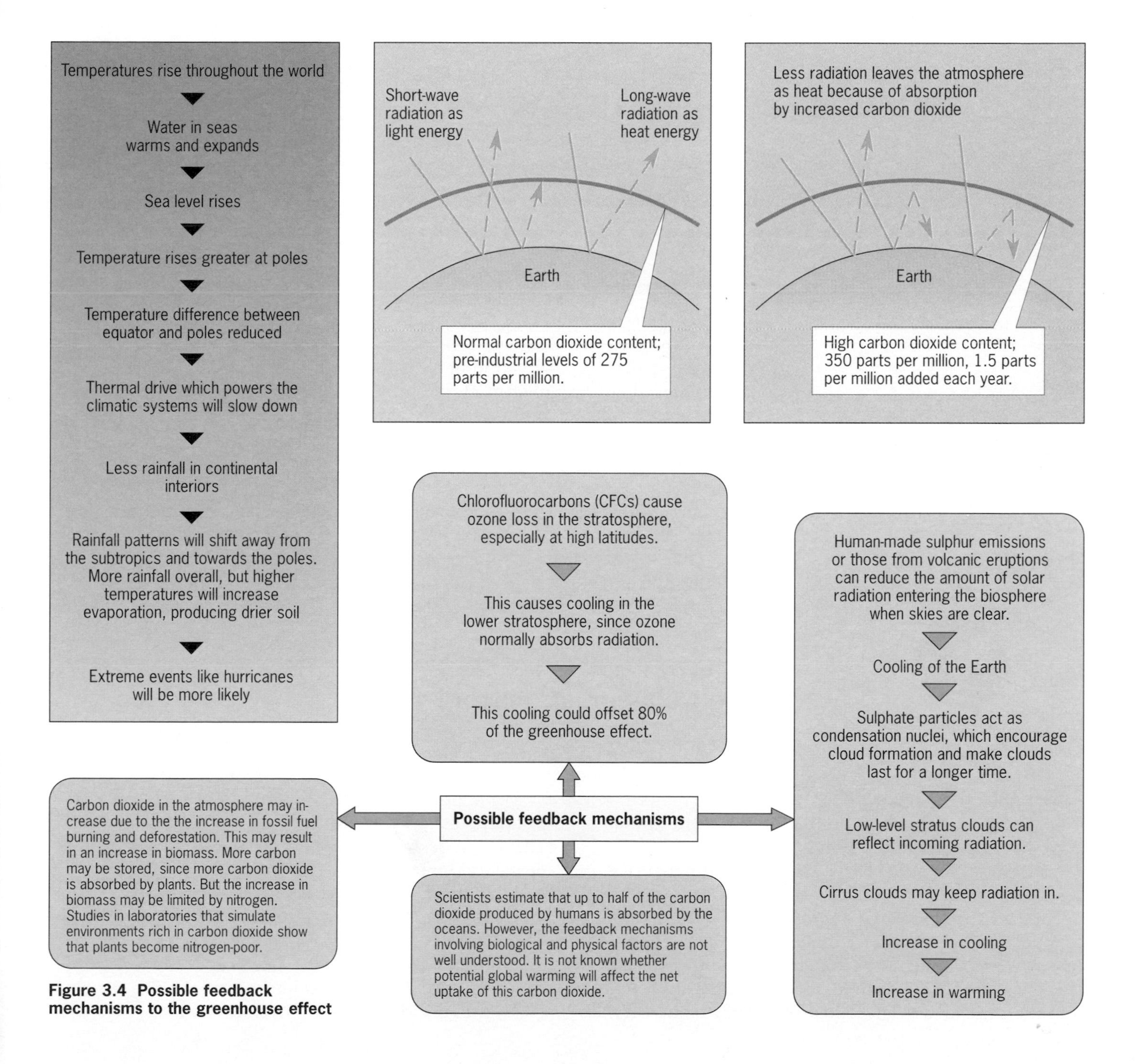

Figure 3.4 Possible feedback mechanisms to the greenhouse effect

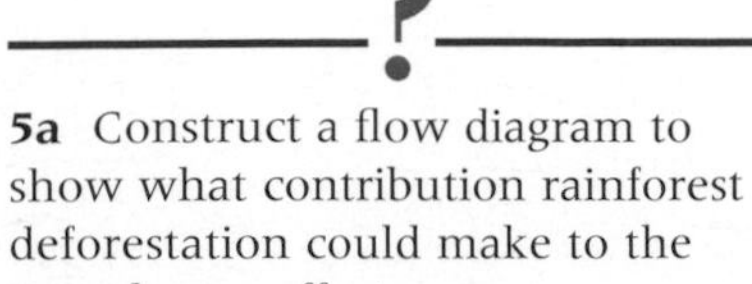

5a Construct a flow diagram to show what contribution rainforest deforestation could make to the greenhouse effect.
b How might the greenhouse effect influence other ecosystems?
c What types of feedback mechanism are shown in Figure 3.4?
d Are the albedo changes shown in Figure 3.5 a form of negative feedback? Give reasons for your answer and explain how albedo changes influence the greenhouse effect.

Albedo

Another variable which may cause change is the amount of energy reflected by a surface. This is called its **albedo**. Different land areas have different albedo levels. Lighter surfaces reflect more **radiation** than darker ones. Rainforests generally reflect 9 per cent of incoming radiation, whereas desert reflects 37 per cent. A change from rainforest to bare soil effectively means a change from a darker surface to a lighter one. The events shown in Figure 3.5 may take place.

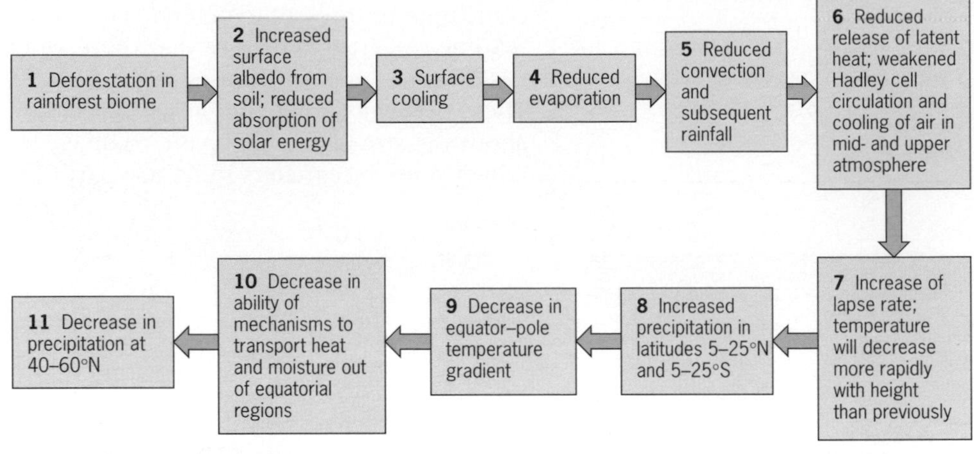

Figure 3.5 The albedo effect

3.3 Managing rainforests

Activities aimed at improving woodland or forest for a particular use are called **management**. They can include tree planting, tree thinning, grazing by animals and the conservation of wildlife. The ultimate aim of forest management may be timber products or leisure activities. This will depend on the owners of the land and their needs – which can include profit, recreation and nature conservation. **Sustainable** management occurs when the level of exploitation of resources is not greater than the ability of the ecosystem to replace itself.

Conservation is the protection of something which people perceive as being of value. This may be a building, an entire ecosystem, a particular species or a way of life. Conservation requires people to take an active part by changing certain inputs to get the desired **outputs**, while preservation means keeping something as it is and not changing the inputs.

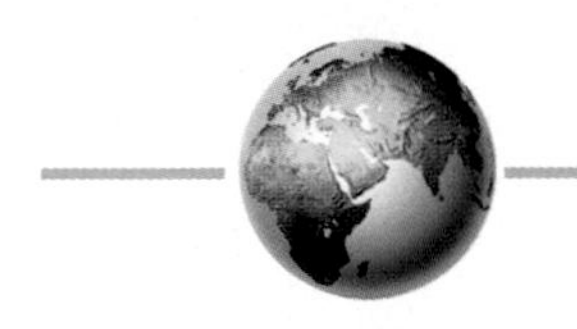

The Gola Forest, Sierra Leone

Deforestation

Agriculture, logging, hunting, mining

The climatic factors are such that Sierra Leone could have over one half of its land surface covered by rainforest but an aerial survey in 1976 estimated that just 5 per cent of the country was closed-canopy forest. Secondary forest covered approximately 3.5 per cent and bush fallow covered 55.2 per cent (Fig. 3.6). Bush-fallow is areas that are burnt, cleared and dug during the dry season. Crops are planted for one or two years and then the area is left fallow for eight to ten years. It is likely this has been further reduced since that date. This forest destruction appears to have had some ecological impact already.

Gola Forest Reserves (East, West and North) (Fig. 3.11) were established in the 1920s and 1930s to supply timber for home and abroad (Fig. 3.8). They are administered by the Forestry Division of the Ministry of Agriculture, Fisheries and Forestry (MAFF). Other forest uses are shown in Figures 3.9–3.11.

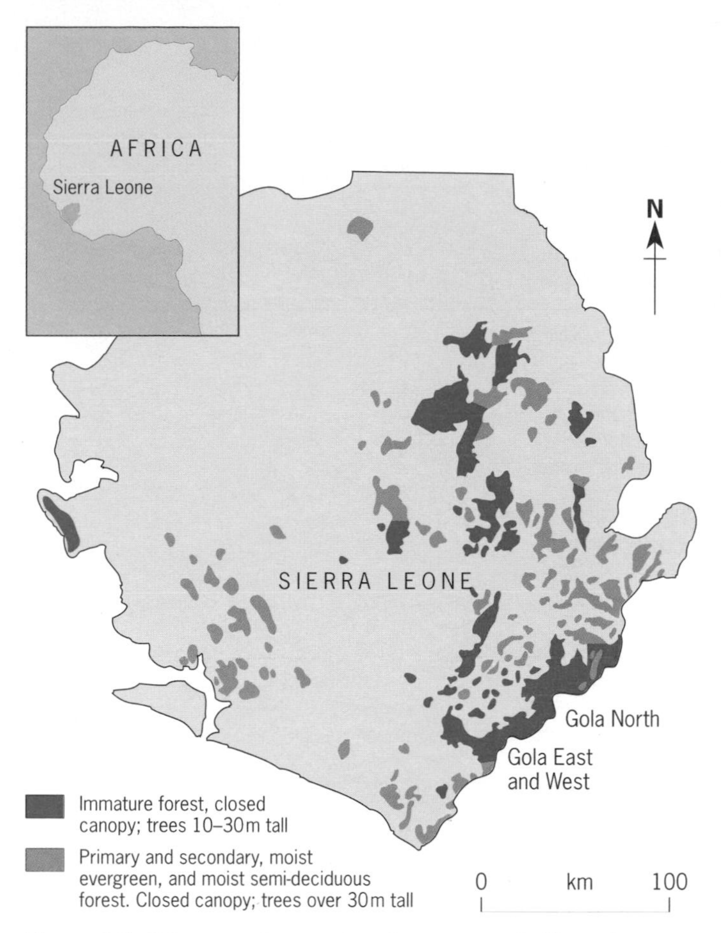

Figure 3.6 Primary and secondary forest cover in Sierra Leone (after Allport et al., 1989)

Figure 3.7 Charcoal production

Figure 3.8 Logging

Figure 3.9 Settlement construction

Figure 3.10 Farming

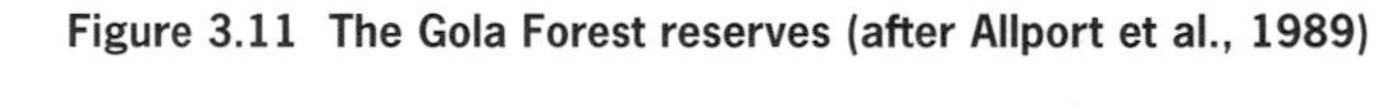

6 Using Figure 3.11 calculate the proportion of the Gola Forest which is used for logging.

Figure 3.11 The Gola Forest reserves (after Allport et al., 1989)

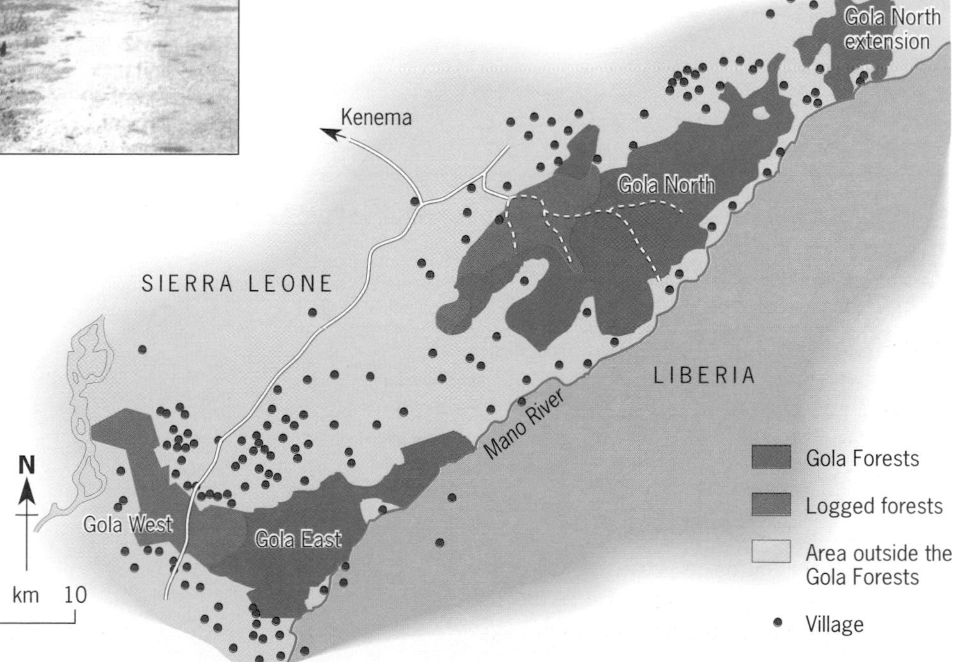

Table 3.3 Views of Western researchers on Gola Forest use

1 Agriculture
Forest has been cleared for rice, mixed crops or cash crops. Rice is the favoured crop, with swamp rice being grown in wetter areas, although hill rice is the preferred variety. Areas of forest covering tens of hectares are burnt and cleared during the dry season. Rice is planted for one to two years alongside root crops such as cassava and yam. The land is then left fallow for eight to ten years (Fig. 3.13).

2 Logging
Up until the start of the war in Sierra Leone, which began in 1991, timber extraction took place within the forest and there was limited management to improve the stand of trees. Damage to the forest is variable. Often more trees are knocked down by trees being felled. Up to 50 per cent of the crown area of the forest can be destroyed to remove 10 per cent of the timber. According to Schminke (1988), if the present scale of deforestation continues, combined with poor logging practices, the forest will only last for 22 more years. The soil can also be compacted by machinery, causing a decrease in its air- and water-holding properties, which reduces the germination of seeds.

Logging increases access to the area and opens it up to hunting, mining and shifting cultivation. These activities create further pressures within the forest.

3 Hunting
Hunting has certainly had a major impact on forest mammals. Many hunters come from outside local communities and sell bush meat at regional markets or export it to nearby Liberia. There is no control of such hunting activities. Shotguns, lures, traps and dazzling (using bright lights to attract and confuse animals) are all used to catch game. Commercial gangs have contributed to the main cause of the reduction of mammal populations in the forest. Only 60 elephants are left in Sierra Leone. Chimpanzees have also been captured for use in medical science in the West. Loss of these animals may lead to an imbalance in forest food webs.

4 Mining
Miners enter the Gola Forest in search of diamonds. They can also cause damage to the forest by adding sediment to streams, altering stream flow and contributing to the area's increase in human population.

?

7a Using Table 3.3, list the causes of forest loss observed by Western researchers.
b Compare your list with Section 3 of Table 3.4. Comment on any similarities and differences.

8 There are differences in perception even among local people living near the rainforest. Table 3.4 shows the results of interviews with rural forest-edge inhabitants (Fig. 3.12) and urban dwellers who live near the forest. Draw suitable graphs to represent the differences between rural and urban views in Sections 2 and 4 of Table 3.4. Comment on your graphs.

9 Suggest possible reasons for the different perceptions of rainforest use.

Table 3.4 Responses to a questionnaire on the Gola rainforest by rural and urban people in Sierra Leone (N = sample population) (*Source:* Amara Lukulay, 1989)

1 Are there noticeable changes in the soils?	Rural N = 100	Urban N = 93
More crop failures	30	27
More dry soils	10	5
More sand soils	8	5
Less manure	12	0
More stones	8	3
More soils eroded	4	4
More red soils	3	2
No changes	22	13
Cannot tell	99	37

2 What are the causes of soil changes?	Rural N = 83	Urban N = 75
Much farming with forest destruction	28	22
God	23	7
Limited fallow periods	13	0
Less rainfall	6	7
Others	2	3
Cannot tell	5	6

3 What are the causes of forest loss?	Rural N = 89	Urban N = 66
Farming	88	50
Logging	1	15
Settlement construction	0	7
Firewood harvesting	0	6
Mining	0	5
Fire setting	0	2
Charcoal production	0	1

4 What are the effects of forest loss?	Rural N = 83	Urban N = 66
Limited forests for farming	22	17
Timber shortages	9	25
Relief from wild animals	19	8
Water shortages	10	8
More areas opened for farming	13	0
Loss of needed animals	3	4
Loss of other forest products	2	1
Rainfall disturbed	0	4
No effect	19	15
Others	6	5

Figure 3.12 People in forest edge settlement

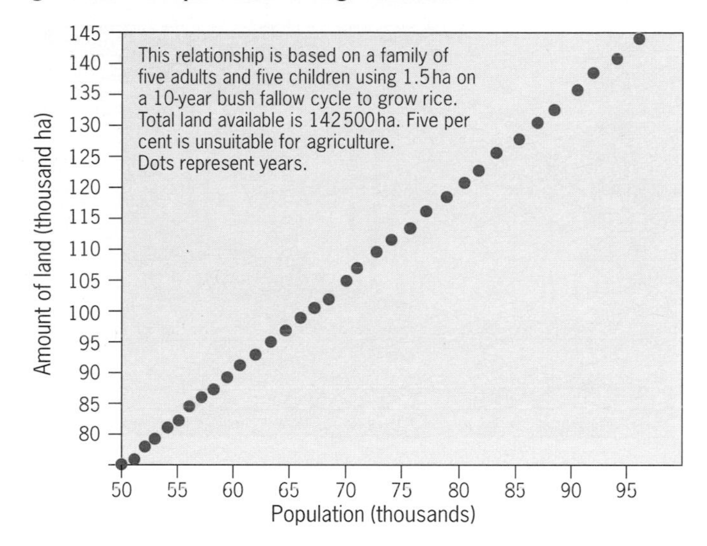

Figure 3.14 The relationship between land availability and population growth in the Gola Forest, Sierra Leone (after Davies and Richards, 1991

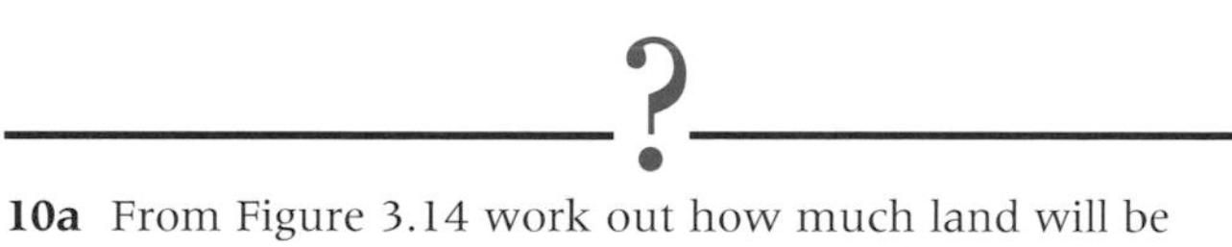

10a From Figure 3.14 work out how much land will be used by 70 000 people.
b How many years will it take to reach this level of use?
c How long will it take to use up the 145 000 hectares of available land?
d Suggest positive or negative feedbacks which might prevent this from happening.

11a Use Table 3.4 and Figures 3.15–3.19 to create a spider diagram showing where the local population gain most of their:

- medicine
- equipment
- bush meat
- income.

b From which of the habitats listed in Table 3.5 do the local people appear to get most of their produce?
c Which other habitat not listed in Table 3.5 would be important to the local people? Give reasons for your view.

Population pressure and habitats

On average there are 32 people per square kilometre in rural districts around the Gola Forest.

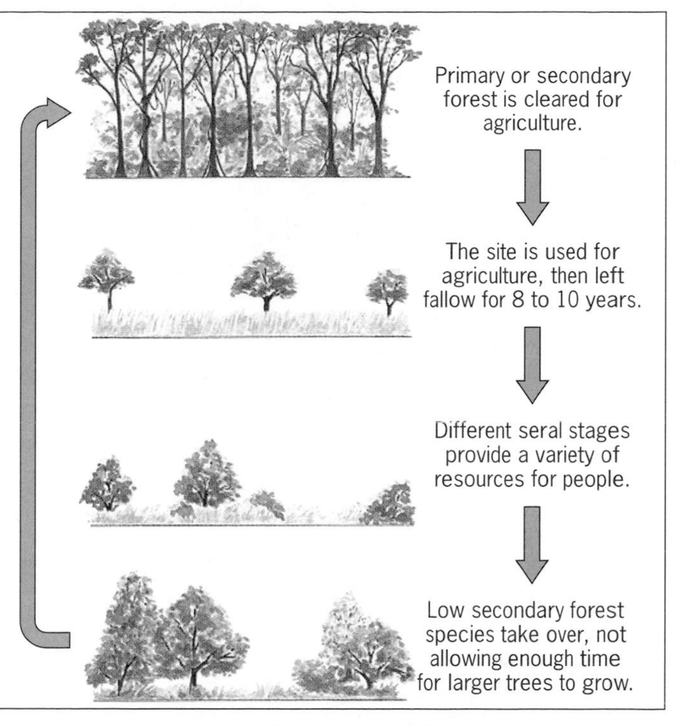

Figure 3.13 Forest succession in a bush-fallow system

Figure 3.15 Primary forest

Figure 3.16 Logged forest

Conservation

Conservationists have studied the Gola Forest in detail to estimate its conservation value. They have defined habitats and recorded the use of each for wildlife.

Figure 3.17 Fallow area of bush-fallow

Figure 3.18 Cocoa/coffee plantation

Figure 3.19 Farm in bush-fallow

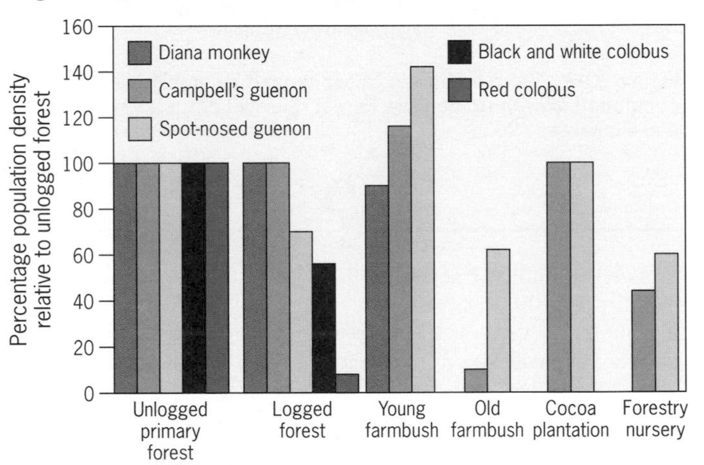

Figure 3.20 The relationship between the amount of unlogged forest and population density of some primate species (after Davies and Richards, 1991)

Table 3.5 Percentage resource use in different habitat areas (*Source:* C. Richards, P. Davies, 1991)

Resource	Forest	Forest edge	Fallow land	Farm land	Swamp	Plantation
Medicine	31	–	48	17	–	2
Household Equipment	10	–	49	8	29	3
Bush meat[1]	26	34	40	–	–	–
Income[2]	3.5	–	–	87	–	–

[1] Although the local people do eat bush meat, this is not the major part of their diet. Vegetable matter makes up 51.9 per cent, and 48.1 per cent is meat, of which 60.2 per cent is meat from fish and reptiles

[2] Miscellaneous income is 9.5 per cent.

Table 3.6 Bird species found in different habitas of the Gola Forest (*Source:* ICBP/UEA, 1989)

Habitat	Farmbush	Cocoa/ coffee plantation	Forest plantation	Logged forest	Primary forest
Total number of bird species	176	126	131	159	169

12 Using Table 3.6 and Figure 3.20, work out two areas where conservationists should concentrate their efforts. Give reasons for your choices.

In addition to developing sustainable logging systems, it is also necessary to develop and improve the sustainable yields of existing bush-fallow agriculture, since 70 per cent of Sierra Leone depends on this form of farming. If the productivity of the bush-fallow was improved, then further destruction of primary forest areas may be prevented in the future.

Future management plans for the Gola Forest area are addressed under four broad headings:

1 Conservation research and reserve design
2 Forest management
3 Rural development
4 Conservation education

1 Conservation research

Additional research is needed to understand the **ecology** of the forest area and to identify the most vulnerable parts. Existing research suggests that Strict Forest Reserves in **primary forest** are essential to the survival of important **species**.

2 Forest management: logged areas

Within Sierra Leone, the Forestry Division has the responsibility for two types of forest area: Forest Reserves and Protected Forests. Protected Forests are small strips of forest, usually along roadsides, approximately 200m wide and are often planted with a single species.

Forest Reserves are managed for timber use, water catchment or habitat protection. Local people have rights of access to collect minor forest products for non-commercial use. Hunting, fishing and firewood collection are therefore permitted, although some species like the *Colobus* monkey are protected from hunting throughout Sierra Leone.

In 1993, the only management plan for a Forest Reserve in Sierra Leone was the one completed for the Gola. However, in 1988 a Forestry Act had begun to encourage the development of such documents.

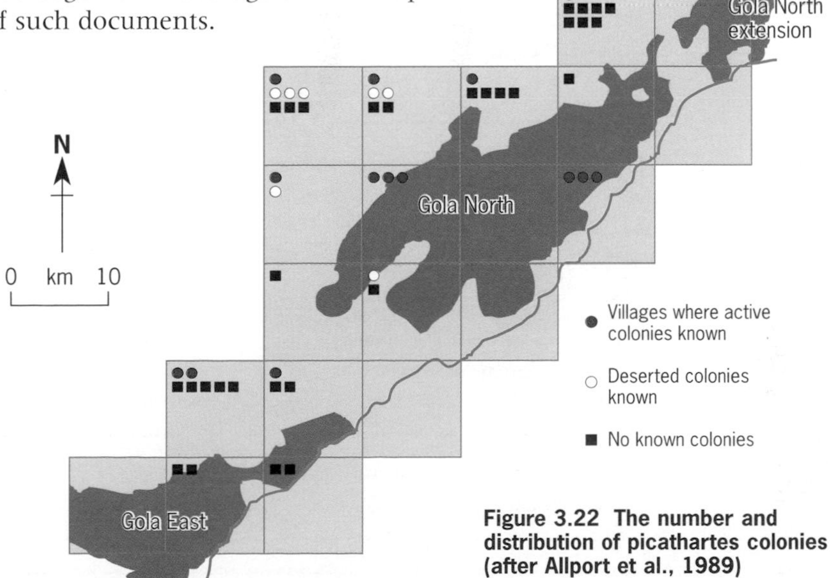

Figure 3.22 The number and distribution of picathartes colonies (after Allport et al., 1989)

Figure 3.21 Hornbill

Non-governmental organisations who are involved in the management, such as the Royal Society for the Protection of Birds and Birdlife International, also recommend that:

a Research must be carried out to determine the rate of extraction permissible to ensure a sustainable yield and the effects of logging on the birds.

b Large fruiting trees should be left in areas of logged forest to ensure food for fruit-eating species such as hornbills (Fig. 3.21).

c Riverside forest should not be logged. This will preserve habitats for water-dependent species. It would also provide corridors along which species could move to recolonise logged forest. Additionally, it will help to ensure maintenance of water quality, by reducing sediment inputs.

d Areas containing the white-necked picathartes, a vulnerable species (Figs. 3.22 and 3.23), should be left unlogged.

e Buffer zones should be created between logged areas and primary forest reserves (see Figure 3.8). There should be no logging in these areas. The best wildlife areas should be left as Strict Nature Reserves.

Figure 3.23 White-necked picathartes

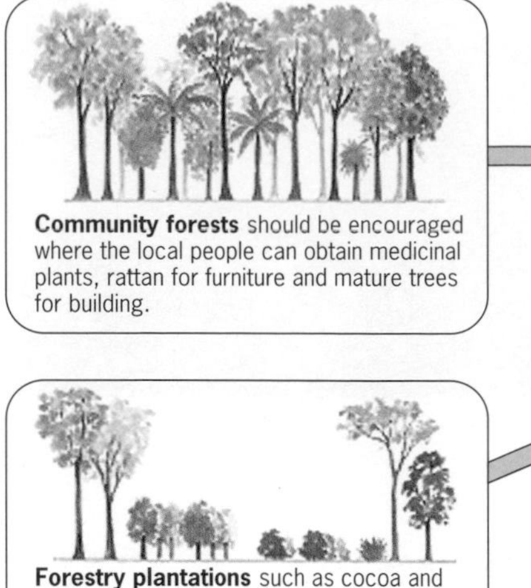

Community forests should be encouraged where the local people can obtain medicinal plants, rattan for furniture and mature trees for building.

Forestry plantations such as cocoa and coffee plantations should be encouraged as an additional buffer zone around the logged forest areas. Bush meat can also be obtained from these areas.

A market for locally-abundant forest products such as rattans, fruit, oil seeds and medicinal plants could be expanded where sustainable development is possible.

The productivity of existing **bush fallow** areas could be increased by small-scale fish farming, continued hunting of 'pest' species or domestication of deer, cane rat and giant rats (which are already used as food).

Agro-forestry can be encouraged to increase the productivity of farm plots. This may reduce the length of the fallow cycle by using soil-enriching trees with nitrogen-fixing nodules in their roots. Timber crops can be grown with food and cash crops. Multi-storied cropping systems are already being used in cocoa and coffee plantations where trees provide shade for cash crops below. This method replicates the layered structure of the original forest.

Figure 3.24 Development schemes in rural communities in rainforest areas

3 Rural development

Measures to develop the sustainable extraction of forest products (Fig. 3.24) should be encouraged.

4 Conservation education

Local people could be involved in conservation at all levels (Fig. 3.25). The costs and benefits of changing their agricultural practices could be assessed. They could also find out why the changes they have observed in their environment are taking place.

People from other cultures who work in the forest areas could educate themselves and others about the values that Mende people place on the forest resource, since they differ from the values of others. Most of the Mende's products come from secondary woodland, not the primary forest. Their social life begins and ends at the edge of a clearing. The Mende people believe that spiritual energy is released as they change forest to farm and this gives them power.

Figure 3.25 Young people concerned about the Gola Forest

Regeneration of the bush following cultivation is a sign that they are blessed by their ancestors and have not angered them by felling forest.

The bush-fallow system is a key to the conservation of the forest, since this is the source of most of the Mende's daily needs. Clear links could be made between any rural development and the conservation of the forest.

It is difficult for the Mende to see themselves as guardians of the forest. Whereas outsiders are able to exploit or protect the forest, the Mende view is that the forest protects and provides for them. They can continue to rely on the forest and revere it in return. Hence, Mende peoples cannot 'Save the Gola Forest' but can 'Get away from behind Gola Forest' (which means stop living under its protection) to prevent forest destruction. The conservation of the area therefore involves managers taking different cultural views on board.

?

13 Devise a plan to manage the Gola Forest. Remember that you have a maximum of 20 years to save the rainforest.
a Using the information about reserve design (page 37) and Figs 3.6, 3.11 and 3.22, trace an outline sketch map of Gola Forest areas.
b Identify the interest groups involved.
c Add zones to your map to account for the requirements of these interest groups.
d Make a list of action points and place them in priority order, taking into account the needs of the inhabitants.

14 Use Figure 3.2 to comment on your plan. What attitudes and values does it represent?

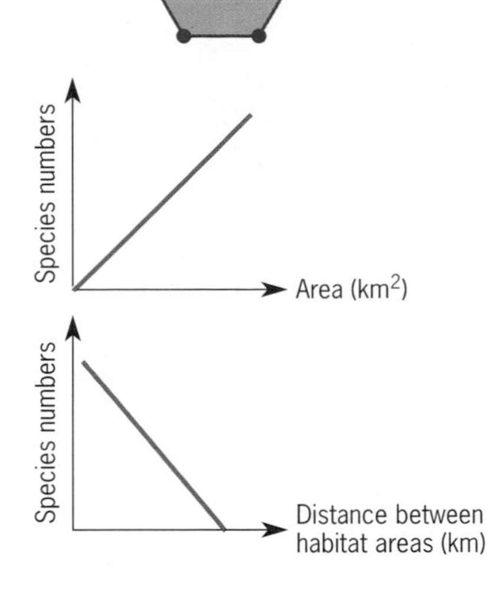

Figure 3.26 A generalised relationship between species numbers and area of habitat and distance from original habitat

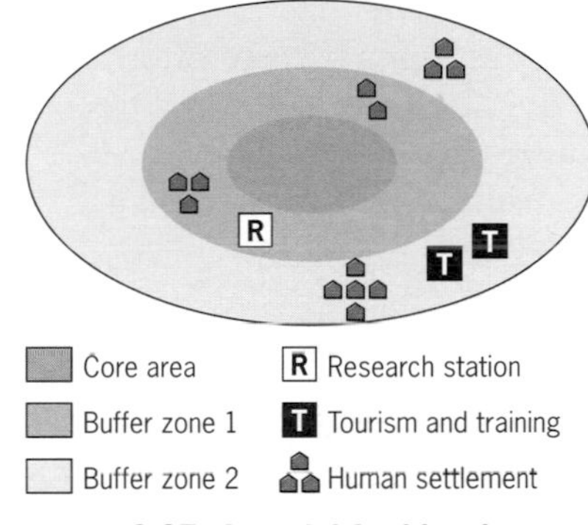

Core area
R Research station
Buffer zone 1
T Tourism and training
Buffer zone 2
Human settlement

Figure 3.27 A model for biosphere reserve design (after UNESCO, 1974)

Island biogeography and nature reserves

One way of protecting rainforests is to establish nature reserves within them. There are ways of making the design of these better for wildlife. The theory of island biogeography is useful in planning the shape and extent of reserves (Fig. 3.26).

The more remote or closed the 'island' (in this case an isolated piece of the rainforest), the lower the immigration rate. The smaller the island, the higher the extinction rate. For equilibrium to be maintained, there has to be a balance between immigration and extinction. If a fraction of a reserve is destroyed, there will be increased pressure on the rest of it to maintain a balance.

Several principles guide conservationists in creating reserves:

1 Larger reserves are better, because:
a A large reserve can hold more species at equilibrium and therefore has lower extinction rates.
b A large reserve is usually better for wildlife than several small ones whose areas add up to a large one.

2 Smaller reserves can be made more viable by carefully placing them near each other and by providing corridors. This will help species movement. Animals also have advantages of being able to cover more habitats, which encourages edge species and offers insurance from disease or fire.

3 A reserve should be approximately circular (Fig. 3.27) so that there is a maximum area-to-perimeter ratio. This avoids peninsular effects where there is a large edge in comparison to the land enclosed. In a circle, there is a long distance to the centre, which therefore prevents disturbance.

Nature reserves are not the only solution to the problems of rainforests, since they will only cover very small areas of the rainforest **biome**. Managing the rainforest and its human community as a whole is probably the most successful option.

3.4 Worldwide management

Tropical rainforests, are not only a local or regional issue within individual countries. They were high on the agenda of the United Nations Conference on Environment and Development in Rio de Janeiro, Brazil, in 1992 (the Earth Summit).

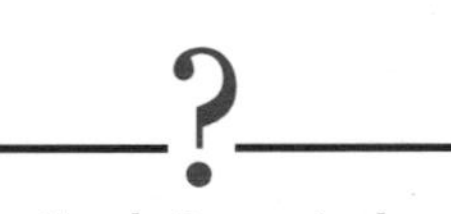

15 At the Earth Summit the observations shown in Figure 3.28 were made about the progress on tropical rainforest discussions. Explain the attitudes of the countries involved. (Remember that economically developed countries were not willing to compensate developing countries adequately for their loss of income.)

> The negotiations on principles for the conversion and rational exploitation of forests were among the hardest of the Summit and produced a feeble document. The developed countries had wanted these non-legally-binding principles to pave the way for a legally-binding forest treaty, but did not achieve that. Developing countries led by India and Malaysia would not accept anything suggesting interference in the way they ran their forests. The outcome is a text saying little more than that forests are important and should be managed sustainably. Nicholas Schon of the *Independent*

> The developing countries themselves seemed split between those who wished to be paid by the North to keep their forests as global carbon sinks (those with more forest than is exploitable in the short term) and those who feared this might interfere with freedom to fully exploit their forests in the short term. Graham Wynn, RSPB

Figure 3.28 Some reactions to the 1992 Earth Summit held in Rio de Janeiro, Brazil

Debt swapping

Economically developed countries have been willing to place money overseas to help rainforest projects. One recent means of achieving this has been through the development of 'debt-for-nature swaps'. By 1993, twenty such swaps had occurred, where rainforest was conserved in return for having debts written off. Non-governmental conservation groups have been able to buy heavily discounted (at lower interest rates) developing world debt, offering to write it off if the country concerned invests in conservation programmes. For example, in 1988 the National Westminster Bank and World Wide Fund for Nature in the United States of America (US) and the US Agency for International Development exchanged a US$3 million debt for nature in the rainforests of Madagascar. Such debt-swapping only accounts for $100 million worth of debts from a total of US$1.3 trillion that developing world countries owe. This represents one ten-thousandth of the total debt.

Sustainability

Western consumers are now being asked by conservationists to consider purchasing only sustainably produced **hardwoods** to safeguard rainforests. However, the International Tropical Timber Organisation (ITTO), set up in 1983 to unite timber exporters and conservationists, estimates that, out of 828 million hectares of productive tropical forest, only about 1 million hectares are managed sustainably.

Although many schemes have been set up, theory and practice are often very different. According to P. F. Burgess (1989), the points in Figure 3.29 are 'golden rules' for effective forest management. However, points 4–9 are inadequately carried out throughout the world in tropical, temperate and **boreal** forests, thereby affecting the true sustainability of the forest.

?

16 Study Figure 3.29. Construct a table to show the advantages to the forest ecosystem of doing the following management tasks: 1, 3, 6 and 8.

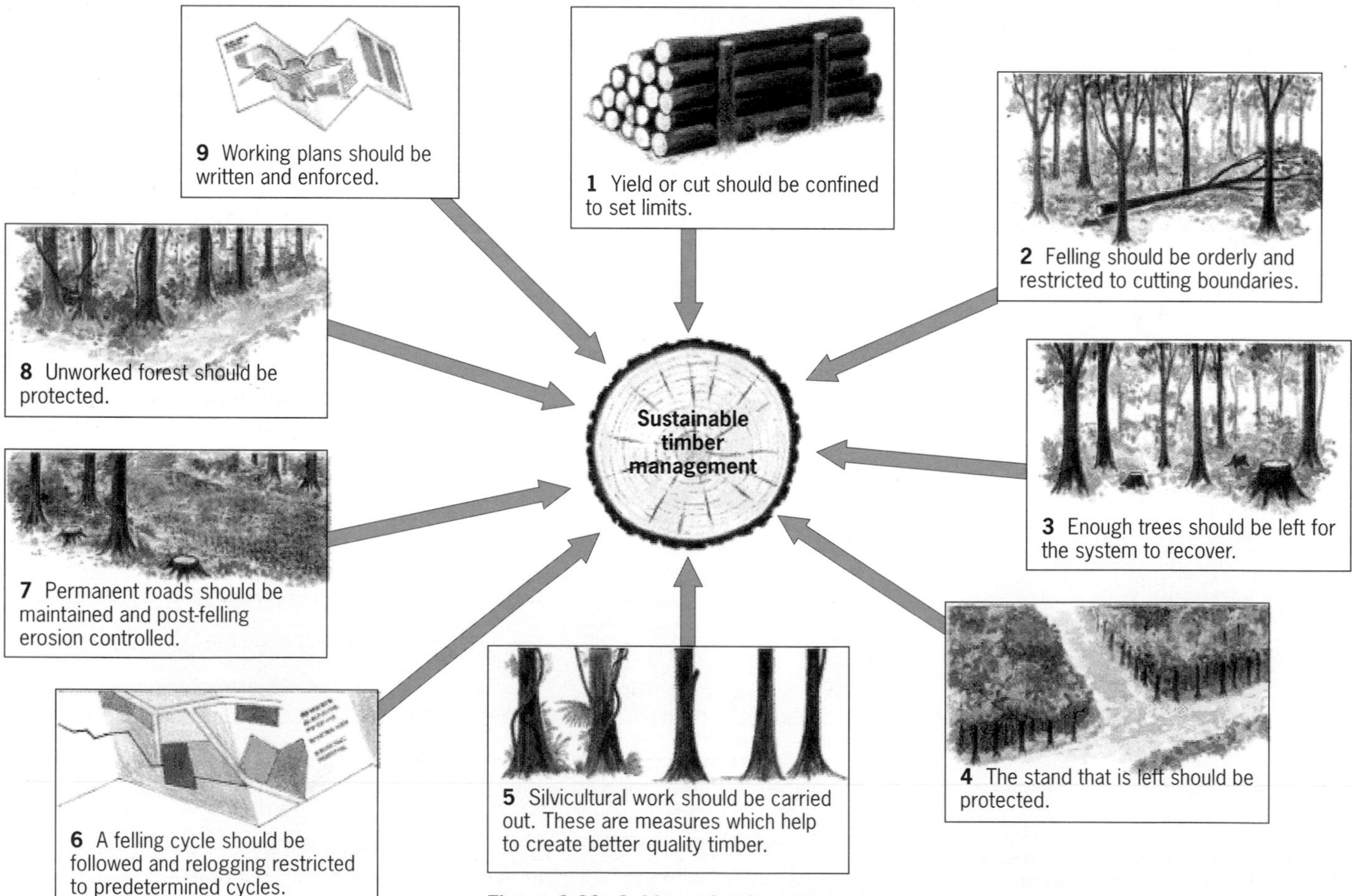

9 Working plans should be written and enforced.
1 Yield or cut should be confined to set limits.
2 Felling should be orderly and restricted to cutting boundaries.
8 Unworked forest should be protected.
Sustainable timber management
3 Enough trees should be left for the system to recover.
7 Permanent roads should be maintained and post-felling erosion controlled.
4 The stand that is left should be protected.
5 Silvicultural work should be carried out. These are measures which help to create better quality timber.
6 A felling cycle should be followed and relogging restricted to predetermined cycles.

Figure 3.29 Golden rules for sustainable timber management

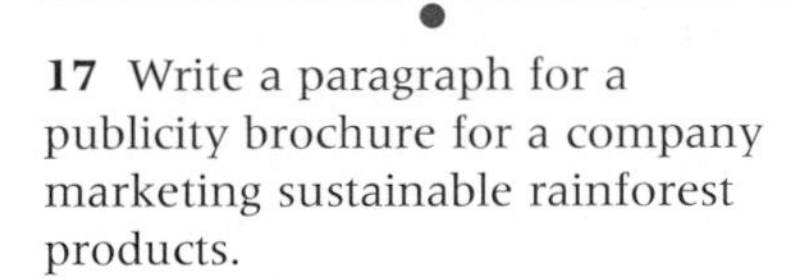

17 Write a paragraph for a publicity brochure for a company marketing sustainable rainforest products.

For years the people of the Hunstein Range in Papua New Guinea have yearned for development, access to the outside world, cash to buy anti-malarial medicines, to pay school fees and to buy household items. The government offered them all these if they would sign their ancestral forests over to a logging company.

The people, torn between their traditions and the call of the market, came together in October of last year to discuss their options. 'If there were only some way that we could earn some money and keep our forests,' they said, 'we would never sign the logging permit.'

They have found an alternative. They now sell carvings and bilum bags to generate an income without destroying their forest home. The bags are made by women from the bark of the tulip tree and are so well made that they sometimes outlive the person who made them.

Figure 3.30 Magazine article on alternative income to forestry for indigenous peoples of Papua New Guinea (*Source: Green Magazine*, April 1992)

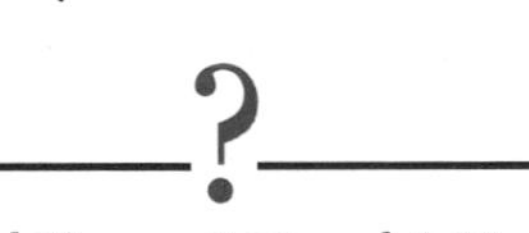

18 Read Figures 3.30 and 3.31.
a For Figure 3.30, draw a simple diagram to explain the pressures on and responses of the local people. Highlight the role of money.
b Do the same for Figure 3.31.
c Choose one of the extracts and describe the attitudes and values of the people involved. Use Figure 3.2 to help.
d Comment on the roles played by international organisations in Figure 3.31.

Sustained yields

There are some instances where sustained yields are promoted. Poore (1989), suggests that in Thailand the Mae Poong Forest is well managed. Additionally, in Malaysia and the Philippines many forestry sites have sustainable yields.

However, strictly speaking, sustained yield is not always sustainable forestry. *Plantations can produce very effective sustainable yields but will not maintain existing wildlife.* In primary tropical rainforests, conservationists are concerned that the ecosystem be maintained as far as possible. In Queensland, Australia, over 160 000 ha (an area of forest scheduled for logging) was, until recently, under a management system for the sustainable production of timber. This was based on research results and was subject to strict environmental guidelines. It ensured that **disturbance** to forest plants and animals was kept to a minimum. This scheme no longer runs, since logging was considered to be an inappropriate activity when the area was nominated as a potential World Heritage site.

Sustainability of rainforests is not, however, limited to logging products. Many other ways exist to use rainforest products for commercial gain (Table 3.7).

Table 3.7 Sustainable use of the rainforest. Not all of these uses are commercially viable at present

Craft items
Craft items are made from sustainable bark and wood (e.g. bilum bags made from tulip tree bark, Papua New Guinea).

Rubber
Harvesting from individual trees.

Brazil nuts
Brazil nuts can not be grown in plantations since pollination relies on rainforest bees.

Fruit trees
A 4.5 m wide fence of fruit trees planted to stake claims on ancestral land stopped invasion by settlers (e.g. Awa Indians, Ecuador).

Butterfly farming
Farming concentrates on largest and most valuable creatures. Needs an intact habitat, therefore favours conservation. Difficult to over-harvest. Small plots of food plants can be grown to farm the butterflies (e.g. WWF community butterfly farm, New Guinea Island, S E Asia). (There are restrictions on exporting rare creatures. Not all countries have signed the agreement.)

Ecotourism
Learning about the rainforest and undertaking conservation work.

Brazil nut oil
Kayapo Indians in Brazil are making this product.

Wax and honey
Bee-keeping, e.g. Zambia, N W Province.

Gene pools
Rainforest communities should own the patent to medicinal plants found in their forests. Merk & Co. have agreed to screen unexplored plant and animal species and will pay US $1 million. This will go to INBIO, a scientific and conservation charity.

Animal trapping
Animals are caught and exported for medical science (chimpanzees) and for private collection. (Restrictions on rare species. Much illegal trading.)

Hunting
Some animals can be culled in a controlled way. (Subject to abuse; Needs careful management. Restrictions on rare species.)

Wild birds
Wild birds taken for pet shops or private collections. (Much illegal trading. Birds are easily killed in transit. Many airlines now refuse to carry wild birds.)

The massive international financing of dam-building projects in Brazilian Amazonia has been halted. The decision follows extended and global protests against Brazilian plans to develop the Amazon, which culminated in an unprecedented rally organized in Altamira, on the Xingu river, by the Kayapo Indians of central Brazil. In a showdown with the Brazilian authorities, and in the full glare of international publicity, the Indians demanded a halt to plans to construct massive dams on their river that would flood them off their ancestral lands.

The decision to halt the funding marks a major victory for the Indians and the international human rights organizations, such as Survival International, that have supported them. It follows a three-year campaign to stop the World Bank, and a consortium of Japanese, German, British and American commercial banks, from investing billions of dollars in Brazil's ambitious dam-building programme.

Figure 3.31 Magazine article about the involvement of indigenous peoples to save Brazilian rainforests (*Source: Geographical Magazine*, June 1989)

Economic pressures and responses
There are enormous financial pressures on governments and local people to extract resources from rainforests. Often, however, local people are not consulted about rainforest exploitation, though they have clear ideas about what they need and want.

3.5 Ecotourism

Ecotourism may be one way of conserving fragile ecosystems and marketing their appeal while providing income for local companies. On the other hand, groups of tourists, however small and well-intentioned, may easily have a negative effect on indigenous people's way of life. It is important to respect the integrity of host communities and follow basic groundrules (Table 3.8).

Table 3.8 Key principles of ecotourism (*Source:* Wright, 1992)

1 Maintain the quality of the environmental resource. Develop it in an environmentally sound way.
2 Produce first-hand participation and inspiring experiences.
3 Educate all parties before, during and after the trip.
4 Encourage recognition by everyone involved of the value of the resource.
5 Accept the resource as it exists and accept that it may limit the number of visits involved.
6 Encourage understanding and involve partnerships between governments, non-governmental groups, industry, science and local people.
7 Promote moral and ethical responsibilities and behaviours in the action of everybody involved.
8 Develop long-term benefits to the resource, local community and to industry.

Possible advantages of ecotourism
Ecotourism may have economic advantages for both tour operators and local communities (Table 3.9).

Sometimes there are opportunities for working with conservation organisations to ensure that some benefits are directed to the environmental resource (Table 3.10).

Table 3.9 Economic advantages of ecotourism (*Source:* Wright, 1992)

1 A greater variety of economic activities in rural and non-industrial regions.
2 Long-term economic stability.
3 Tourists involved in ecotourism tend to spend more and stay for longer.
4 Demand for local goods and services.
5 Development of infrastructure such as roads, airports and bridges.
6 Increase of foreign-exchange earnings.

Table 3.10 Benefits of links with conservation organisations (*Source:* Wright, 1992)

1 Donation of portion of tour fees to local groups.
2 Education about the value of the resource.
3 Opportunities to observe and take part in scientific activity.
4 Involvement of local people in providing support services and products.
5 Involvement of local people in explaining cultural activities or relationship with natural resources.
6 Promotion of a tourist and/or operator code of ethics for responsible travel.

Measuring the benefits

With the growing increase in ecotourism there has been a need to measure its environmental benefits. Green Flag International is a non-profit-making company formed in response to the growing demand for conservation advice from tour operators and the travelling public. It rates holidays, tour operators and resorts using the following checklist:
1 Consideration to landscape, wildlife, cultural heritage
2 Efficiency
3 Waste disposal and recycling
4 Interaction with local communities in terms of goods and services
5 Sympathetic building and architecture.

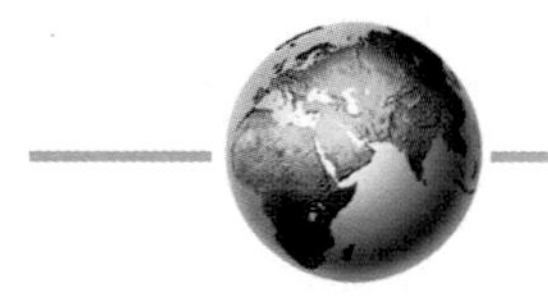

The Cuyabeno Nature Reserve, Ecuador

Cuyabeno is one of the areas in the Ecuadorian rainforest (Fig. 3.32) with facilities for ecotourism. It is remote: access involves several hours' drive or a flight from Quito, followed by a motorboat journey down river. Ecuadorian tour companies have built jungle lodges from local materials in selected areas, or there is a 'flotel' (floating hotel) on the Rio Aguarico. Some of these facilities were provided with the help of the World Wide Fund for Nature.

The lodges are run by staff from outside the area, but other basic services are provided by local Siona and Secoya Indians. Food and drinking water are brought in from outside the Reserve. All tourist groups (maximum size 16 people) have to be accompanied by trained guides, who are usually graduate biologists. Visitors may not hunt or fish in the Reserve or otherwise endanger the wildlife or pick plants. Tourists may explore the rainforest on foot using existing trails, or in canoes on the lakes, creeks and rivers (Fig. 3.33).

Wildlife is varied and abundant: the Ecuadorian rainforest is one of the most species-rich ecosystems in the world. Cuyabeno has many species of monkeys, birds, insects, reptiles and fish, as well as freshwater dolphins.

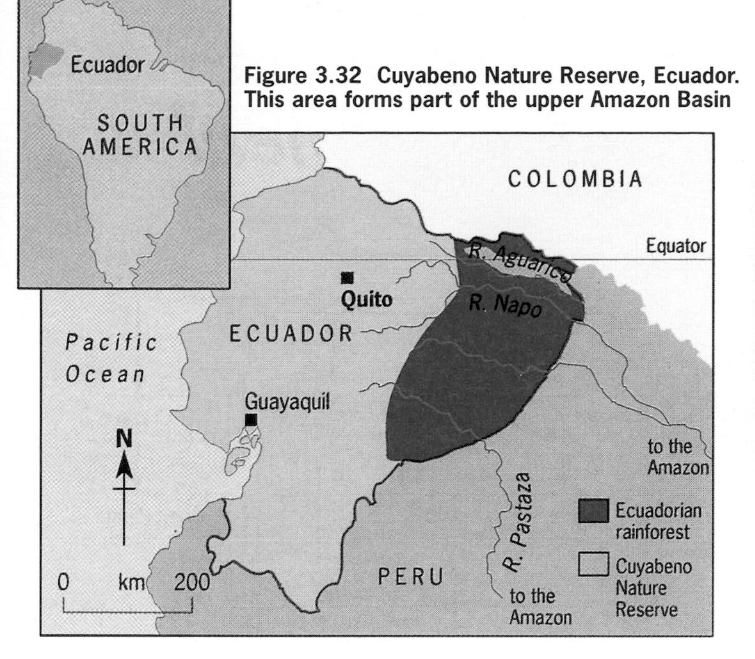

Figure 3.32 Cuyabeno Nature Reserve, Ecuador. This area forms part of the upper Amazon Basin

This area of Ecuador contains oil deposits and, even though Cuyabeno has protected status, it is severely threatened by oil exploitation. New oil towns have appeared, roads and pipelines have been built, settlers have followed and cleared many hectares of rainforest for settlements and farms. Several oil spills have been recorded, with consequent damage to the flora and fauna. The government employs only a few wardens to patrol the area. Local and international agencies are trying to protect Cuyabeno; part of the tourists' education is becoming aware of the threat posed by oil.

Figure 3.33 A group of ecotourists leave their river lodge for a guided trip through the rainforest

?

19a Write a list of guidelines that you could give to tour operators in Cuyabeno to make sure they get a high rating from Green Flag International for their tours to the tropical rainforest.
b In the role of a tour operator, write a speech asking local people to become involved in ecotourism in the rainforest.
c Write a paragraph to the people in Figure 3.33 as an introduction to their rainforest tour. Use the guidelines in Table 3.6 to comply with best practice in ecotourism.

?

20 Essay: Evaluate the advantages and disadvantages of ecotourism as part of a strategy for sustainable rainforest management.

Summary

- Rainforests are sites for HEP, mineral extraction, plantations, agriculture, settlement and logging. All these put pressure on the environment.
- Deforestation and removal of biomass lead to a breakdown in the essential cycling systems and leave the ground unprotected from heavy rain and solar radiation.
- Climatic changes on global, regional and local scales follow deforestation, due to a contribution to the greenhouse effect, changing albedo and evapotranspiration rates. The hydrological cycle is adversely affected, as are the soil systems on which the forest grew. Such changes prevent the forest from attaining a climatic climax community again.
- The way that the environment is 'managed' depends on people's values and attitudes.
- Economic problems within countries with rainforests present barriers to successful management. Short-term emergencies restrict long-term policies.
- The perceptions and needs of local people should be understood in order to manage the rainforest on a local scale.
- Sustainable yield is not necessarily sustainable forestry.
- Rainforests can be managed sustainably. This depends on a market for sustainably produced goods on all scales.
- Ecotourism is one method of rainforest management. However, certain guidelines must be followed to ensure sustainability.

4 Temperate deciduous woodland: management

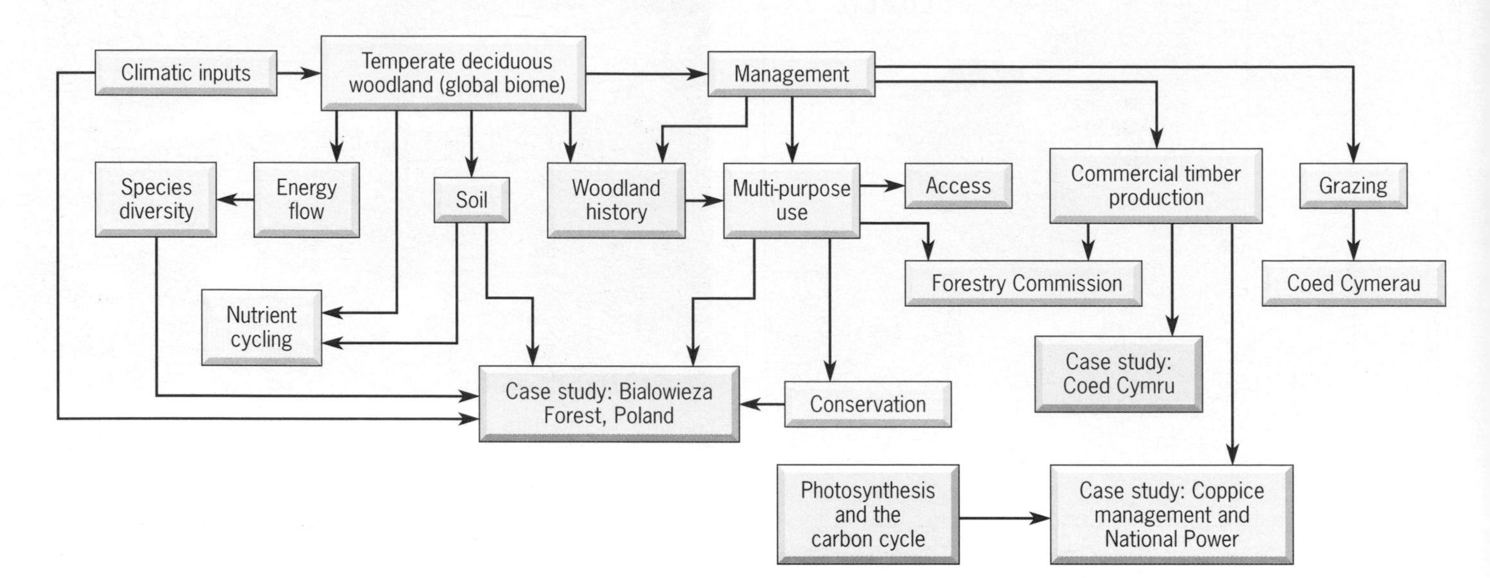

4.1 The nature of the ecosystem

Climatic inputs

Trees need more sunlight, warmth and moisture than many smaller plants. They require a long growing season for the growth and ripening of new shoots. Temperature therefore limits the distribution of trees. Roots must be able to take up water to balance **transpiration**. The soil must therefore remain moist. In temperate latitudes, forests only occur where there is an annual rainfall of at least 350 mm. When the soil is cold, the tree roots can only take up small quantities of soil water, and growth almost stops. The low temperatures cause a **physiological drought**. **Deciduous** trees lose their leaves at one season of the year. Their leaves drop before winter low temperatures start. Losing leaves during winter helps to reduce transpiration.

?

1 Using Figure 4.1, explain why temperate deciduous forests will not receive as much solar **radiation** as tropical rainforests.

2 Using Figure 4.2 describe:
a the temperature range
b the yearly total of rainfall
c the distribution of rainfall.

Figure 4.1 Distribution of temperate deciduous forests

Figure 4.2 Climate graph for a temperate deciduous forest, London

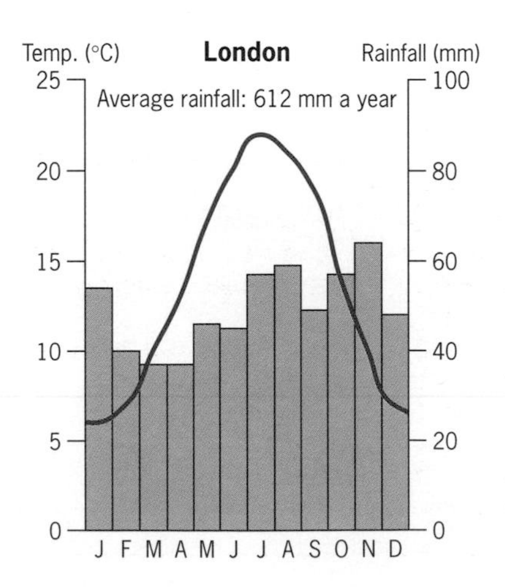

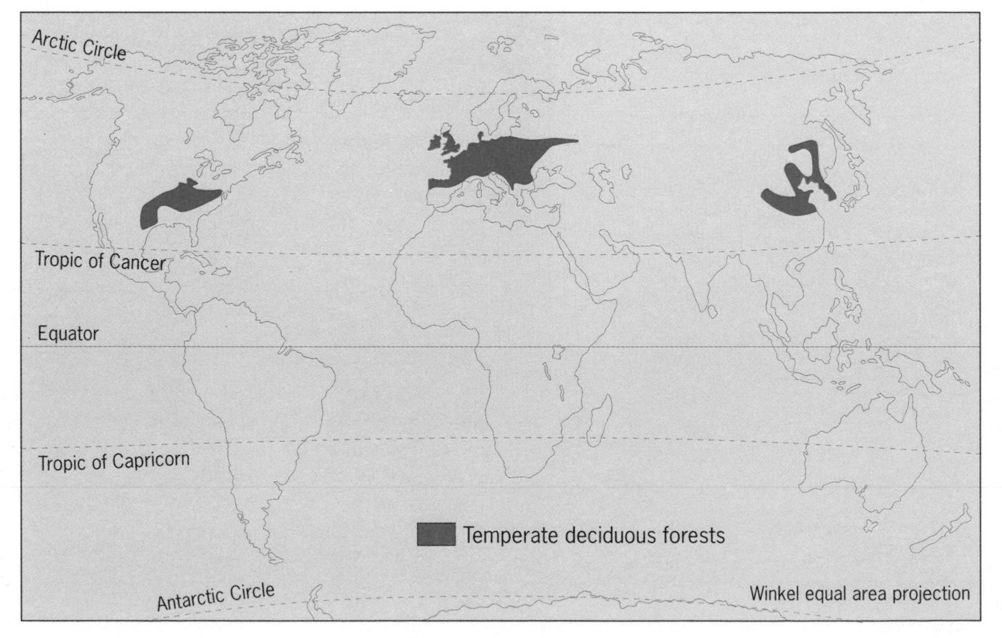

Figure 4.3 Spring in a temperate deciduous forest, Bedfordshire

Figure 4.4 Summer in a temperate deciduous forest, Gloucestershire

Figure 4.5 Autumn in a temperate deciduous forest, Kent

Figure 4.6 Winter in a temperate deciduous forest, Kent

Western and central Europe has a rainfall maximum in summer, in contrast to the British Isles. Higher evaporation in summer means that less water is available. Beech is able to grow under drier conditions. Beech can dominate oak because it is taller and has denser foliage, which allows it to **compete**. Although temperate deciduous forests grow across the area shown in Figure 4.1, they will have different dominant species.

Energy flow

Inputs of energy vary with seasons, so the total **biomass** is limited by this in temperate deciduous forests. The density of undergrowth is related to the amount of light which passes through the canopy. The undergrowth is much denser where trees are well spaced. Most trees and shrubs have broad leaves, which absorb much of the radiation as it hits the canopy.

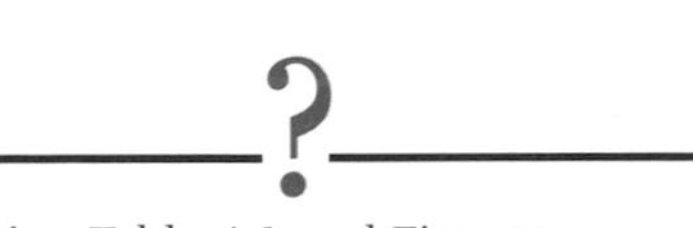

3a Using Table 4.1 and Figures 4.3–4.6, list the seasonal changes affecting plants in a temperate deciduous woodland.
b Explain how the changes affect the productivity of woodland throughout the year.
c Use annotated diagrams to explain how climatic characteristics (Figs 4.1, 4.2) may affect the yearly growth cycle of temperate deciduous forests.

4 Using Table 4.1 and Figure 1.11, explain how seasonal changes affect a selection of woodland fauna.

Table 4.1 Seasonal changes in a temperate deciduous woodland

Autumn – winter	Spring – summer
• Fungi thrive in the damp, mild conditions of early autumn. • Trees and shrubs lose their leaves (except some evergreen species such as holly). • More light reaches the woodland floor, but the temperatures are too low for plants to grow. • Ground flora dies down and survives under the ground as bulbs, roots, rhizomes or tubers. • Many mammals (e.g. dormice), reptiles, amphibians and insects hibernate. • Some insects survive at a dormant stage (eggs, larvae or pupae). • Some birds migrate south or move to gardens in search of food.	• Temperature rises. Ground flora (e.g. violets) grows and flowers before trees fully develop leaves. • Trees develop leaves. Less light reaches the woodland floor. • Birds return to the woodland. • Insects emerge from the dormant stage and hibernation. Mammals, reptiles and amphibians emerge from hibernation. • Birds and mammals nest and rear their young. Reptiles and amphibians produce young. • Plants flower and produce seeds. • Nuts, acorns and berries provide food for many animals. Some animals (e.g. squirrels) store food for the winter and early spring.

?

5a Using Table 4.2, construct a pyramid to show transfers of energy. Use Figure 1.9 (a histogram) as a model. Draw the area of the bars to scale.
b Use Figure 1.10 to explain your pyramid.

6a Study Figures 4.7 and 4.8. Which woodland matches Figure 4.9 most closely? Give reasons for your decision.
b Explain why the vegetation structure of the two woodlands differs.

7 Using Figure 4.10 describe where most nutrients are stored in a beech forest. Explain why this may be so.

Table 4.2 Energy (in biomass equivalent) transferred between trophic levels in a temperate deciduous woodland

Level	Average productivity (g/m²/year)
I (Producers)	1200
2 (Primary consumers)	20
3 (Secondary consumers)	5
4 (Tertiary consumers)	1.25

Nutrients

Nutrient cycling slows down during colder months. Lower temperatures limit the activities of **decomposers.** During these months trees take up nutrients in smaller quantities, since they lose their leaves.

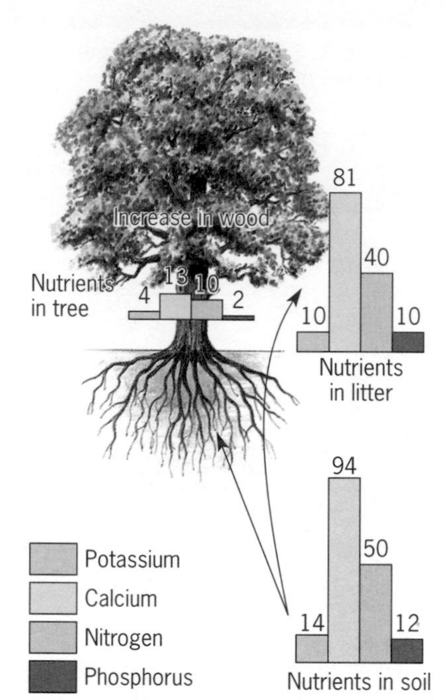
Increase in wood
Nutrients in tree
4
13
10
2
81
40
10
10
Nutrients in litter
94
50
14
12
Nutrients in soil
Potassium
Calcium
Nitrogen
Phosphorus

Figure 4.7 A beech woodland, England

Figure 4.8 An oak woodland, England

Figure 4.9 Generalised structure of a temperate deciduous woodland (after Smart and Andrews, 1985)

Figure 4.10 Cycling of four nutrients in a temperate deciduous forest in kilograms per hectare per year (after Duvigneaud and Denaeyer De Smet, 1970 in O'Hare,1988)

Canopy
The uppermost layer – the branches, twigs and leaves of larger trees.
Shrub layer
Beneath the canopy – smaller trees, shrubs and some of the climbers.
Field or herb layer
The taller non-woody plants, including flowering plants and herbs.
Ground layer
Mosses and other small or creeping plants growing on or close to the ground.

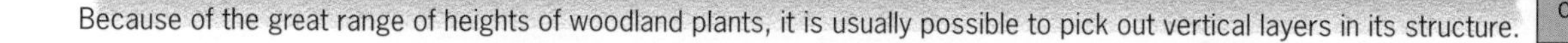

Because of the great range of heights of woodland plants, it is usually possible to pick out vertical layers in its structure.

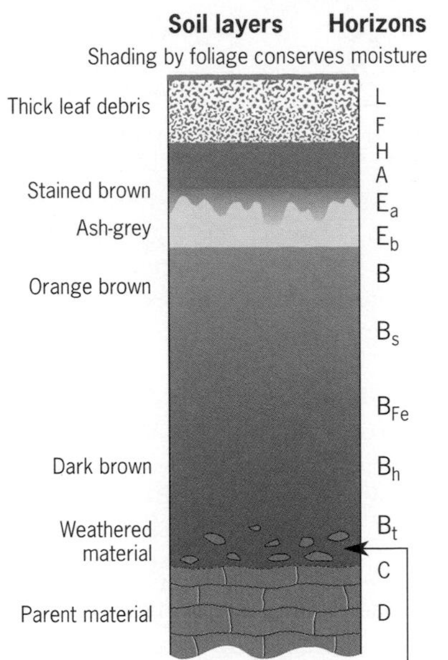

L–B horizons may be one metre deep. Soil fauna in the more basic soil and help to mix the A,E and B horizons.

Nutrients and minerals

L	Many bases returned
F	Partially decomposed litter
H	Well-decomposed humus
A	Mixed mineral-organic horizon
E	Less rapid decomposition of clays and less leaching
B	Iron and aluminium oxides deposited

Figure 4.11 Profile of a brown earth soil

Brown earths are good agricultural soils; their leaf litter gives **mull** humus. Earthworms mix this into the soil, giving a dark brown colour to the upper horizon.
Lower horizons are coloured by iron compounds that move downward through the soil. Water washes out some clay particles from the upper horizons and deposits them lower in the profile. Lower horizons are often of a heavier texture.
Brown earths are usually weakly acidic due to leaching.

Soil

The soil profile reflects the inputs of weathered material, precipitation, mixing by organisms and nutrient cycling (Fig. 4.11).

Climate also has an impact on soil-forming processes. Precipitation can affect the movement of nutrients and minerals through soil **horizons**.

Soil profile

The soil-forming processes and the nature of the climate, relief and parent rock lead to the development of distinctive layers or horizons within the soil. These can often be distinguished because of variations in colour, texture and structure, and together they constitute the soil profile.

The main soil horizons that occur in most soil profiles are shown by letters:

L the top organic horizon made up of decomposing organic matter
A the topsoil, containing both mineral and organic matter
E the eluviated horizon, which has nutrients and other substances washed out of it by downward percolating of water
B the subsoil, generally illuviated, having nutrients and other substances washed in by downward percolating water
C the bedrock.

Soil texture

The type of texture is determined by the proportions of different particle sizes:

- sandy soil has lots of sand
- loam has equal proportions of sand, silt and clay
- clayey soil has a large proportion of clay.

The organic matter in the soil resulting from the decomposition of plant debris by fungi and bacteria, is called humus (a dark brown colloidal substance with a greasy feel). A colloid can exist as a flexible solid or gel, or as a mobile fluid or sol. Colloids influence the character of soil in many ways:

- waterholding capacity
- adhesion between the larger sand and silt particles
- the aggregates or peds produced by the cementing vary; they give rise to particular types of soil structure, which influence the aeration and drainage of the soil and affect the ease with which plant roots can penetrate
- when the peds are small or contain air spaces, they are called crumbs; they are well-drained and good for roots
- peds can form angular or subangular blocks; these occur in clay-loam soils and are well drained and aerated
- vertically elongated peds in columns give a prismatic structure in clayey soils or limestones, allowing for water movement and root development
- where the vertical axis overlaps like a series of plates on top of each other, the soil is said to be platy; if machinery, people or animals compact platy soil, it hinders water and air movement, and root system development.

Soil acidity

The greater the concentration of hydrogen ions in the soil, the more acid it is. The soil's acidity is determined largely by its **humus** content. Plants (e.g. oak) that contain large quantities of metals such as calcium are likely to form a humus which is weakly acid to alkaline (mull). When these metals dissolve in water, they form bases such as calcium carbonate, and give rise to an alkaline humus. Soils with lower concentrations of these minerals are likely to form a strongly acid humus (mor). Coniferous trees tend to encourage the production of mor humus. In addition to acids made from the humus, soil water also contains dissolved carbon dioxide plus soluble salts formed from weathering of bedrock. Salts exist in solution as ions.

Soil fertility

Clay and humus molecules join together to form the clay-humus complex. It carries a negative charge and so attracts the positive ions, such as potassium and ammonium. This helps to prevent the ions from being washed out of the soil, making it more suitable for plant growth. Plants take mineral ions out of the soil in solution through their roots. They return to the soil when the plant decays.

Leaching

When water moves through the soil, some base minerals (substances that dissolve in water to form hydroxide ions) are replaced by hydrogen ions. This process is known as leaching. If minerals are replaced in the clay-humus complex, the soil will become more acidic. The greater the input of water into the system, the more leaching will occur.

?

8a Study Figure 4.12 and refer to Table 4.3 to answer the following questions.
a Which trees were found in the boreal and pre-boreal periods?
b Refer to the growth strategy, soil and shade preferences of the species in Table 4.3. How does the appearance of each tree species in the pre-boreal and boreal periods relate to the climatic changes?
c Which physical factors caused oak and elm to compete with pine and eventually to succeed as the dominant tree species during the Atlantic period?
d Why do you think the elm died out to make way for the ash?

9 Study Figure 4.13 and calculate the percentage of Great Britain's woodland that has disappeared since 2000 BC.

4.2 Woodland history

Table 4.3 Characteristics of some temperate deciduous and coniferous tree species

Species	Method of dispersal	Growth strategy	Soil preference	Shade tolerance	Use by people
Aspen	Wind and suckering	Pioneer species (coloniser)	Heavy soils, acid or alkaline	Does not like shade	Matches, paper
Birch	Wind	Pioneer	Well-drained to moist mountains, moorland soils. Will grow in Arctic conditions, acid soils	Light-demanding	Brooms, shoes, plywood, shingles
Scots pine	Wind	Pioneer	Well-drained acid soils	Light-demanding Withstands cold winters	Timber
Hazel	Animals	Long-term shrub	Rich, heavy alkaline soils	Shade-tolerant, will grow under oak and Scots pine	Nuts, leaves used by early people for fodder, branches for hurdles, wattle and daub
Alder	Wind and water	Pioneer and long-term competitor	Damp and swampy ground, alkaline on wet land		Window frames, clogs, charcoal
Oak	Animals	Long-term competitor	Heavy loams, alkaline; acid	Shade-tolerant	Acorns for pigs; timber for building
Elm	Wind	Slow coloniser, suckering. Problem of elm disease - fungal attack aused by wood-boring beetle	Heavy alkaline soils	Shade-tolerant	Leaves for fodder, wood for water pipes, coffins, furniture. Declined 4000–5000 years ago. 'Elm decline' associated with Neolithic people
Lime	Wind	Long-term competitor	Heavy alkaline	Shade, prefers warm climate	Bark fibres (bast) used for rope, string, coarse cloth, leaves for fodder
Ash	Wind	Long-term competitor	Moist soil, alkaline	Shade-tolerant	Leaves used as fodder, branches for fires, tool handles
Norway maple	Wind	Long-term competitor	Heavy alkaline	Shade-tolerant	Not native, ornament
Beech	Animal	Long-term competitor	Well-drained alkaline	Shade-tolerant	Nuts for food, timber, furniture, building
Hornbeam	Wind	Long-term competitor	Heavy soils alkaline	Shade-tolerant	Cattle yokes, tool handles
Norway spruce	Wind		Well-drained, acid soils	Shade-tolerant	Not native, timber

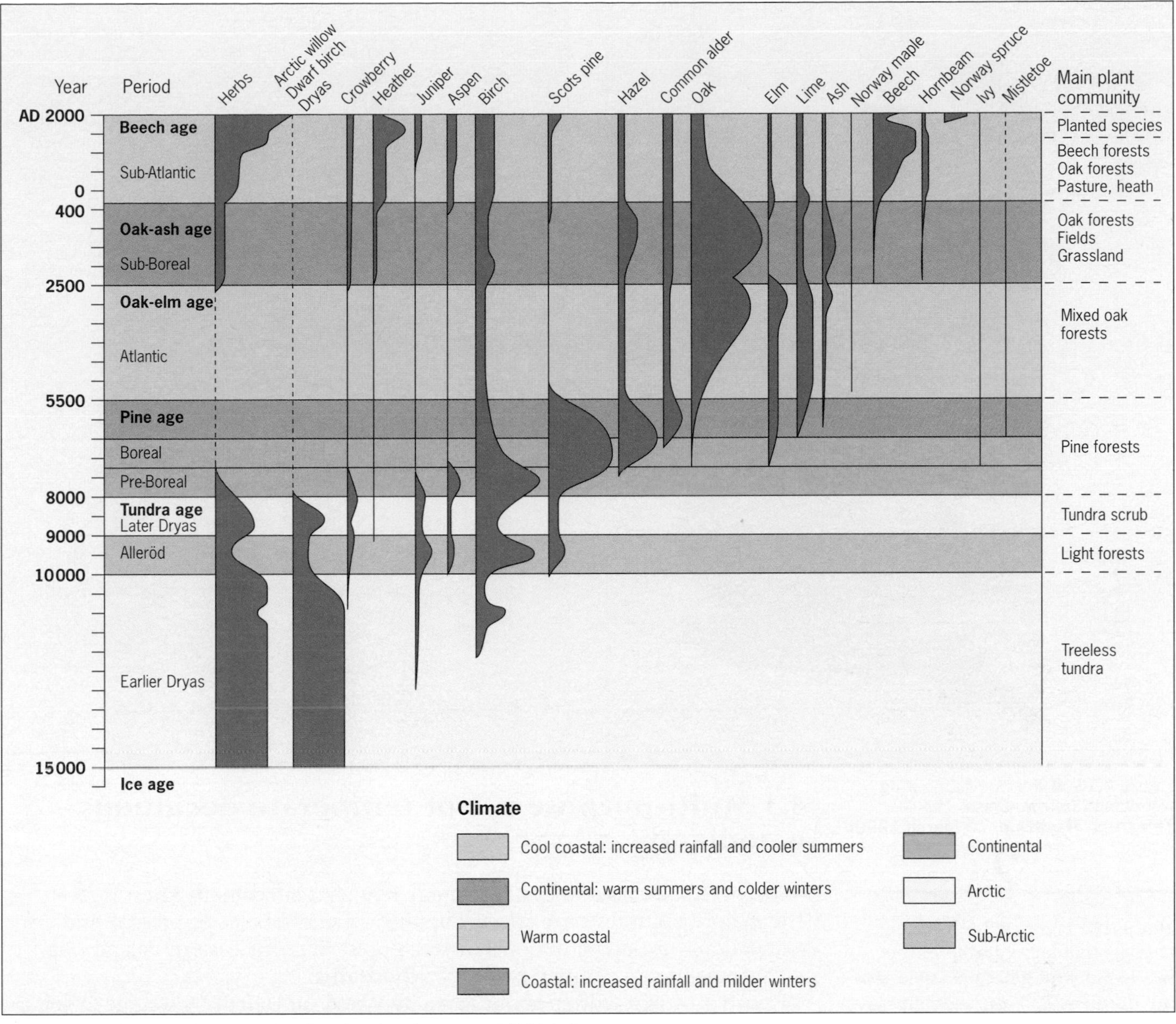

Figure 4.12 A pollen diagram showing the development of British forests from prehistory to modern times (after Vedel and Lange, 1960)

Before the ice age

Fifty million years ago, Europe had a tropical climate and was covered with evergreen forests, similar to those that grow today in tropical Asia, Australia, Africa and South and Central America. **Plate tectonics** states that Europe moved further away from the Equator and into cooler climatic zones. By about 2 million years ago plant communities similar to present-day **climatic climax communities** grew where Great Britain is today.

During the ice age

In Europe this pattern of vegetation was severely disrupted by a series of glaciations (ice ages). The most recent, the Devensian, ended approximately 17 000 years ago. Each glaciation removed much of the cover on land, destroying many plant **species**. As the ice advanced and retreated, so did zones of plants and **fauna**.

After the ice age

Relict communities were gradually re-colonised; species also **colonised** new areas. Evidence of climatic change and its effects on vegetation since the last glaciation has been preserved in the layers of ancient peat bogs and lake muds. Core samples from these deposits have been radio-carbon dated. An analysis of the pollen grains found in these layers shows us the kinds of woodland that grew in the surrounding areas (Fig. 4.12).

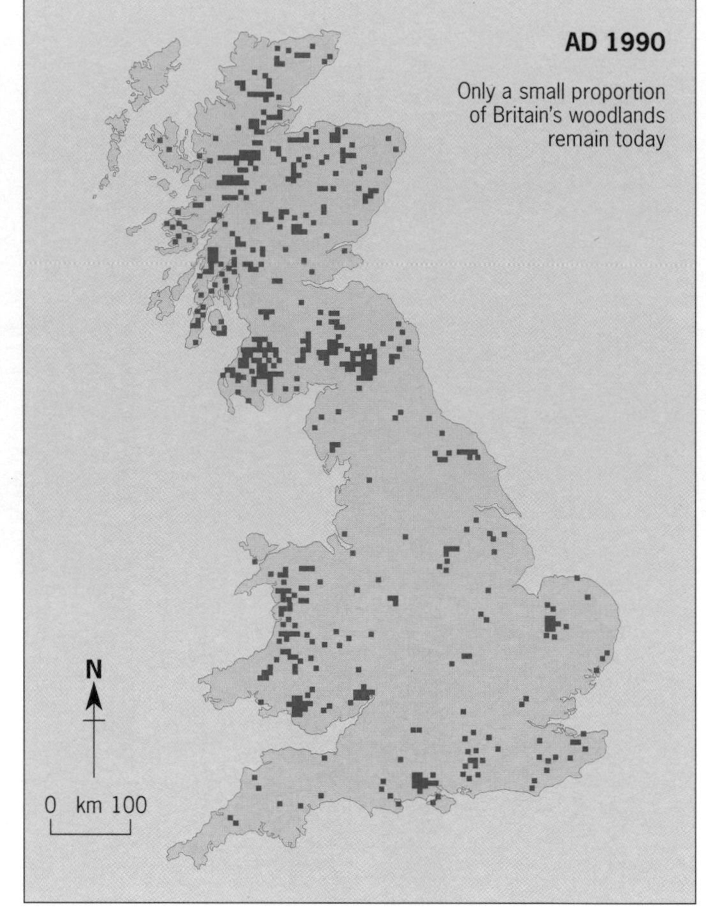

Figure 4.13 Britain's disappearing woodland (*Source:* Swan, *The Telegraph Magazine*, 10 March 1990)

?

10a From Figure 4.14, list the animals that were grazed in the woodland, and where possible give the numbers and any seasonal restrictions to the grazing regime.
b Study Table 4.3. Find the species listed in the account and suggest what each of them was used for.
c What type of soil and light demands did each species have?
d What does the information in **c** tell you about the physical environment of Penkelly Wood?
e To what extent do you think that the woodland was being managed **sustainably** (see Chapter 3)?

4.3 Multi-purpose use of temperate deciduous woodlands

In Great Britain, natural climatic climax temperate deciduous forest is rare. The majority of forests are affected by the activities of past or present human populations. In Europe only Bialowieza Forest in north-eastern Poland (Fig. 4.16) remains as a genuine **primary woodland**.

Figure 4.14 was written about Penkelly Wood (in Dyfed) by George Owen in 1594. The use of Penkelly Wood allowed people to integrate timber production, grazing and at the same time provided a rich wildlife **habitat** and opportunities for game (hunting and fishing).

Historically the **management** of woodland for wood production was an important part of the rural economy. We have the remnants of that managed woodland, and conservationists aim to recreate sustainable management of it.

To manage for several objectives it is important to understand how woodland can be managed for each use and how compatible they are with each other. This is known as multiple-use management.

> ... the forest of Penkelly... contyneth of the usual measure of that contrey aboute 500 acres of woodds and is enclosed with quicksett and pale rownde about the under locke it is all growne about with greate okes of 200 yeres growth and more and some younge wodde of 50 yeres growth and most of it well growen with underwooddes, the herbage whereof... will somer 300 breeding mareas and winter 300 sheepe and 200 cattell well and Sufficiently, beside swyne which may be kept there. Allso there is... great store of woodcockes taken yearly...

Figure 4.14 A description of Penkelly Wood, Dyfed, in 1594. This account goes on to mention other features of the woodland such as bees and the honey that is collected from them and the rich harvest of underwood and timber that was taken for all kinds of local use.

?

11 Why do conservationists want to encourage sustainable management? (See Section 3.4.)

12 Take the items from Figure 4.15 and divide them into three categories. You may find that many of the activities fall into more than one category.
a Those which could be commercially profitable (if the market exists).
b Those which in your opinion are for leisure.
c Those which may provide opportunities for wildlife.

Game management
Woodland can be used by farmers or let out to shooting syndicates for hunting game birds and animals. It can also be managed for traditional fox hunting.

Habitat for birds and animals
Woodland can be managed to increase their value as habitats for certain plants and animals. Formal designations such as SSSIs aim to protect or increase the wildlife interest of a site.

Stock shelter and grazing
Woodlands are used to reduce exposure and provide grazing land among the trees.

Timber production
A wide variety of wood products can be harvested or have been in the past.

Recreation
This can include activities in which the landowner uses the area for personal recreation or other activities such as walking on public footpaths, pony trekking through private land and off-road driving on private tracks (run as a source of income).

Shelter belts
A line of trees can be used to protect crops from the wind and help prevent soil erosion.

Landscape
Woods that are in scenic places may have a value as part of the landscape itself. This may include a formal designation such as inclusion in a National Park or an Area of Outstanding Natural Beauty.

Figure 4.15 Significant uses of woodland

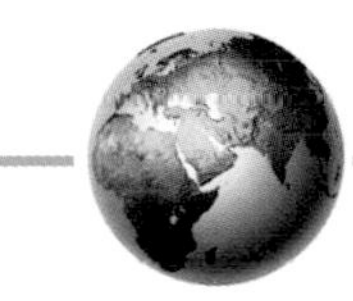

Bialowieza Forest, Poland

The Bialowieza Forest contains 580 square kilometres of mixed forest and is situated in north-eastern Poland (Figs 4.16 and 4.17) This includes some nature reserves and a protected area which makes up the Bialowieza National Park. It is a remnant of the primary woodland that once covered all of Europe. The forest used to be the hunting ground of kings, but is now state-owned. The park is a World **Biosphere Reserve** and is listed by UNESCO as a World Cultural and Natural Heritage Site.

Figure 4.16 The Bialowieza Forest in eastern Poland

Figure 4.17 Primary woodland in the Bialowieza Forest

Climate

Table 4.4 Climate of Bialowieza Forest

Average January temperature	- 4.7°C
Average July temperature	17.8°C
Average annual precipitation	641 mm
Average duration of snow cover	92 days per year, varying from zero to 132 days
Spring comes five days later than in central Poland, autumn 15 days earlier	

Wildlife

Table 4.5 Wildlife in the Bialowieza Forest

Species	Number of species	Notes
Trees	26	Six indigenous species of mammal have become extinct and three have been introduced. Bears were hunted to extinction in the 19th century, but a few appear from time to time from Byelorussia - Wolves, otters and ermine are rare, but foxes, badgers, marten, weasels and polecats are more common. Beavers have returned to the forest. The most common hoofed mammals are the wild boar and red deer. The European bison, which used to be common, became extinct in 1919 but was reintroduced in 1929. There are now over 200 bison, of which about 30 live in reserves (Fig.4.18) used for breeding, scientific research and showing to visitors.
Vascular plants	990	
Fungi	1000	
Insects	Over 8500 (including 2000 species of beetle and 1000 of butterflies)	
Invertebrates	Vast, as yet unknown	
Mammals	62	
Birds	228	
Reptiles	7	
Amphibians	12	
Fish	24	

Figure 4.18 A bison reserve

Soils

The soils consist of brown earths, podsolised brown earths as well as some peat and chernozem.

Management

Area outside the National Park

Outside the National Park areas, the forest is intensively used. The proportion of mature trees is rapidly decreasing and the proportion of young trees and cultivated areas is increasing.

For centuries the forest was used to supply trees for construction and for ships' masts. Currently about 150 000 cubic metres of timber are extracted each year. Other resources extracted by people include bark for tanneries, resin, berries, wild mushrooms and medicinal herbs, as well as game. There are 21 villages on the Polish side of the forest.

The National Park

The National Park itself is a strict conservation area, with the aim of protecting the vegetation and wildlife, including the use of animal reserves and tree nurseries. Many species of animals and plants are rare and protected by law. However, there is an illegal trade in some species. The Park contains a number of 'monumental' trees: some of the oaks and a few other species with a trunk diameter of at least 1.2 metres have been designated as 'natural monuments'. Their age is hard to calculate, but oaks live up to 400-500 years, and Bialowieza pines up to 300 years.

Forest resources are not extracted from the Park: it is a centre for scientific research, education and tourism. The Park includes a former palace in its own grounds and a small museum. Horse-drawn vehicles and bicycles are permitted on roads within the Park. Tourists have to keep to designated trails and be accompanied by an official guide. Groups are limited to 25 people, and access for very young children is restricted. There are hotels, a youth hostel and a camping site.

?

13 How do the figures in Table 4.4 compare with Figure 4.2? What differences may lead to Bialowieza being a mixed forest rather than completely deciduous?

14 Under which species would you expect to find brown earth soils?

15a Using Figure 4.19 and Table 4.3, make a list of the soil preferences of species found in the Bialowieza Forest.
b Describe the locations of each type of species, noting the elevation of the land and the distance from water courses and swampy areas.
c Explain the location of each of the species using the information you have collected in **a** and **b**.

16 Using Table 4.3, identify the tree species in Bialowieza which will support most insects.

17 Suggest why Bialowieza contains more species of plants and animals than other forests within Europe.

18 Using evidence from Figure 4.19 give grid references for evidence of the different uses of the forest in the areas outside the National Park.

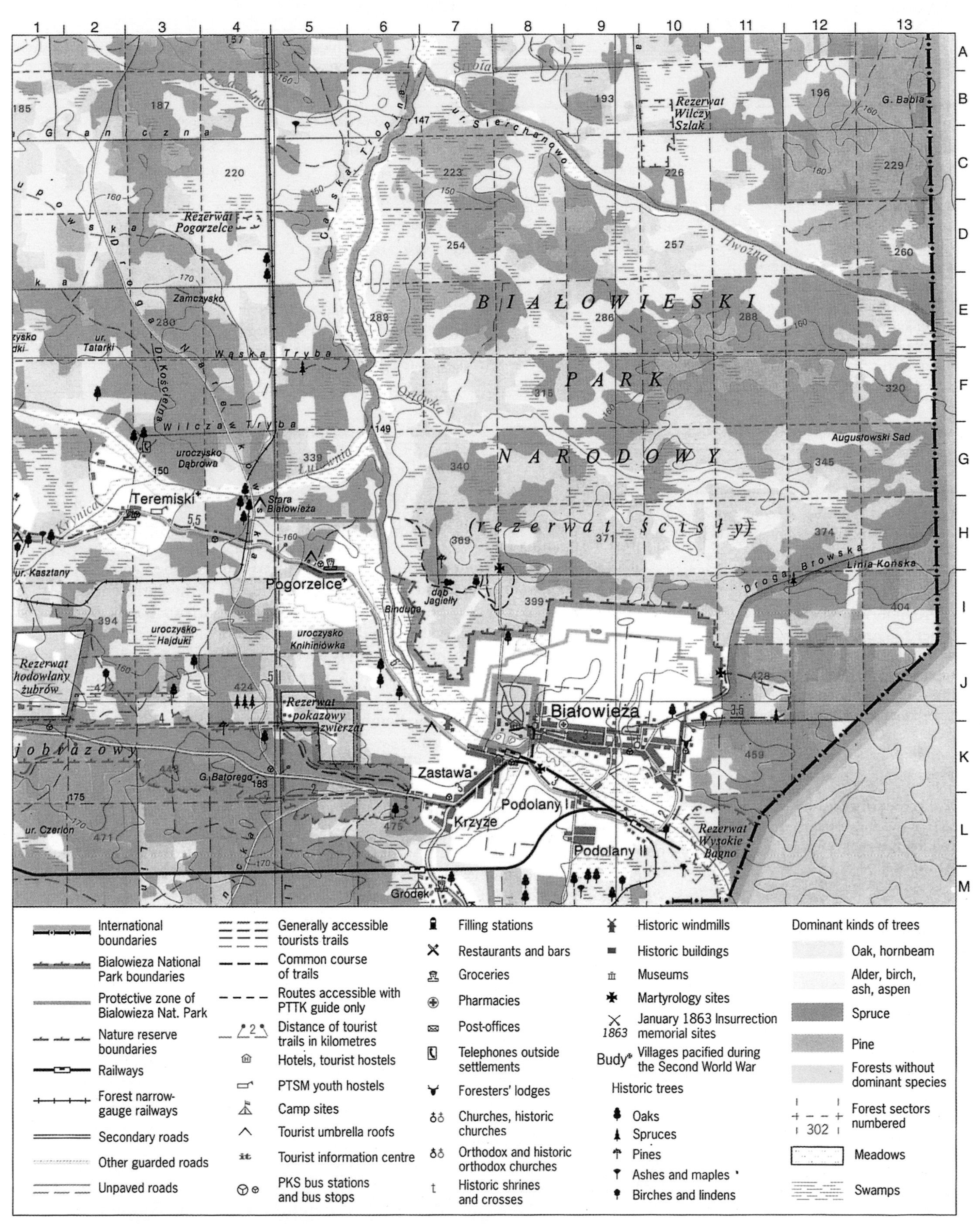

Figure 4.19 The National Park of the Bialowieza Forest © PPWK Co., Warsaw, 1989

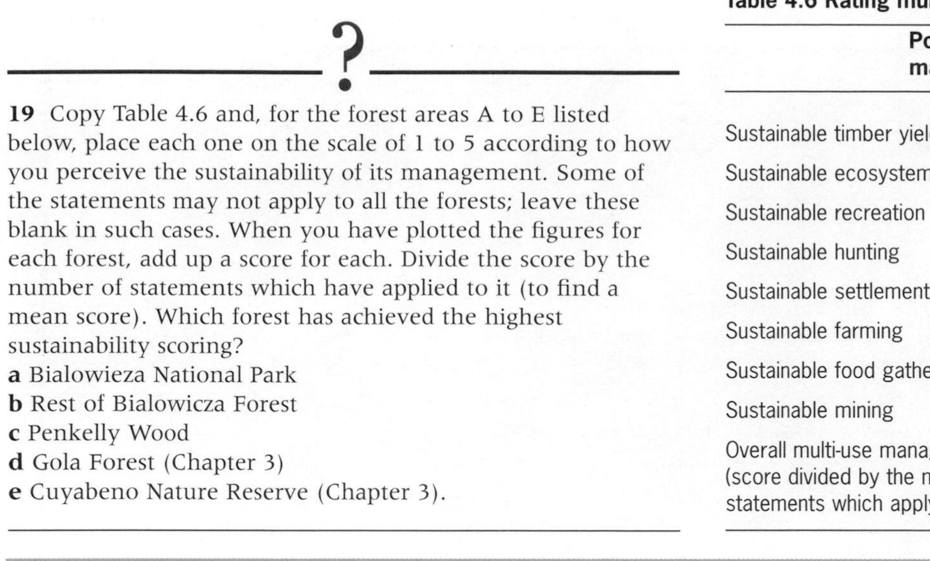

19 Copy Table 4.6 and, for the forest areas A to E listed below, place each one on the scale of 1 to 5 according to how you perceive the sustainability of its management. Some of the statements may not apply to all the forests; leave these blank in such cases. When you have plotted the figures for each forest, add up a score for each. Divide the score by the number of statements which have applied to it (to find a mean score). Which forest has achieved the highest sustainability scoring?

a Bialowieza National Park
b Rest of Bialowicza Forest
c Penkelly Wood
d Gola Forest (Chapter 3)
e Cuyabeno Nature Reserve (Chapter 3).

Table 4.6 Rating multi-use management

	Poor management	1	2	3	4	5	Good management
Sustainable timber yield		1	2	3	4	5	
Sustainable ecosystem		1	2	3	4	5	
Sustainable recreation		1	2	3	4	5	
Sustainable hunting		1	2	3	4	5	
Sustainable settlement		1	2	3	4	5	
Sustainable farming		1	2	3	4	5	
Sustainable food gathering		1	2	3	4	5	
Sustainable mining		1	2	3	4	5	
Overall multi-use management rating (score divided by the number of statements which apply to the forest)							

?

20a Assess your own reactions to a forest trip using a questionnaire based on Appendix 1. Then collect information from others in your group or more widely if possible.
b From the information that you gather you can rank the benefits and emotional responses in order of their importance. Remember that it is impossible to be objective when surveying feelings and reactions.
c What facilities would landowners need to provide to cater for these responses?

21 Write a list of pros and cons for each of the suggested future woodland management options from the RICS.

Access

Woodlands are popular places for trips and other recreational activities. People may perceive a range of benefits and experience different emotions in relation to a trip to a forest (Appendix 1).

In 1989 the UK government announced that a quarter of a million acres of **Forestry Commission** woodland would be sold by the end of the twentieth century. At this time, 60 per cent of the nation's forestry was already in private hands. Malcolm Rifkind, the Scottish Secretary, then said, 'We are concerned, however, that the general public should also continue to enjoy access to the forests to be disposed of by the Commission in a way that is compatible with the management for forestry and other purposes.'

Chris Hall, the Chairman of the Ramblers Association, in 1989 cited the case of Penwood, a few kilometres south-west of Yeovil, which was sold in 1983 and then lost its car park, picnic places and trails.

Commercial use of woodland does not necessarily mean limited access. The Royal Institute of Chartered Surveyors (RICS) provides a list of things for owners to do to improve income from woodland access. This includes car parks, permissive paths, observation posts, trail bike paths, car rallying and war games. Chris Hall worries about these developments: 'Woods are going to be planned as recreational areas and we are going to pay to visit them. The wild wood, the free wood will disappear.'

4.4 Commercial timber production

We can use a variety of management practices to harvest timber from woodland. The main techniques are shown in Figure 4.20.

The Forestry Commission

It is impossible to consider commercial timber production within the UK without an understanding of the role of the Forestry Commission. This government department was established in 1919, following difficulties of timber supply during the 1914–18 war. Its initial aim was to provide a strategic timber reserve for the nation. State woods therefore grew in number from nil to some 1 250 000 acres between 1919 and 1957, largely in the form of **softwood** plantations with species such as sitka spruce, larch and Norway spruce. By 1957 their strategic role was diminishing and the Forestry Commission's objectives focused on commercial and social aspects, including a recognition of its role in helping rural economies. The FC aimed to harmonise forestry with agriculture and the environment.

In 1988, the UK government announced an annual target of 33 000 ha of forestry to be planted in addition to existing areas. The government's forestry policy is aimed at the sustainable management of the existing resources and a steady expansion of tree cover to provide multiple benefits.

The Forestry Commission encouraged more **broadleaved** planting after its review, *Broadleaved Woodlands Policy*, in 1984–5. This followed pressure from conservationists to improve the conservation value of commercial plantations.

The Forestry Commission has continued to encourage private investment in broadleaved woodlands with a system of grants through the 'Woodland Grant Scheme' and its associated 'Farm Woodland Premium Scheme'. These provide significantly higher payments to farmers and land owners who plant broadleaved rather than **coniferous** species.

The Forestry Commission has a dual role, advising on and implementing government policy for all the UK's forests, in addition to managing 40 per cent of the country's forests. Managing forests includes responsibility for timber production, recreational opportunities and conservation.

Figure 4.20 Techniques of harvesting timber

Coppice
Most broadleaved tree species produce shoots from a stump. Coppicing involves cutting stems on a rotation of 7 to 25 years. A labour-intensive method, it produces wood for firewood, charcoal, fencing and chair legs.

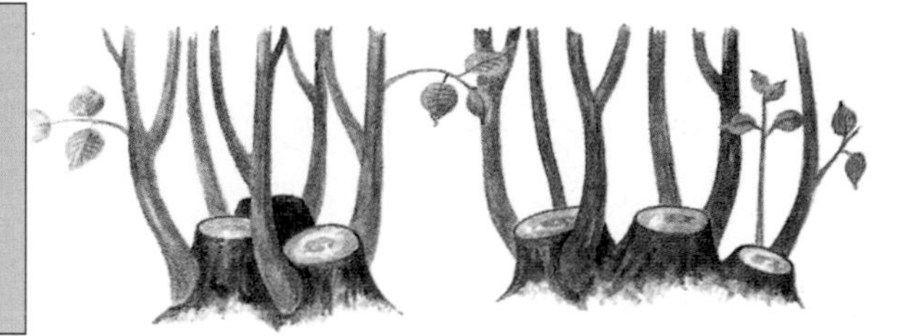

Clearfelling
This method is used when larger-sized timber is needed. The whole woodland or a larger area is felled at one time and a new crop planted to replace it.

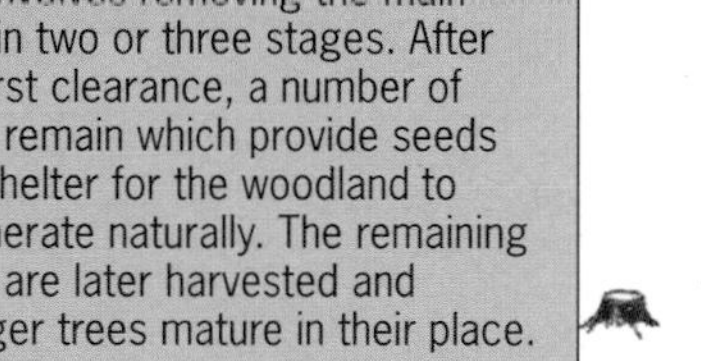

Shelterwood
This involves removing the main crop in two or three stages. After the first clearance, a number of trees remain which provide seeds and shelter for the woodland to regenerate naturally. The remaining trees are later harvested and younger trees mature in their place.

Selection
This involves woodland with a variety of ages of trees. When a tree reaches a suitable size, it is felled and the woodland continues to regenerate naturally in the space created.

Table 4.7 Advantages and disadvantages of five timber harvesting

Management techniques	Advantages	Disadvantages
Coppice		
Game management for pheasant/deer	Pheasants require a sheltered woodland floor with areas of low vegetation for breeding cover. These conditions are available in coppice. A mosaic of habitat and cover at various stages is available for deer.	
Birds and animals (not game)	Cover encourages some species of birds such as nightingales and small mammals. Grazing for deer and other browsers. Good way of keeping open spaces in woods; ideal for some insects.	Seed crop can be temporarily reduced. Red squirrel and small mammals often affected. Little or no mature timber or dead wood for insects.
Landscape	Constant feature in the landscape	Temporary bare patches when harvested.
Shelterbelt	N/A	N/A
Grazing	None	Grazing animals must be excluded from coppiced areas.
Clear felling		
Game	Good for woodcock	No cover, total habitat removal
Birds and animals (not game)	Sheltered and sunny open space important for some insects. Good for woodcock, nightjar, short-eared owl	Destruction of woodland community. Native trees lack diversity of structure to support very rich insect fauna
Landscape	None	Loss of amenity and aesthetic value
Shelterbelts	N/A	N/A
Grazing	Renewed ground flora	No shelter
Shelterwood		
Game	Excellent cover	None
Birds and animals (not game)	Excellent, more nesting sites for species	Mature trees removed; dead wood reduced
Landscape	Constant visual amenity. Age variety	Some elimination of mature species
Shelterbelts	Very effective	None, as understorey encouraged
Grazing	Incompatible	Lack of regeneration
Selection systems		
Game	Excellent	None
Birds and animals (not game)	Excellent	None
Landscape	Excellent	None
Shelterbelts	Excellent	None
Grazing	Incompatible	Lack of regeneration

N/A = information not available or not applicable

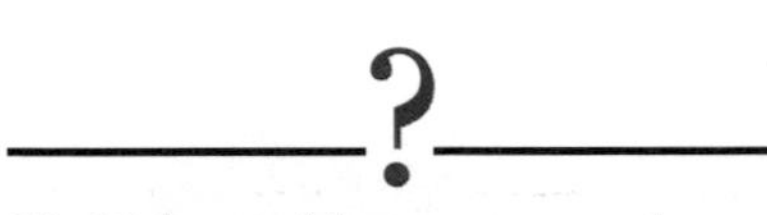

22 Make a table to compare the advantages and disadvantages of each method of timber production highlighted in Figure 4.20 and Table 4.7 for:
a a landowner who wants to produce timber
b a walker.

In 1992 the Commission split these roles to form the Forestry Authority and Forest Enterprise. The Forestry Authority acts as a government department, administering grants, research and monitoring all aspects of the industry. Forest Enterprise is responsible for the multiple-purpose management of the Commission's own forests and woodlands. Its main aim is to develop and maintain attractive and productive woodlands for public benefits.

Many organisations are responding to the Forestry Commission's aim of multi-purpose management in woodlands. In many areas this management relies on economic forces. Without some form of income, landowners may be reluctant to manage their land for a variety of uses.

?

23a From the figures in Table 4.8 construct pie charts to show the information.
b Suggest an explanation for the differences between the use for recreation of Forestry Commission and private woodland.
c Suggest an explanation for the difference between the area of land planted with conifers and the area planted with broadleaved species.
d Explain the fact that coppicing was once carried out in a significant proportion of Great Britain's deciduous woodlands but has now dropped to the low levels shown in Table 4.8.
e With reference to the aims of the Forest Authority and Forest Enterprise, predict what may happen to the statistics in the future.

Table 4.8 Amount of land (1000s of ha) used for different types of forestry (*Source:* FC, 1993)

Woodlands	Conifer	Broadleaved	Coppice	Amenity and recreation
Forestry Commission	793	51	1	37
Private	727	551	39	162

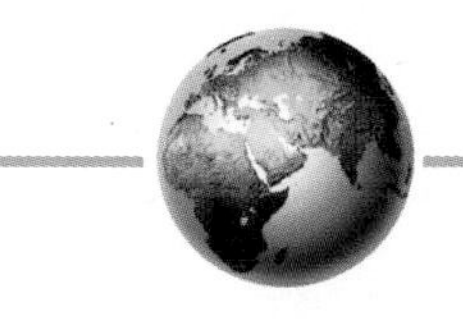

Coed Cymru

Coed Cymru is an organisation that was established in 1986 with the financial support of the Countryside Council for Wales to promote the good management of broadleaved woodlands in Wales:

The aims of Coed Cymru
- To re-establish the economic cycle that sustained woodlands in the past
- To produce wood while increasing its structural and species diversity
- To reduce the neglect of many Welsh woods, including overgrazing
- To make Welsh woods an asset to landowners and to the landscape and wildlife of Wales

Figure 4.21 A valleyside in Wales

Figure 4.22 Trees in a Welsh wood

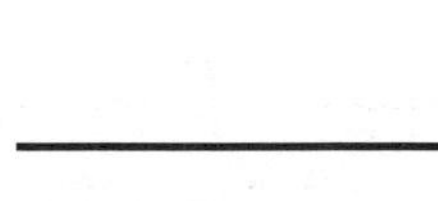

24a What factors will be of major importance in harvesting wood commercially from the woodland shown in Figure 4.21?
b What factors will have contributed to the condition of the trees shown in Figure 4.22?
c If this timber were to be harvested, what uses do you think could be made of it?

A market for timber

Much of the timber is taken from hillside woodland on poor soils which cannot be used for other purposes. Coed Cymru have highlighted the potential for these woodlands of Wales (Fig. 4.23).

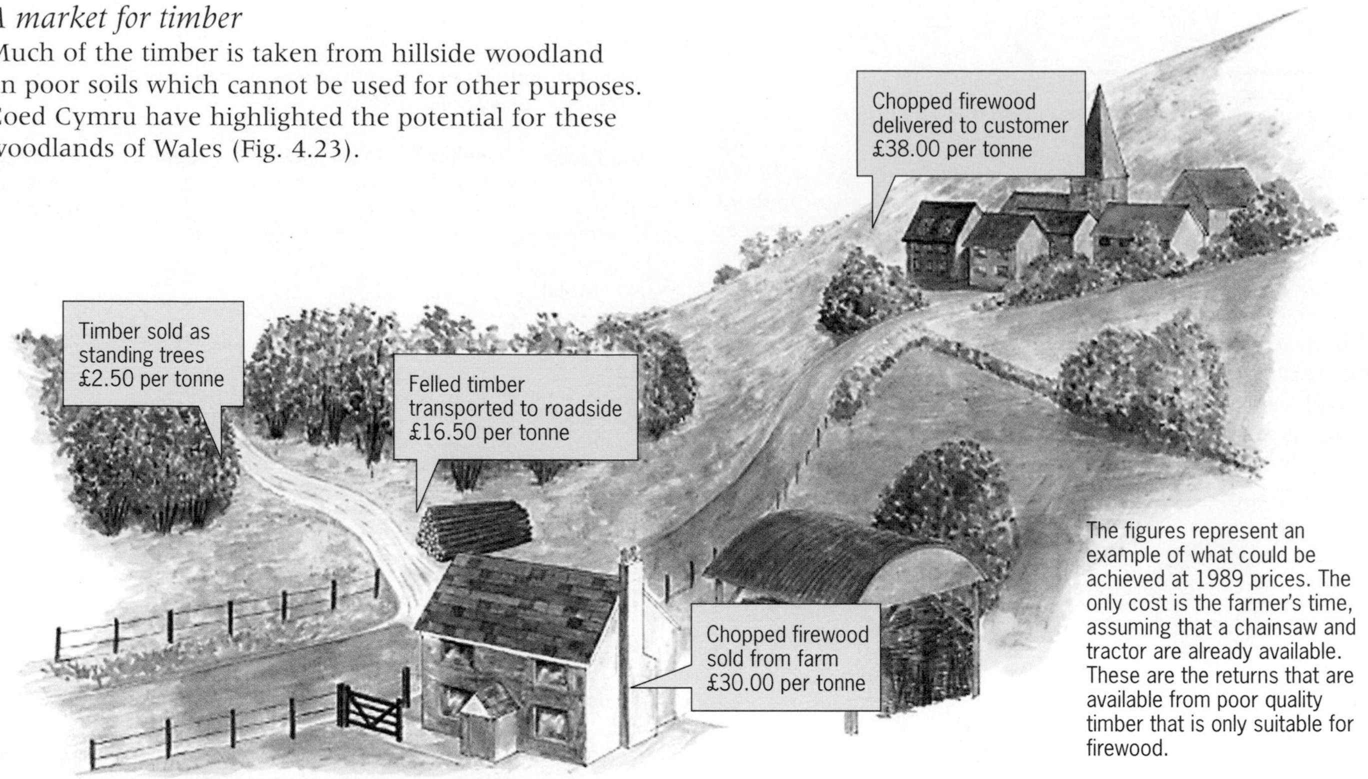

Chopped firewood delivered to customer £38.00 per tonne

Timber sold as standing trees £2.50 per tonne

Felled timber transported to roadside £16.50 per tonne

Chopped firewood sold from farm £30.00 per tonne

The figures represent an example of what could be achieved at 1989 prices. The only cost is the farmer's time, assuming that a chainsaw and tractor are already available. These are the returns that are available from poor quality timber that is only suitable for firewood.

Figure 4.23 Firewood production in Wales

Marketing better-quality timber

On average a 1-hectare block of trees would yield about 100 tonnes of timber. On average 10 per cent would be better-quality timber. The roadside value of better-quality timber is £22 per tonne at 1989 prices. This timber could be further processed with the hire of additional basic equipment such as a band-saw for wood processing for £259.

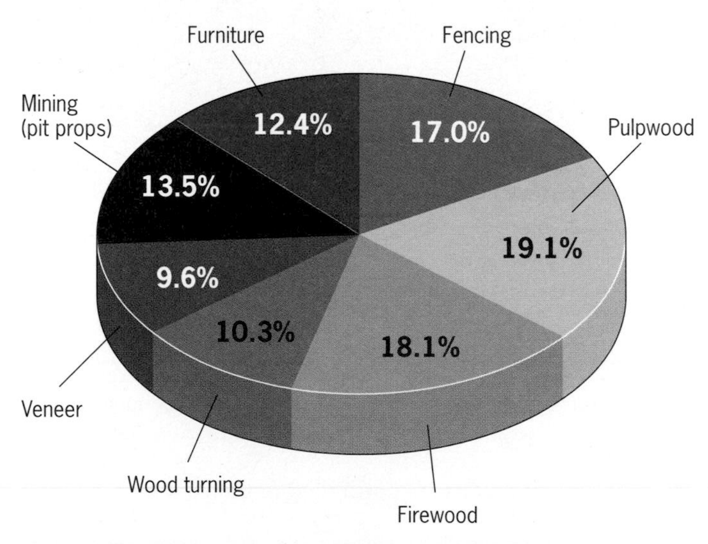

Figure 4.24 Markets used by contractors selling Welsh timber (after Hughes et al., 1991

?

25a For each of the stages in the firewood operation shown in Figure 4.23, calculate the percentage increase in value of the crop compared with its original standing value.
b In the context of wood harvested from Welsh mountains, suggest why the potential added value is so great in the first stage of the operation.

26 Ten tonnes of better-quality timber could be converted to:
- 244 fencing stakes at £1.10 each
- 37 gate posts at £30 each
- 4 heartwood planks at £25.70 each (1989 prices).

Use these figures and the information in Figure 4.24 to calculate the income that could be achieved from a 1 hectare woodland block with 10 tonnes of better quality timber and 90 tonnes of poorer quality when:
a It is sold as standing timber.
b It is sold at the roadside.
c It is sold fully processed for maximum return.
d What percentage increase in return is achieved by processing the wood, compared with selling the wood standing?

27 Which of the markets shown in Figure 4.24 require high-quality timber and which require a lower-quality material?

28 What factors in the growing and harvesting processes limit:
a the amount of good-quality timber available?
b the economic value of harvested timber?

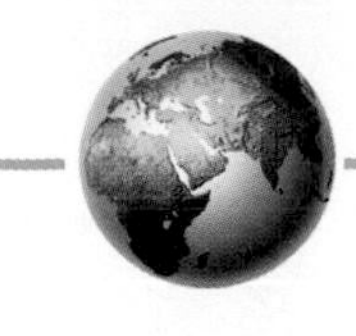

Coppice management and National Power, England and Wales

In the UK, the practice of **coppice** management has been neglected due to the lack of demand for the timber produced by this method. It is only used for craft items at a local scale.

National Power, an electricity-generating company, is developing a modern version of the traditional coppice technique to produce wood as a commercially profitable fuel for burning directly. It is designed to be a fast-growing system suitable for mechanical harvesting instead of high labour inputs (Figure 4.25).

The economics of energy coppice

One of the major costs of energy coppice is planting the trees initially. This is done by planting cuttings, which then take root and grow. Buying the cuttings costs £1000 per hectare. At the end of the first year, the shoots that have grown are cut off to leave a stump, and the cut-off material can then be used to plant another area at no extra cost. If the farmer plants five blocks to allow for a five-year rotation, this reduces the overall cost of cuttings to £200 per hectare.

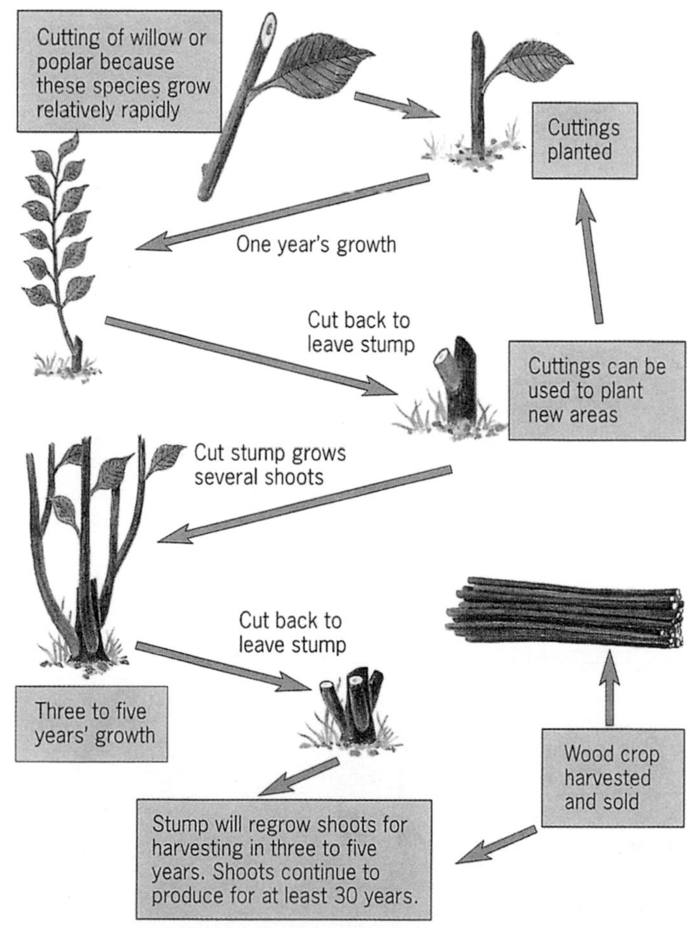
Cutting of willow or poplar because these species grow relatively rapidly
Cuttings planted
One year's growth
Cut back to leave stump
Cuttings can be used to plant new areas
Cut stump grows several shoots
Cut back to leave stump
Three to five years' growth
Wood crop harvested and sold
Stump will regrow shoots for harvesting in three to five years. Shoots continue to produce for at least 30 years.

Figure 4.25 The process of growing coppice wood for fuel (after RSPB, 1991)

The other costs involved in establishing a plantation are for herbicides to clear the site initially and a further treatment with herbicide at the end of the first year.

The returns from the crop are dependent on a number of inputs such as climate and land quality. However, an average cropping rotation might be expected to produce 12 tonnes of dry matter per hectare per year, and this would provide a gross profit of £252 per hectare.

Gross profit

Gross profit is the financial return from selling the crop, less the costs of such things as fertilisers, contractors, etc. Gross profit does not take into account the fixed costs on a farm such as the cost of buying farm equipment which is used in all aspects of farm work.

In comparison, a winter wheat crop would produce a gross profit of about £470 per hectare, and spring barley £300 per hectare. However, coppice will grow not only on very good quality cereal land but also on poorer-quality land which would not support cereals. Such land, however, may have conservation value. For example, wetlands which have not been able to be cultivated are very important habitats for some rare plants and birds.

Set-aside grants

The European Union has been producing surpluses of agricultural crops, particularly grain. To reduce these surpluses the EU has been paying farmers since 1988 to take land out of agricultural production using **set-aside grants**. Energy coppice is classified as a 'non-agricultural' use and so in 1993 qualified for an annual grant of £150 per hectare.

?

29 Explain why the trees are cut back after their first year of growth.

30a From the point of view of a farmer, could energy coppice be an economic alternative as a crop?
b Would energy coppice be an economic proposition if coppice was classed as an agricultural crop?

31 Suggest some advantages and disadvantages of energy coppice for wildlife conservation.

32 Essay: Wood is classed as a renewable fuel. Explain the meaning and the importance of **renewable** resources.

Photosynthesis and the carbon cycle

The energy in the wood that is being used as fuel originally comes from the sun. **Photosynthesis** transfers this energy into chemicals in trees (Fig. 4.26).

In the process of burning wood, carbon dioxide gas is released. This is one of the **greenhouse** gases. Wood-burning also creates some fly ash, soot and carbon deposits.

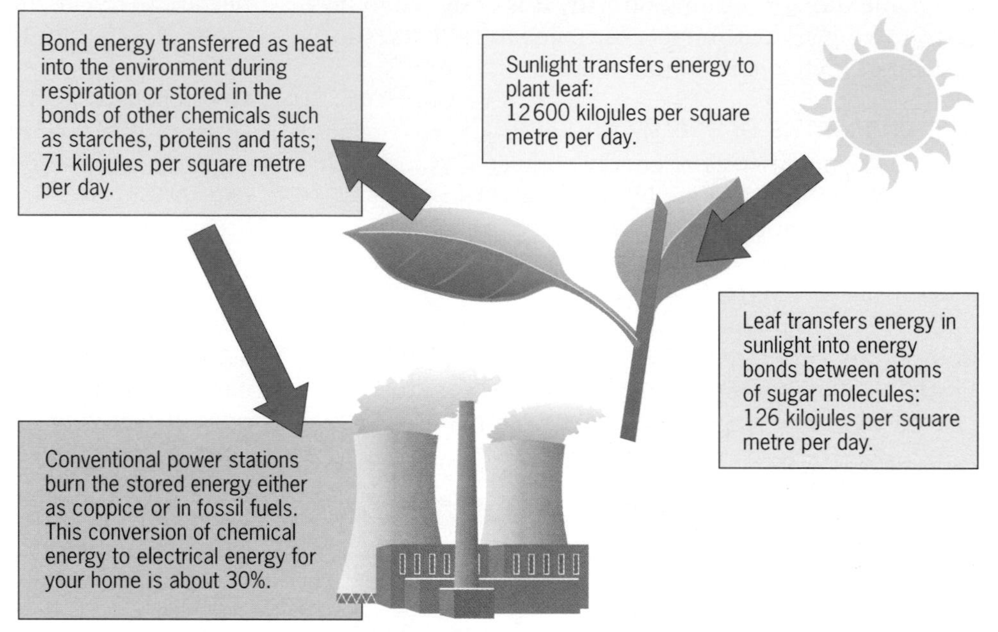

Bond energy transferred as heat into the environment during respiration or stored in the bonds of other chemicals such as starches, proteins and fats; 71 kilojules per square metre per day.
Sunlight transfers energy to plant leaf: 12600 kilojules per square metre per day.
Leaf transfers energy in sunlight into energy bonds between atoms of sugar molecules: 126 kilojules per square metre per day.
Conventional power stations burn the stored energy either as coppice or in fossil fuels. This conversion of chemical energy to electrical energy for your home is about 30%.

Figure 4.26 Energy flow during photosynthesis

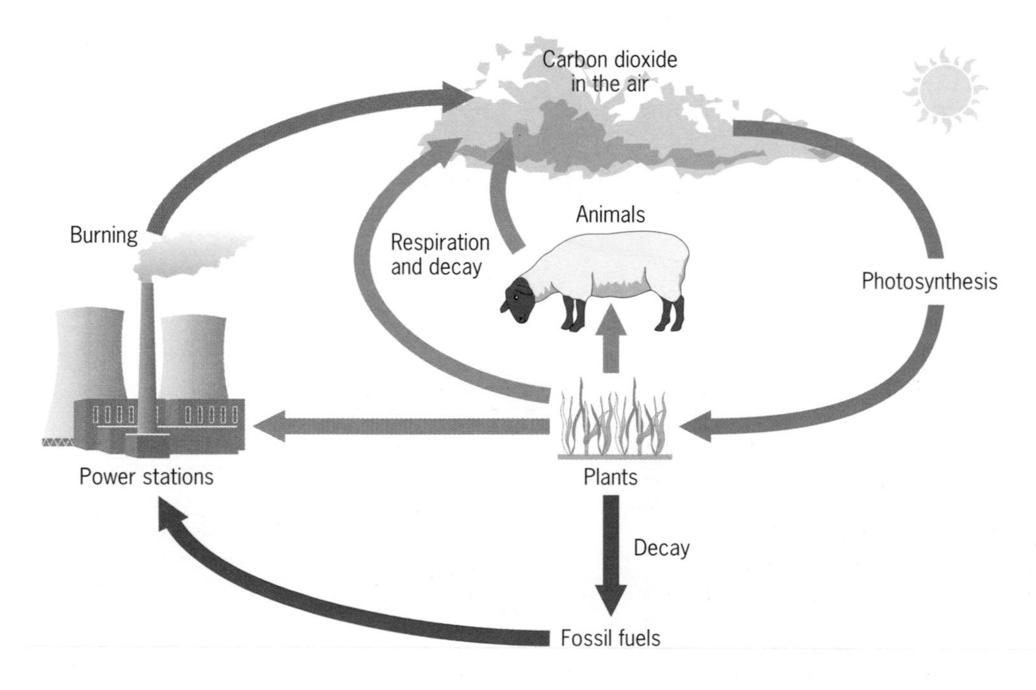

Figure 4.27 The carbon cycle

?

33a Use the figures given in Figure 4.26 to calculate the percentage of the sun's energy absorbed by a plant and used in photosynthesis.
b Calculate 30 per cent of 71 kilojoules per square metre per day to find out how much energy would be available for electricity in the home.

34a Use Figure 4.27 to identify where the carbon in the plants comes from.
b How will wood as a fuel affect the level of carbon dioxide in the atmosphere, compared with the use of fossil fuels?
c What other acid-rain-producing gases are given off by fossil fuels but not by wood burning? (see Chapter 5)

35 Essay: The efficiency of thermal power stations is about 30 per cent. This means that 70 per cent of the heat produced by burning fuels such as wood is wasted. Examine the environmental advantages and disadvantages of coppice management for electricity production.

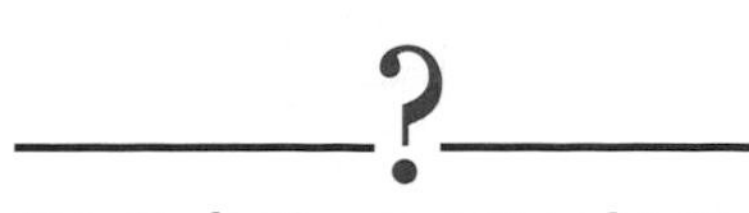

36a Study Figures 4.28 and 4.29 and describe the difference in structure of the two woodlands.
b Suggest reasons how fencing and grazing could create the differences that you described in **a**.

Table 4.9 The difference in the number of saplings in Coed Cymerau

Year	Number of saplings
1964	195
1979	5316

4.5 Grazing in upland broadleaved woodlands

Woodlands have been important in agriculture for centuries as part of the annual grazing cycle. Traditionally they were of particular importance in upland areas. The animals were put into woodland in bad weather, especially in the winter months, because the trees provide shelter. So, during the summer period, seedlings could become re-established.

Modern upland farming practices, encouraged by EU support (see Chapter 10), mean that the numbers of sheep have greatly increased. With the poor nutrition levels available to sheep on marginal land, farmers have had to provide extra feeding from the woodland. The purpose of this farming system is to keep the animals off the enclosed grazing and hay fields. This avoids reducing the growth of the grass which will yield the single summer hay or silage crop.

Woodlands which are heavily grazed have an ecological character of their own. The animals selectively graze and trample the ground flora. This usually results in a grassy sward with poor species diversity between the trees. They also add manure to the **ecosystem** (Figure 1.10).

Figure 4.28 Grazed oak woodland, Wales

Figure 4.29 Ungrazed oak woodland, Wales

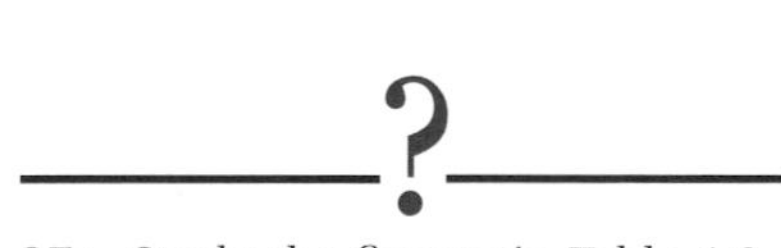

37a Study the figures in Table 4.9. Is it possible to relate these totals to grazing? Give reasons for your answer.
b If grazing had continued over a long period, how would this have affected the age structure of trees in the woodland?
c What might have been the long-term effects on the woodland if the original pattern of grazing had continued unchanged?

Coed Cymerau woodland

Coed Cymerau is a woodland in North Wales. Historically it was part of a farm, and the woodland was grazed by both cattle and sheep. The woodland was designated a National Nature Reserve in 1962, and a study area was fenced in 1963 to prevent grazing by farm animals. For 30 years natural regeneration processes have been allowed to operate.

As vegetation in the fenced areas of the reserve moved through **successional stages**, brambles became abundant. They formed a dense growth in places, completely shading the soil beneath. them. This affected the growth of herbaceous plants, some of which were rare species. Such herbaceous plants are important for insects, which in turn are part of the food web of the woodland.

An appropriate livestock density in a woodland is important for maintaining some herbaceous plants. However, farmers' knowledge of correct stocking levels has been lost. The knowledge which has been used to manage woodlands sustainably for hundreds of years can only be gained back gradually by trial and error.

?

38 Should woodlands be managed with livestock to protect rare herbaceous plants or should the wood be left to follow the natural process of succession? Give reasons for your view.

39 From the information provided in Figure 4.30, describe the relationship of regeneration to grazing level in the wood.

40 Draw a diagram using Table 4.10 to show the changes that will take place within the woodland if grazing takes place.

41 Assume that you are an agricultural adviser visiting a farm to advise on management of the woodland. The farmer has noticed that his woodland is ageing, with no young trees. However, he is definite about the fact that he must continue grazing his stock in the woodland for both shelter and for the grazing that it provides. Using Table 4.11 suggest two ways to help to regenerate the woodland in these circumstances. Give reasons for your choice.

42 After taking the advice that you provided on his farm in question 41, the farmer dies and his child sells the woodland to the County Trust for Nature Conservation. You are asked to revisit the woodland. Advise the Trust on the management policies likely to be beneficial to their interests.

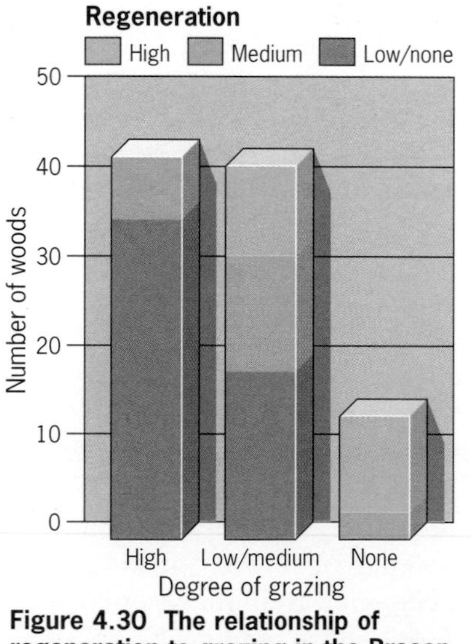

Figure 4.30 The relationship of regeneration to grazing in the Brecon Beacons (after Hughes et al., 1991)

Table 4.10 Some differences between grazed and ungrazed woodland

Grazed	Ungrazed
Overgrazing reduces the number of perennial grasses which tend to be replaced by annuals.	More perennial grasses
Grazing by sheep will reduce the ground flora and prevent regeneration	Denser ground flora and many small seedlings
Existing large trees which are not browsed sufficiently at a high level may compete for light and nutrients	The canopy may be denser, with less light reaching the floor, since fewer ground and understorey species compete for nutrients.
Some seeds of plant species are spread by being carried in guts/wool	
Passage of matter through the gut and out as faeces changes the nitrogen cycle, so the soil could be richer in nitrogen than in an ungrazed area. This may have an affect on the invertebrates	The floor will not be enriched by nitrogen and so species may not be the same.
Grazing can increase species diversity by opening up the community and creating more niches	
Excessive grazing may reduce the size of soil particles and plant litter to a point where they can be eroded. Puddling the soil surface by trampling can speed up erosion, since infiltration capacity will be reduced	The infiltration rate should be faster
Heavy grazing can kill plants and reduce their level of photosynthesis	
Trampling and grazing resistant plants may dominate, as palatable or fragile plants are damaged. Bracken may expand, since it is slightly poisonous to young stock and will be avoided. Similarly elder, gorse, broom and rag weed will thrive in grazed areas	Plants not resistant to trampling and grazing may dominate, such as bluebells. Plants with opposite characteristics to those listed for grazing-tolerant species will be found.
Plants resistant to grazing tend to have the following characteristics: • small size, since the plant will get more protection from the soil around it • plant stems and leaves spread close to the ground. • small leaves (less easily damaged by treading) • firm tissue and bending ability • strong vegetative increase characteristics, e.g. by spreading stolons. • small, hard seeds and a large number of seeds per plant, since treading will damage many.	

Table 4.11 Tree protection against sheep

Type	Longevity	Efficiency	Cost
Timber guards	Long-term, 20 years. Can be damaged by animals rubbing against them.	At 0.25 metres away from tree, they must be 1.10m high; wire must have 50mm vertical and horizontal spacing. 0.90m if further away, 150mm wire.	Expensive
Tree shelters	5 years, but may be damaged by animals rubbing	Growth rate rapid, acts as greenhouse. 5 times normal rate. 1.2m high for sheep	If less than 1ha, cheaper than fencing
Chemical	2–3 months	Applied to trees in winter not when active growth occurs.	Cheaper than timber guards

Figure 4.31 Woodland landscape, Scotland

Figure 4.32 Woodland landscape, Scotland

The use of a set of pairs of opposed adjectives is one way of assessing landscape. This method is known as the bipolar semantic scale, in which you can use different sets of adjectives.

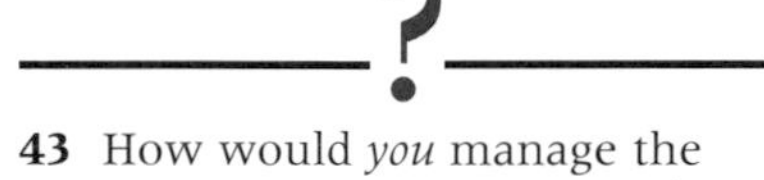

43 How would *you* manage the woodland? Give reasons for your decisions.

44 Essay: Broadleaved woodlands are under pressure. How can these pressures be reduced?

45 Look at Figure 4.31 and make your assessment of the landscape using the scale below. Note the number closest to the adjective that best describes your assessment. Three is neutral.

Enclosed	1 2 3 4 5	Exposed
Harmonious	1 2 3 4 5	Discordant
Random	1 2 3 4 5	Ordered
Interesting	1 2 3 4 5	Boring
Beautiful	1 2 3 4 5	Ugly

46a Figure 4.32 also shows woodland. Carry out the same landscape assessment on this scene. Add other adjective pairs of your choice.
b Comment on your perceptions of the two landscapes.

Summary

- High latitudes and reduced solar radiation inputs cause productivity, biomass and species diversity to be lower in temperate deciduous forest than in tropical rainforests.
- Temperate deciduous forest is adapted to seasonality. Compared with tropical rainforests it has fewer species, more balanced distribution of nutrients throughout its stores and shuts down nutrient transfers temporarily during the non-growing season.
- Since human activity in Europe accelerated, woodland has declined. Few areas of primary forest remain. Climate has affected the distribution of some species but the decline is associated with deforestation by humans and the rise of agriculture.
- Woodlands have traditionally been managed on multi-purpose, sustainable principles.
- As the use of woodlands has intensified, many have lost their sustainability.
- Sustainable use of woodland is once again being encouraged, along with an integration of timber production, grazing, a rich wildlife habitat, opportunities for game and recreation, plus aesthetic appeal.
- Some forests have a primary objective such as timber production or recreation. Other interest groups will gradually be integrated into these aims.
- The Forestry Commission aims to use multi-purpose principles to run its Forest Authority and Enterprise divisions.
- Markets for small-scale woodland timber products are being pioneered and will encourage multi-purpose use on private land.
- Funding is currently available which favours planting of broadleaved species.

5 Native and non-native boreal forests

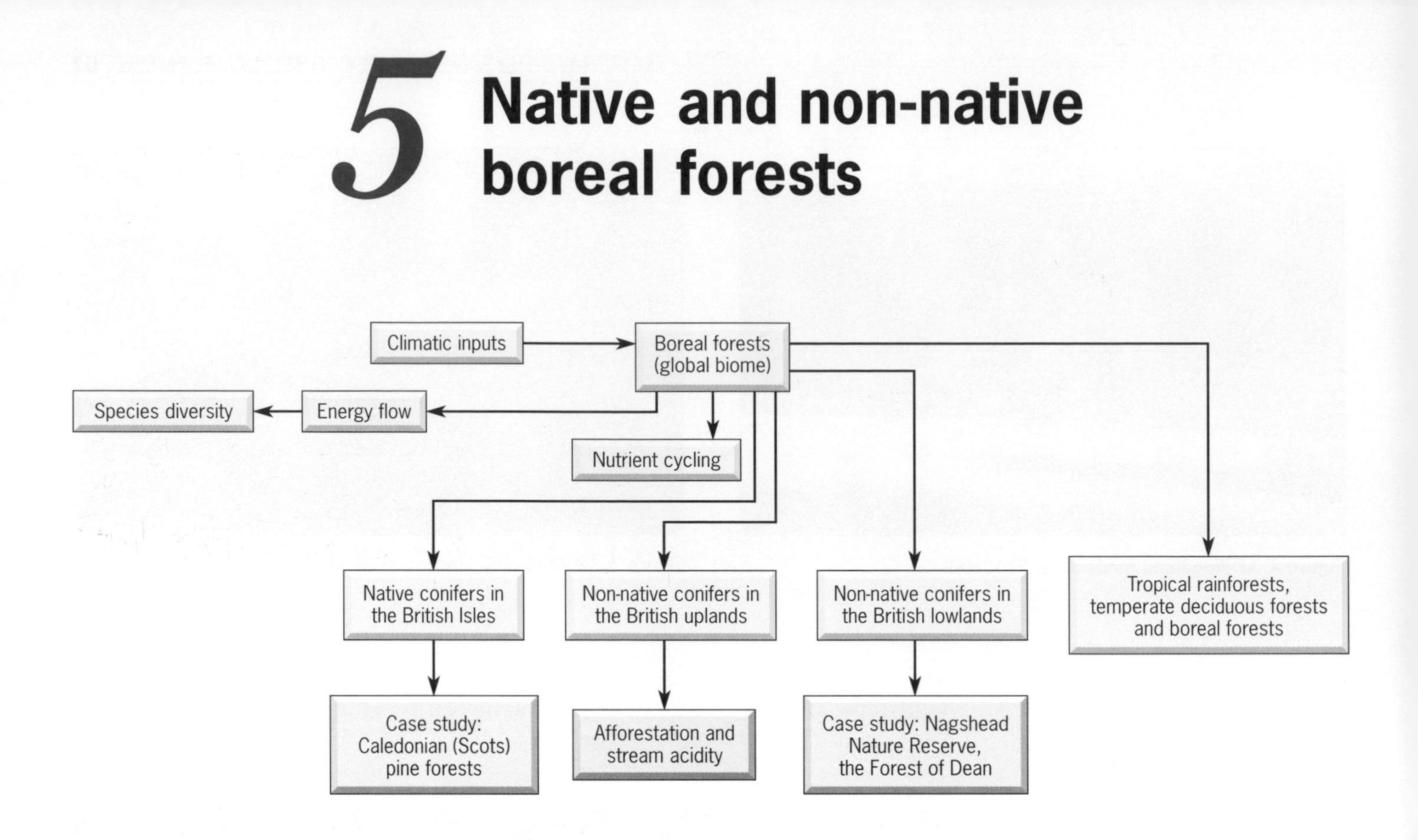

5.1 The nature of the ecosystem

Climatic inputs

Most plants cannot grow below 5–6°C. Plants have various **adaptations** to survive in such an environment (Table 5.1).

Figure 5.1 Primary boreal woodland, Sweden

Table 5.1 Adaptations of advantage in colder climates

Temperate deciduous trees	Boreal trees (Fig.5.1)
1 Thick bark to protect woody tissues against frost damage.	1 Thick bark which helps to protect woody tissues against intense cold.
2 Leaf loss to reduce **transpiration.**	2 Needle-like leaves, thick cuticles, aromatic oils and stomata sunk in pits all slow down transpiration levels.
3 Rise in sugar concentration in the sap to prevent water loss.	3 Sugar concentration in the sap increases in winter allowing water to enter the cells but it cannot move outwards. This offers further protection against water loss,
4 Deciduous trees produce leaves and flowers very soon after the start of the growing season. Sometimes flowering begins before leaf growth.	4 Most trees are evergreen which allows transpiration and **photosynthesis** to begin immediately when warm weather arrives.
5 When the soil is cold, tree roots can only absorb small quantities of water. Growth almost ceases and the low temperatures cause a **physiological drought**. The trees lose leaves to reduce transpiration.	5 Larch can grow in very cold environments. It is deciduous to reduce transpiration even further.
	6 Shallow root systems enable the tree to start taking up water as the surface layer thaws, even if the subsoil is still frozen.

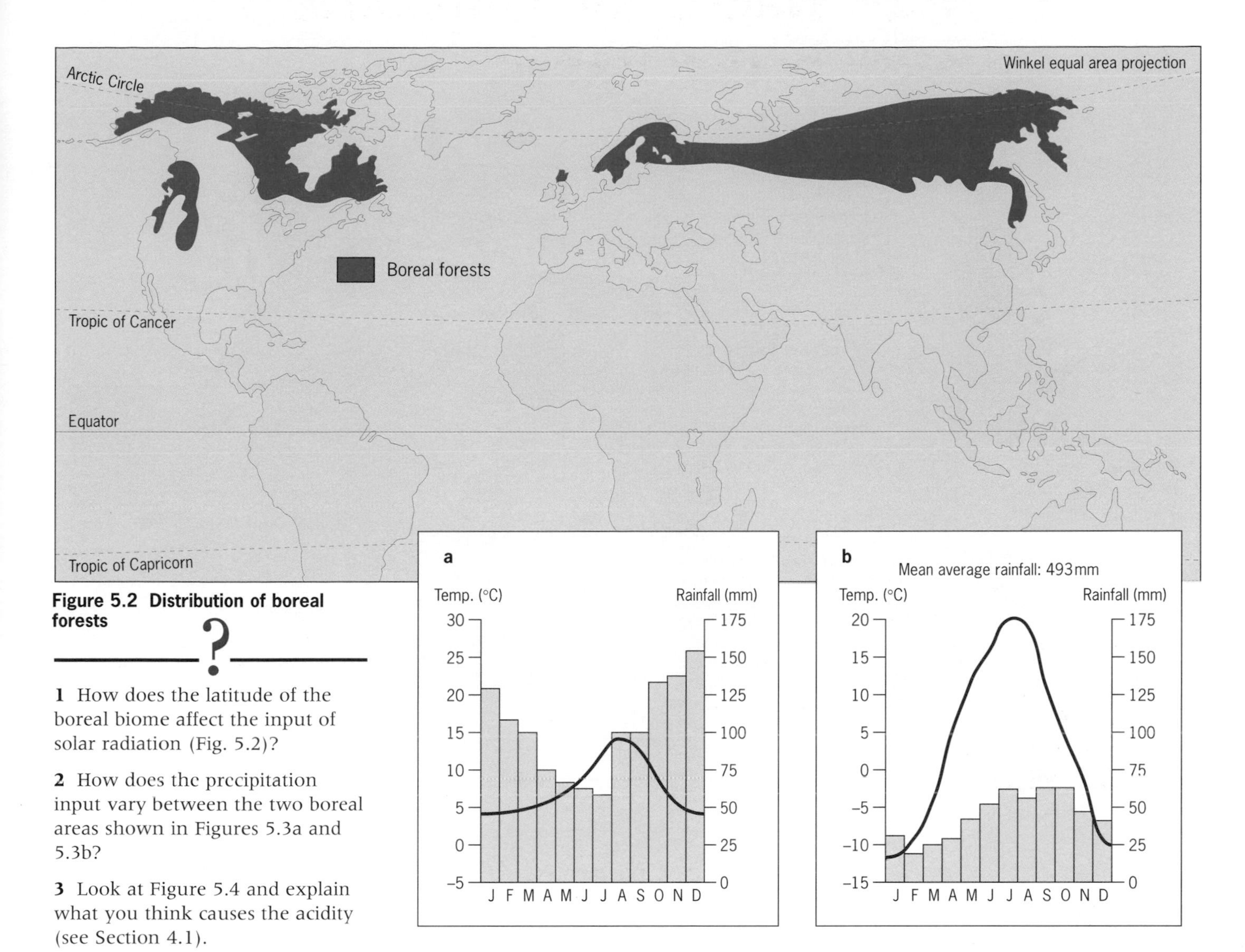

Figure 5.2 Distribution of boreal forests

?

1 How does the latitude of the boreal biome affect the input of solar radiation (Fig. 5.2)?

2 How does the precipitation input vary between the two boreal areas shown in Figures 5.3a and 5.3b?

3 Look at Figure 5.4 and explain what you think causes the acidity (see Section 4.1).

Figure 5.3 Climate graphs for (a) boreal site in Scotland, (b) Arkhangel'sk, Russia

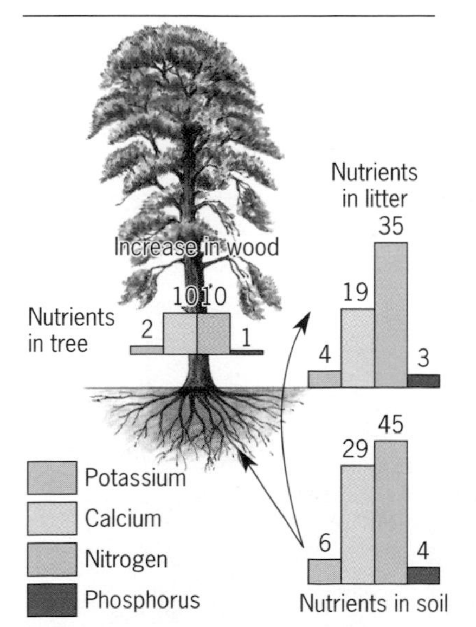

Figure 5.4 Nutrient cycling in a boreal pine forest in kilograms per hectare per year (after Duvigneaud and Denaeyer De Smet, 1970 in O'Hare, 1988)

Boreal forest replaces temperate **deciduous** trees as the dominant vegetation where the growing season falls below six months and the frost-free period below four months (Fig. 5.1). Most of these trees are **conifers**. Such trees have seeds which are not in enclosed seed cases. Generally conifers are associated with long, cold winters, but other **species** are also adapted to these areas associated with physiological drought (conditions where water is reduced by physical factors other than drought). Higher rainfall can contribute to leaching and soil acidity in some boreal areas. However, less rainfall in some boreal areas means that the characteristic **podsol** soil **horizon** is acidic for another reason.

Energy flow

Energy inputs decrease as latitudes increase. Therefore the high-latitude boreal biome receives a lower total energy input.

Nutrient cycling

Decomposers within the litter layer are fewer than in temperate deciduous forests or tropical rainforests because of the lower temperatures. This means that the recycling of nutrients is slower. The task of mixing acidic (mor) **humus** into horizons below is very slow, particularly since earthworms (which help to mix humus into lower mineral horizons) do not thrive in acidic conditions. This extra acidity leads to rapid leaching, called podzolisation. Iron moves downwards to a new layer, leaving behind a horizon of mainly silica. The iron layer is broken up by the roots of conifers and does not form an iron pan.

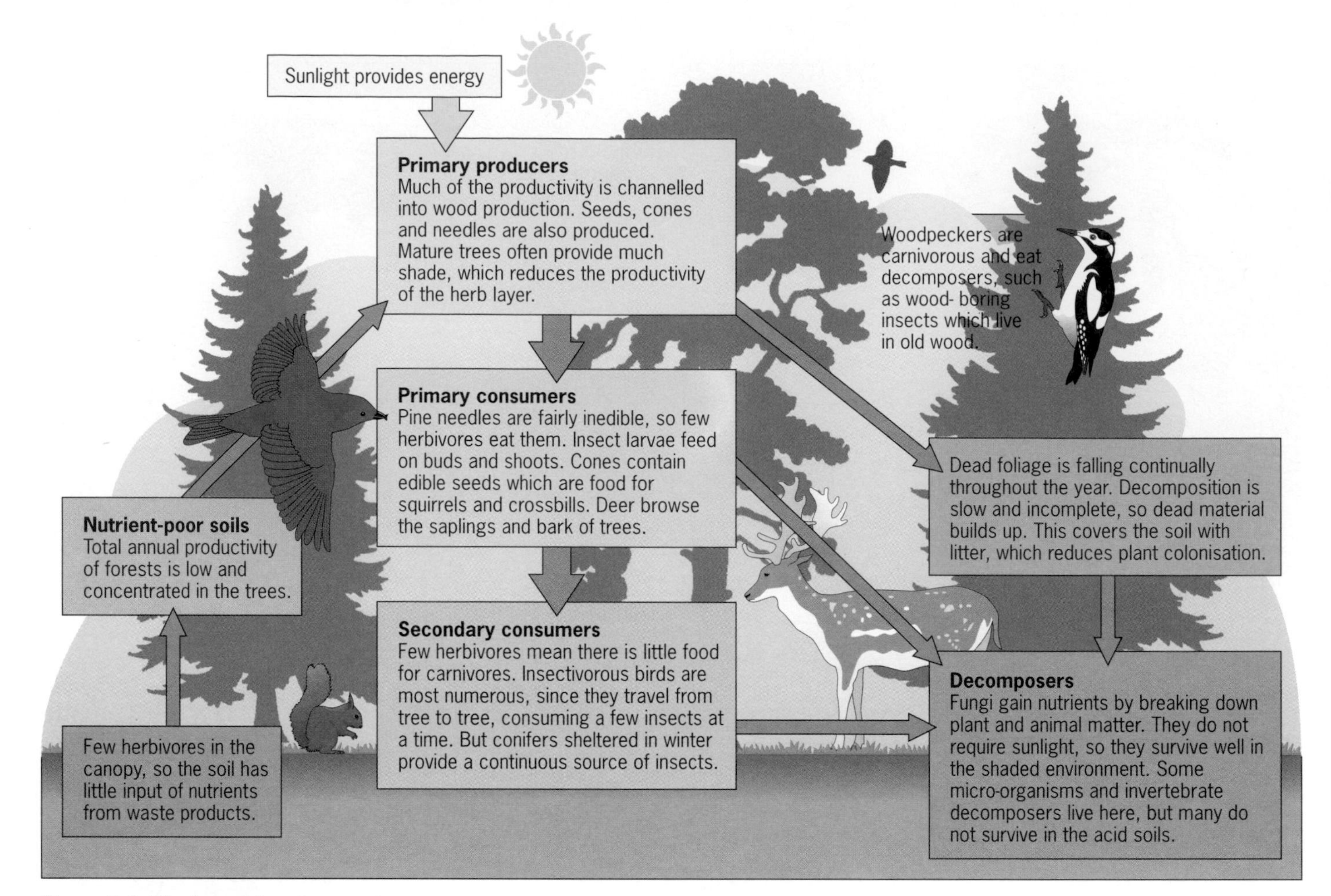

Sunlight provides energy
Primary producers
Much of the productivity is channelled into wood production. Seeds, cones and needles are also produced. Mature trees often provide much shade, which reduces the productivity of the herb layer.
Woodpeckers are carnivorous and eat decomposers, such as wood- boring insects which live in old wood.
Primary consumers
Pine needles are fairly inedible, so few herbivores eat them. Insect larvae feed on buds and shoots. Cones contain edible seeds which are food for squirrels and crossbills. Deer browse the saplings and bark of trees.
Dead foliage is falling continually throughout the year. Decomposition is slow and incomplete, so dead material builds up. This covers the soil with litter, which reduces plant colonisation.
Nutrient-poor soils
Total annual productivity of forests is low and concentrated in the trees.
Secondary consumers
Few herbivores mean there is little food for carnivores. Insectivorous birds are most numerous, since they travel from tree to tree, consuming a few insects at a time. But conifers sheltered in winter provide a continuous source of insects.
Decomposers
Fungi gain nutrients by breaking down plant and animal matter. They do not require sunlight, so they survive well in the shaded environment. Some micro-organisms and invertebrate decomposers live here, but many do not survive in the acid soils.
Few herbivores in the canopy, so the soil has little input of nutrients from waste products.

Figure 5.5 The boreal forest ecosystem

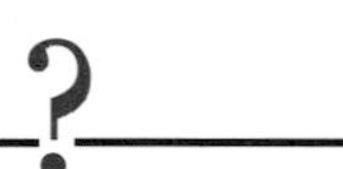

4 From Figure 5.5 write down three food chains.

5 Why do you think that the boreal forest has a relatively simple food web?

6a Find out the net primary productivity and biomass for boreal forests from Table 1.1.
b Explain how you think the inputs into the boreal forest ecosystem might affect its primary productivity and biomass.

7 Explain the distribution of nutrients shown in Figure 5.4.

8 From Figure 5.6, explain:
a Why the layers in the upper soil horizon are not decayed.
b Why so many metals are moved downwards (leached).
c What might happen to water when it reaches the iron pan. What effect could this have on soil acidity?

5.2 Conifers in boreal areas

In some boreal zones of the British Isles, conifers are native species and therefore have their own distinctive communities associated with them.

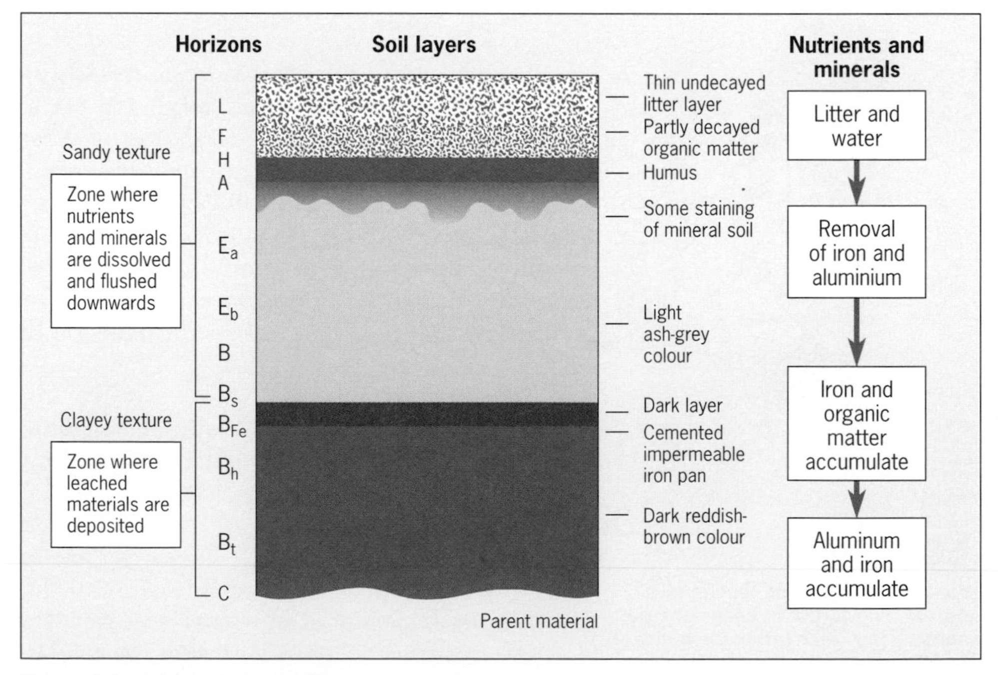

Figure 5.6 A podsol soil profile

Caledonian (Scots) pine forests

Figure 5.7 A plantation of Scots pine

Figure 5.8 A Caledonian (Scots) pine forest, Abernethy

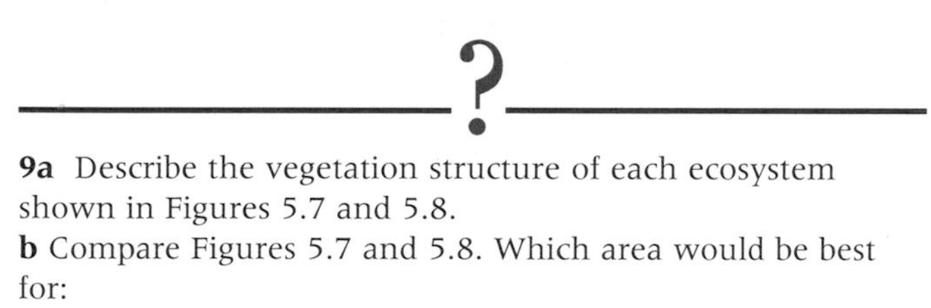

9a Describe the vegetation structure of each ecosystem shown in Figures 5.7 and 5.8.
b Compare Figures 5.7 and 5.8. Which area would be best for:
- timber production
- wildlife?

In native pine woods, Scots pine is the dominant species, with birch, rowan, aspen and juniper as shrub flora. It has a wide distribution in the boreal zone of the northern hemisphere on the poorer soils to which it is well adapted. Natural climatic factors, and other factors such as fire (caused by lightning) and windthrow, produce a forest with open glades, wet and boggy areas and a varied age structure with a proportion of dead and dying trees. The Great Wood of Caledon which was once extensive in Scotland (Fig. 5.9) has developed its own characteristics in isolation since the British Isles became separated from the European continent.

Native pine forest, the remnant of the Great Wood of Caledon, supports a wide variety of plants and animals, including a ground flora of rare plants such as one-flowered wintergreen, twinflower, and an orchid – creeping ladies' tresses. It has a characteristic group of **invertebrates**, many uncommon elsewhere. New species are still being recorded. No mammals are entirely dependent on this **habitat**. However, the red squirrel, a declining species, favours it, and it is also the natural home of wildcat, pine marten and red deer. Red deer also eat young pine.

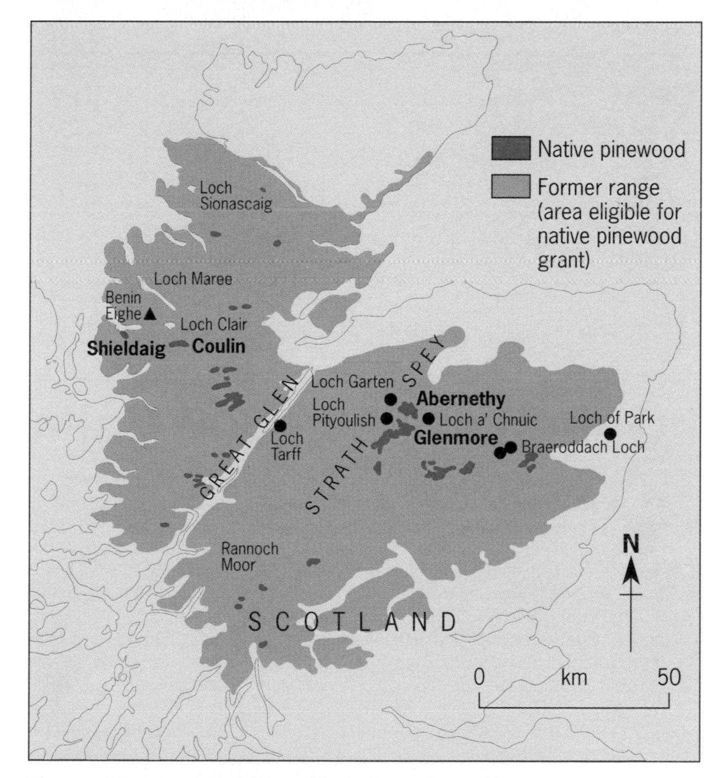

Figure 5.9 The Great Wood of Caledon (after Sullivan, 1977)

Figure 5.10 A pair of Scottish crossbills

Seventy species of birds regularly breed in the pines, three of which – crested tit, capercaillie and the unique Scottish crossbill (Fig. 5.10) – are found nowhere else in the world.

Twenty five per cent of the area remaining in 1957 had been lost by 1990, due mainly to clearfelling and underplanting with non-native species. Estimates suggest that the current area of native pine is 12 000 ha, while 47 000 ha of land area are capable of regeneration from native seed stock. Some of the 12 000 ha of native pine are plantation pine. The remaining pine forest is now largely confined to:

- private land and reserves
- steep-sided valleys
- soils unsuitable for alternative uses
- areas of low grazing density
- areas where grants have been used to replace non-native stock.

?

10a Why do you think that conservationists are worried by the loss of native pine woods?
b Why have native pine woods been lost?
c Suggest why these figures add up to more than 100 per cent.
d How much of the total area is protected pine forest?

Table 5.2 Areas in the Great Wood of Caledon with native pine

Status	Percentage of total area of native pine
SSSI	80
Candidate SPA	30
RSPB reserves	13
Other private ownership	<10
Forestry Commission	20

Table 5.3 The Woodland Grant Scheme, 1992

A Establishment grants (£ per hectare): for planting, restocking and natural regeneration. Grants for native pinewoods are paid at the broadleaved rate.

Grant band	Coniferous	Broadleaved
Less than 1 ha	1005	1575
1–2.9 ha	880	1375
3–9.9 ha	795	1175
10 ha and over	615	975
Plus better land supplement	400	600

B Management grants (£ per hectare): for looking after woodlands

Size of wood	Standard Conifers	Standard Broadleaved	Special
Less than 10 ha	15	35	45
10 ha and over	10	25	35

C Plan preparation grants: for woodlands eligible for management grant and not previously grant-aided by the Forestry Commission.

Grant per woodland: £100.

Queen in £300,000 Balmoral row

The Queen is seeking £300,000 of public money to save an ancient pine forest on her Balmoral estate. The move will provide fresh controversy over whether Britain's richest family should expect the taxpayer to contribute to work on the royal family's private holdings.

Opposition MPs yesterday accused the royal family, which does not pay taxes, of sponging off the state. Conservationists attacked the plan, involving fencing off the forest to prevent deer intruding and causing destruction, as environmentally misguided.

The row follows last week's revelation in parliament that public scrutiny of the £10m of civil list money paid each year to the Queen and other members of her family out of the public purse has been barred until 2000. Labour and Liberal Democrat MPs are planning to demand an emergency debate this week on why the government agreed to keep the 1990 deal secret.

The Balmoral grant application has been made to the Forestry Commission. The Deeside estate is seeking financial assistance under the commission's native pinewood grant scheme to encourage the regeneration of Ballochbuie forest, one of the largest remaining fragments of the ancient Caledonian forest that once swathed the Highlands. The forest is dying out because there are too many red deer eating young trees.

Balmoral's plan is to erect a huge fence around the 1,000-acre site to keep out the animals and allow trees to grow. The £300,000 application, which is by far the biggest made by the estate, is likely to be granted in the next few months.

'Many people will find it remarkable that the royal family is so assiduous in claiming grants for absolutely everything that they do, much of it with no public benefit,' said Labour's rural affairs spokesman, Brian Wilson. 'If you don't pay taxes you should not be able to claim grants out of the taxes which other people pay.'

The leader of the Scottish National Party, Alex Salmond, suggested that they would always be controversy so long as the tax question remained resolved in favour of the Queen and her relations. 'You can't have the bun and the penny,' said Salmond. 'My preference is for the royal family to become subject to the same taxes as everyone else.'

The Balmoral plan is also being criticised by conservationists who claim the fencing scheme will damage the landscape and wildlife.

Dr Adam Watson, an authority on Deeside's native pinewoods, argued that fences would make the forest 'completely unnatural'. It also presented a danger to rare birds like the capercaillie and black grouse, which could be killed by flying into the fencing wire.

Fencing would not restore Ballochbuie to full ecological health, according to Dave Morris, Scottish officer of the Ramblers' Association. 'Surrounding areas will continue to degenerate while an artificial, patchwork forest will appear inside the fences. If massive sums of public money are to be spent on Balmoral to help the Caledonian forest, they should be directed towards increased culling of red deer, not on ineffective, intrusive fences.'

In the past the Balmoral estate has been given a series of much smaller grants from other public bodies for footpath repairs and small birchwood enclosures.

The application to the Forestry Commission follows the breakdown of prolonged negotiations with the government's Scottish Natural Heritage agency, which had been hoping to persuade the estate to cull more deer. Balmoral is factored by Martin Leslie and overseen by the Duke of Edinburgh.

Details of the grant request are regarded as confidential by Buckingham Palace, which simply confirmed yesterday that one had been made. A spokesman for the Forestry Commission said the application was within the terms of the woodland grant scheme, and was under consideration.

Figure 5.11 Newspaper report about the controversy to save an ancient pine forest (*Source: Scotland on Sunday*, 12 July 1992)

?

11a According to Figure 5.11, what is considered to be the main cause of loss of pinewood in 1992?
b What percentage of remaining native pine did this area represent?
c What government incentives were available for improving native pine?

12 From the information in Figure 5.11 and from the rest of the case study, list the arguments for and against fencing part of the Ballochbuie estate.

13 You have inherited 100 ha of pine wood in Strathspey. Twenty per cent is true native pine forest, and part of this 20 per cent has been designated as an **SSSI**. An SSSI agreement has been made after you were approached by Scottish Natural Heritage. You have developed an agreed **management** policy with them, on the understanding that an annual payment for management of the land will reflect any income you lose as a result of the agreement. The remaining 80 per cent was planted with Scots pine 50 years ago and has received no management of any kind. Details of the Woodland Grant Scheme are given in Table 5.3.
a Identify the value positions involved.
b Outline the management agreement made between you and Scottish Natural Heritage.
c List the steps you would take to improve the remaining area as pinewood wildlife habitat and at the same time obtain some money from its management. Include the sources of any advice you would seek.

5.3 Non-native conifers in UK uplands

In 1984 the **Forestry Commission** reached a target of 2 million hectares of productive woodland. Most of this total is in upland areas. Almost half of this is in Scotland, where 12 per cent of the land is **afforested.** This is small in comparison to other EU countries, where 22 per cent is the average. The Forestry Commission estimate that another 1.7 million hectares of Scottish uplands are capable of growing timber.

Forestry does not have to observe statutory planning controls, but bodies such as planning authorities are consulted when private owners and the Forestry Commission make grant applications. However, there is no neighbour notification scheme, so the first that someone knows about a forest on adjacent land may be its actual planting!

The Forestry Commission's duties include sympathetic multi-purpose management, but local communities have identified some disadvantages of forestry in addition to benefits (Figs 5.12 and 5.13).

Figure 5.12 The community and forestry

Figure 5.13 A forester engaged in plantation work

Figure 5.14 Afforestation in the middle zone of mountains, Loch Laggan

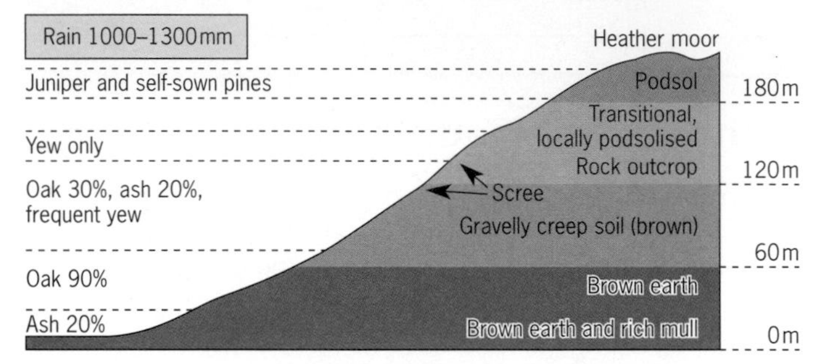

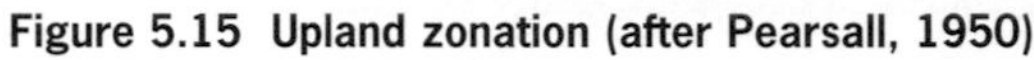

Figure 5.15 Upland zonation (after Pearsall, 1950)

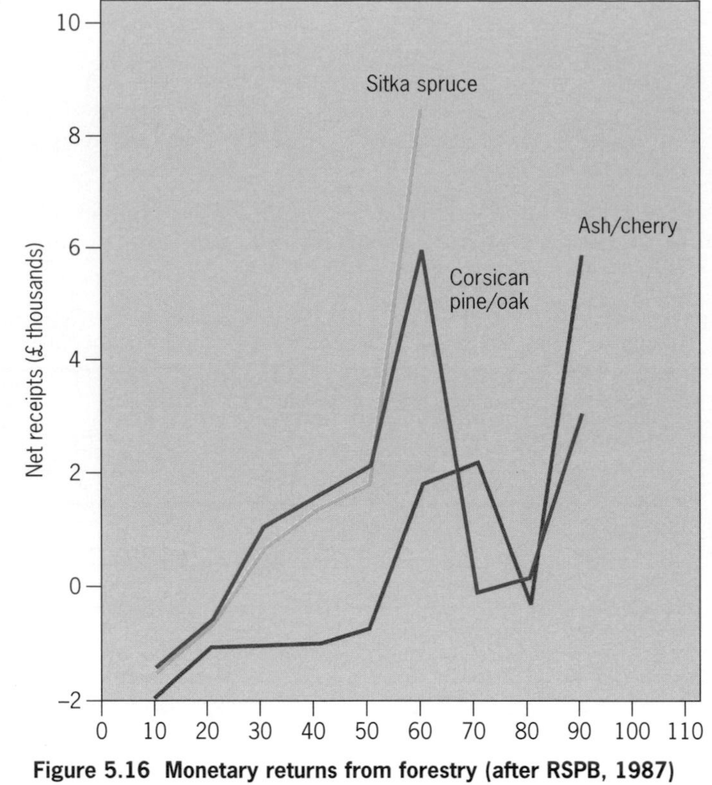

Figure 5.16 Monetary returns from forestry (after RSPB, 1987)

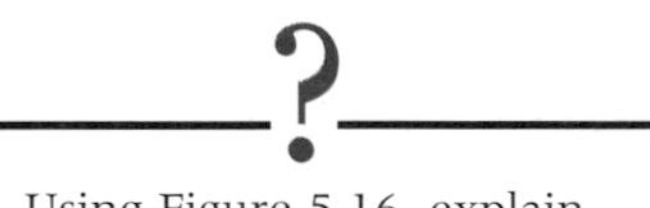

14a Using Figure 5.16, explain why people may favour growing conifers.

b Suggest why some people are opposed to this.

Upland afforestation usually takes place in upper middle zones, where growth is quicker, leaving mountain tops bare (Figs 5.14 and 5.15).

The tree line extends from 350 to 550 metres and does not descend on to better improved hill pasture and arable farms. The system of planting is generally of even-aged uniform plantations, with periodic clear felling and replanting. There is therefore little structural or species diversity. Most of the post-1945 planting was of sitka spruce from north-west USA. This species reaches maturity between 40 and 60 years of age, which is twice as fast as an oak tree. Other common trees include lodgepole pine, Douglas fir, larches, Norway spruce and Scots pine.

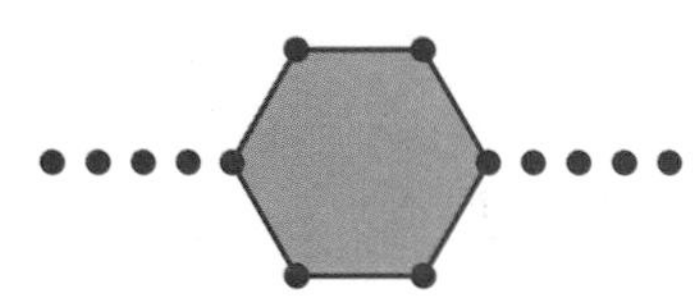

Afforestation and stream acidity

It is widely accepted that some rivers and lakes have become increasingly acid in the twentieth century because atmospheric pollutants, especially sulphur dioxide, are deposited as **acid rain**. This has been noted particularly in Scandinavia, North America and Great Britain.

In Great Britain the acidification has been linked with the planting of conifer forests of non-native species. Some ecologists think that afforestation may have the following effects:

- The tree crowns can trap acid pollutants and salts easily because of their shape.
- Forests evaporate more water than moorland, and so acid pollutants become concentrated.
- Chemical processes take place in the tree crown, particularly the uptake of ammonium and release of hydrogen.

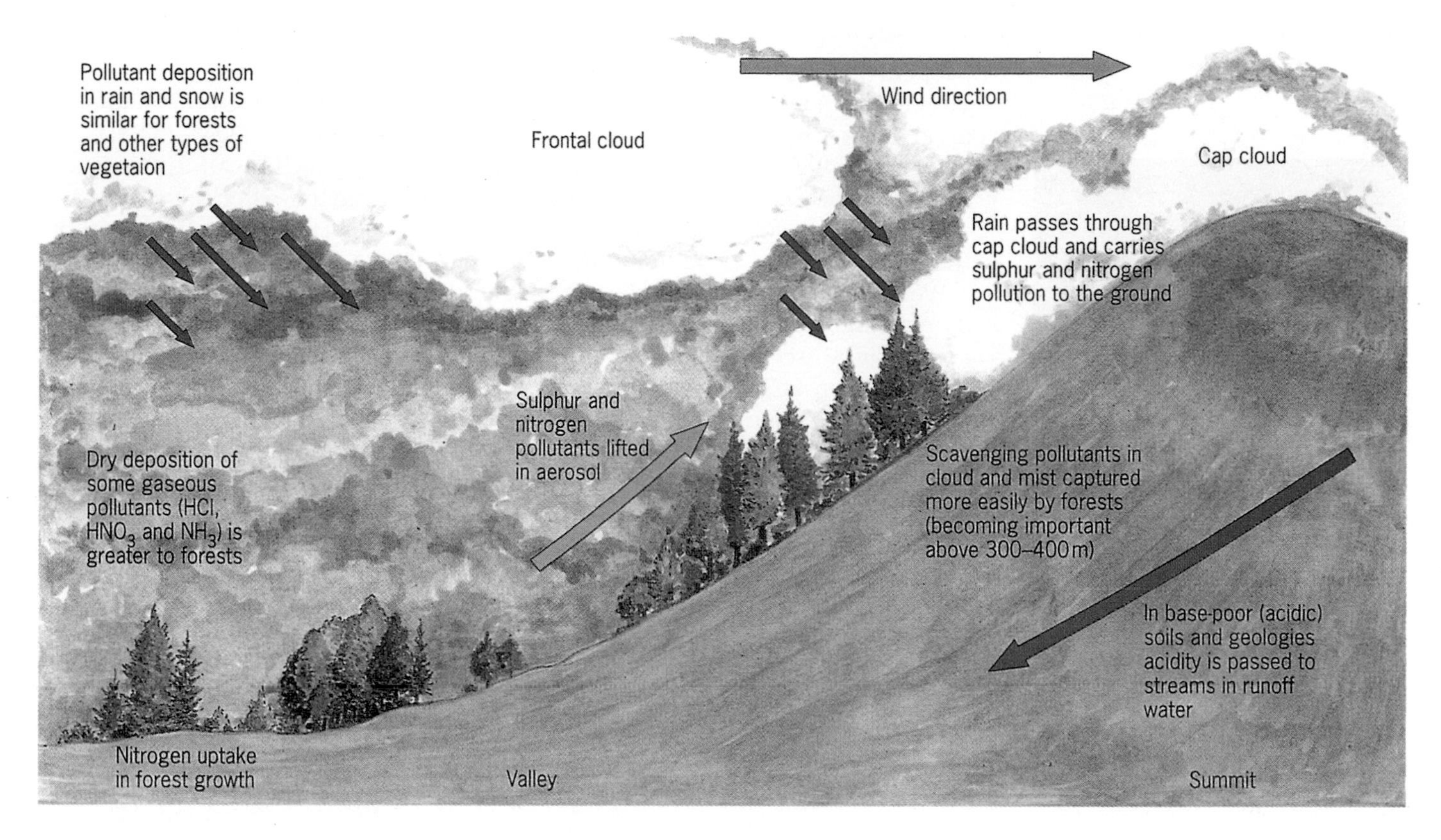

Figure 5.17 Acid rainfall in coniferous forests (after Forestry Commission, 1993)

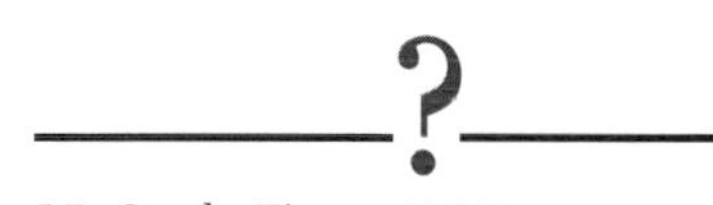

15 Study Figure 5.17.
a Describe how the stream increases its acidity.
b Explain why acidification is likely to be worse where there are acidic rocks.

16a Using Table 5.4, plot a scattergraph of pH against the level of afforestation.
b Describe any relationships that you can infer from your scattergraph.

- Planting trees causes the soil to dry out. This causes greater oxidation of sulphur and nitrogen compounds in the soil.
- The drying of soil causes cracks in the peat. The cracks and forest drains (which are dug when planting takes place) mean that water moves in an unnatural way through the forest.
- Acidic humus produced by pine needles encourages increased soil acidity. Water moving through the soil will cause leaching. Both of these processes combine to increase soil acidification.

Table 5.4 Effect of afforestation on streams in Scotland, 1987

Waterway	Mean summer pH	Areas afforested (%)
Big Water of Fleet	5.7	90
Big Water of Fleet	6.6	30
Dalwhat Water	7.2	20
Garpel Burn	7.0	5
Garrouch Burn	7.1	5
Little Water of Fleet	5.5	100
Little Water of Fleet	6.1	40
Moneypool Burn	7.0	5
New Abbey Pow	6.7	10
Palnure Burn	6.7	80
Penkiln Burn	6.3	65
River Girvan	7.5	10
Scaur Water	7.5	10
Shinnel Water	7.5	35
Skyre Burn	8.1	0
Southwick Burn	7.1	10
Water of Cree	5.7	55
Water of Minnoch	6.3	85

The effect of stream acidity on dippers

Dippers are birds which are closely associated with fast-flowing streams. They nest under bridges and in cavities in stonework. They feed on stream invertebrates by swimming under water to catch their prey (Fig. 5.18).

The breeding areas of dippers have been greatly changed by afforestation. Data have been collected to investigate the various factors affecting dippers.

Seventy-four streams in mid-Wales were surveyed. They were divided into those with dippers present and those without (Table 5.5).

We can compare the difference between the streams with and without dippers and assess statistical significance of the various factors. In this case a *t*-test is used (see Appendix 2). The results show that the probability of the differences in forest cover between streams with and without dippers being purely due to chance is very small indeed.

Figure 5.18 A dipper

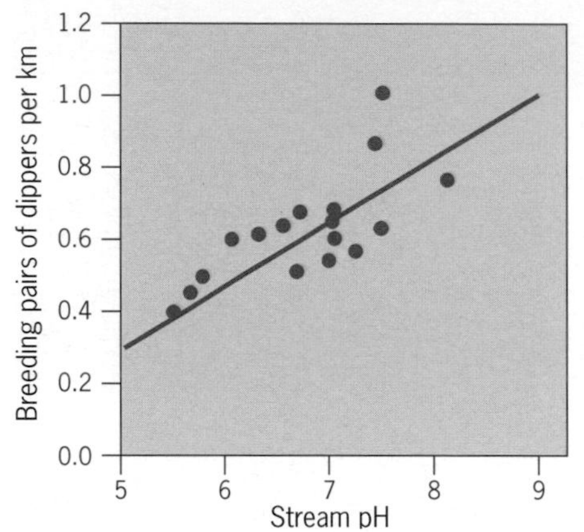

Figure 5.19 The relationship between stream pH and breeding density of dippers (after Tyler, 1985)

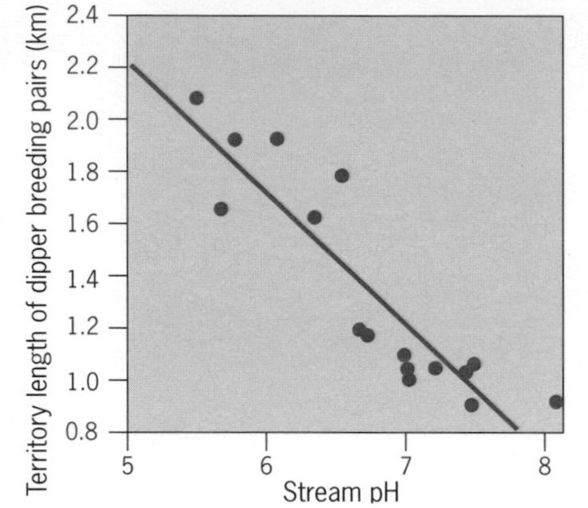

Figure 5.20 The relationship between stream pH and territory length of breeding pairs of dippers (after Tyler, 1987)

Figure 5.21 The relationship between the abundance of caddis larvae and the abundance of breeding pairs of dippers (after Tyler, 1985)

Table 5.5 Results of a dipper survey in mid-Wales

	Streams with dippers		Streams without dippers	
	X (mean)	SD (standard deviation)	X (mean)	SD (standard deviation)
Area of water catchment afforested %	8.4	26.9	36.3	26.9
Mean pH	6.4	0.4	6.0	0.4
Caddis larvae (abundance index)	12.5	2.8	4.8	2.8
Number of sites	$N_1 = 21$		$N_2 = 53$	

?

17a Carry out a *t*-test for the differences in pH and the abundance of caddis larvae (Table 5.5) (see Appendix 2).
b State the conclusions that you can draw from these data about factors connected with dipper distribution.

18a What relationships are shown by Figures 5.19, 5.20 and 5.21?
b How are these relevant to dippers?
c There is no direct causal link between dipper numbers and the acidity of the water. Suggest factors that could lead to decreasing numbers of dippers in a particular area.
d Explain what the numbers of dippers indicate about the food webs and therefore the **biodiversity** of the ecosystem.

5.4 Non-native conifers in lowland areas

Most afforestation has taken place on upland moorland. There, moorland and peat bog communities have been replaced by those of the forest. Very often such landscapes are perceived as 'waste ground' (see Chapter 9). However, afforestation has also taken place in lowland areas.

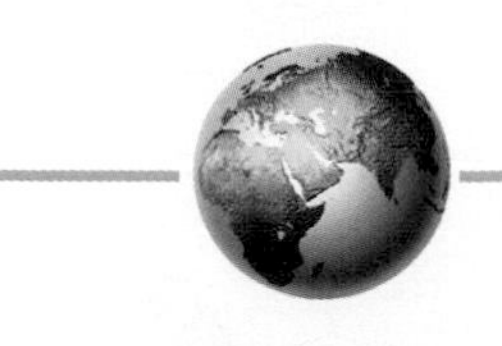

Nagshead Nature Reserve, the Forest of Dean

History of the Dean

One lowland area where non-native conifers and **broadleaved** trees are both grown is the Forest of Dean, Gloucestershire (Fig. 5.22). The Dean has produced timber for centuries. In Norman times (AD 1066–1100) the main aim of the 'Royal Hunting Forest' was to provide cover for animals of the chase. In Tudor times (AD 1485–1603) the **hardwood** was used for charcoal to supply the iron industry. The Civil War (1642–9) demanded increased timber, and new hardwood plantations aimed to supply this. From Henry VIII's day (1509–47) until the late nineteenth century, oak was grown for wooden naval ships. Once the Forestry Commission was established in the twentieth century, conifers were introduced to the area and these are now managed along with hardwood areas to produce a **sustainable** yield of timber.

Nagshead Nature Reserve

Nagshead Nature Reserve is an area of the Dean which is owned by the Forestry Commission and is used to aid conservation (Fig. 5.22). It is managed by the Royal Society for the Protection of Birds. Within the reserve there are contrasting oak wood and Douglas fir plantations (Figs 5.24 and 5.25). The oaks were planted in 1814 at the time of the Napoleonic Wars to produce oak for warships. The Douglas fir were planted in 1962 as a commercial **softwood** timber crop (Fig. 5.26).

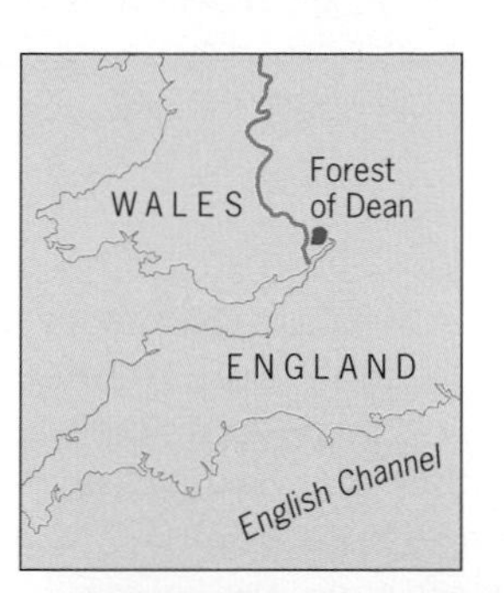

Figure 5.22 The Forest of Dean

Figure 5.23 Nestbox and pied flycatcher

Figure 5.24 A Douglas fir plantation

Figure 5.25 An oak plantation at Nagshead

To establish the effect that coniferisation has on the ecosystem, a sample site was set up and the following variables were studied: light (an important measure of energy input), soil moisture, soil pH, soil texture, infiltration (important for nutrient supply), leaf litter, soil invertebrates, plant cover, girth and height of trees (a measure of the productivity and diversity of the ecosystem).

Sampling strategy

A day-long field study investigation was carried out by the RSPB during November 1992. Sample sites 100mm long and 20 metres wide were set up in two adjacent areas of woodland: one in the conifer plantation, the other in the oak. Care was taken to choose similar topography and aspect so that direct comparisons could be made. Both areas are open to grazing by sheep, so this has a marked effect on the flora and ecological **succession**. Each site was chosen as being representative of each type of woodland. Within each site a systematic sampling strategy was used: sampling with a metre quadrat every 10 metres. Samples or tests were always taken at the centre of each quadrat.

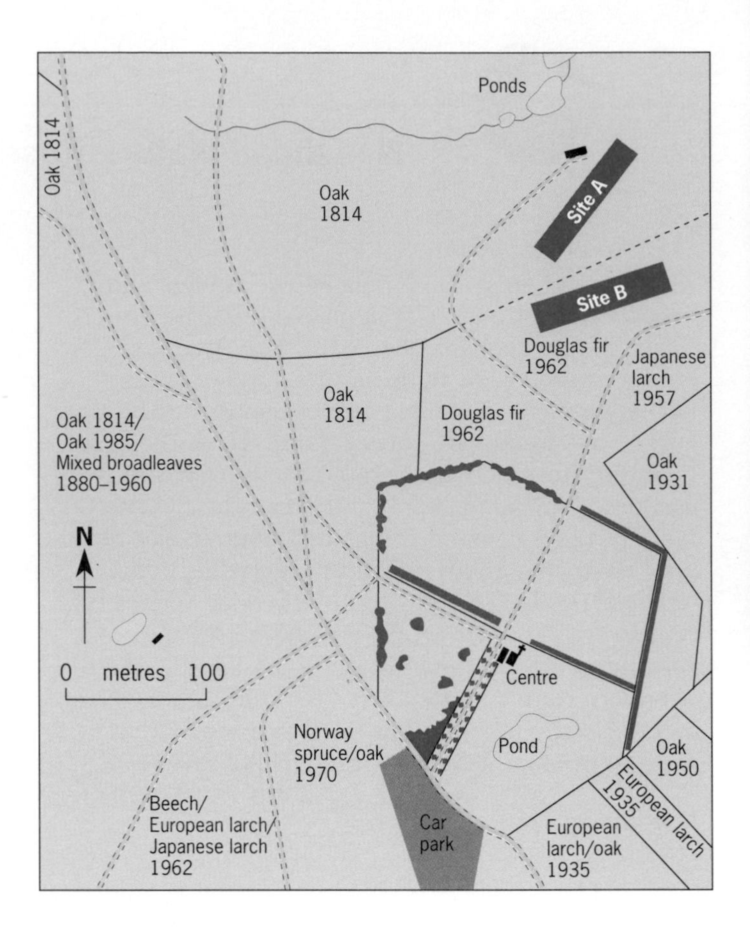

Figure 5.26 Plantation dates and sample site location on Nagshead Reserve (after Cave, 1992)

Table 5.6 pH reading taken at the litter/mineral soil interface

Site A										
Quadrat	1	2	3	4	5	6	7	8	9	10
pH	5.7	5.7	5.7	5.7	6.0	6.0	5.5	6.2	5.5	5.2
Site B										
Quadrat	1	2	3	4	5	6	7	8	9	10
pH	5.5	4.5	4.5	5.0	5.0	4.5	5.0	5.5	5.0	5.0

Table 5.7 Leaf litter

Site A										
Quadrat	1	2	3	4	5	6	7	8	9	10
Leaf litter (cm)	1	1	3	3	2	2	2	< l	< 1	3
Brash (twig cover) (cm)	–	–	–	–	–	–	–	–	–	
Site B										
Quadrat	1	2	3	4	5	6	7	8	9	10
Leaf litter (cm)	3	5	7	5	5	4	4	3	4	5
Brash (twig cover) (cm)	5	5	3	5	10	40	10	30	15	40

Table 5.8 Number of invertebrates: rough search from three handfuls of leaf litter sorted in bowls (H = high, M = medium, L = low)

Site A										
Quadrat	1	2	3	4	5	6	7	8	9	10
Number	M	M	M	H	M	M	H	M	H	M
Site B										
	1	2	3	4	5	6	7	8	9	10
	L	–	L	L	–	–	L	L	–	–

?

19a Refer to Tables 5.6, 5.7 and 5.8. Describe the differences between the two sites for each of these variables.
b Which site contains samples taken in the conifer plantation? Give reasons for your answer.
c Look at the species and numbers of leaf litter invertebrates that were collected in each site. What effect might this have on higher trophic levels? Give reasons for your answer.

Soil

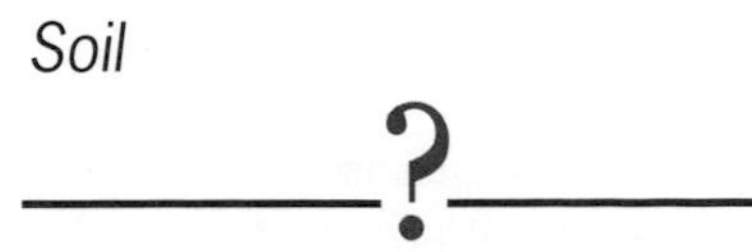

20 Using Tables 5.9 to 5.12, write a report as a soil analyst. Include the following information:
a Describe how the texture, pH and moisture readings vary.
b Suggest reasons for the different readings.
c Describe how infiltration varies. Using Table 5.12, explain why this might be so.
d Using data from Tables 5.10–5.13, suggest what effects the differences in invertebrate numbers may have on the soil profile.
e Describe any relationship between litter depth and moisture levels. Explain your findings.
f How does the pH vary? Why is this so?
g How does the soil moisture vary? Why is this so?

Table 5.9 Soil texture (Scale 1–10: coarse sand 1, silty clay 4, clay–silt 5, pure clay 10)

Quadrat	1	2	3	4	5	6	7	8	9	10
Oak wood	4	4	4	4	4	4	4	3	5	4
Quadrat	1	2	3	4	5	6	7	8	9	10
Douglas fir	4	3	4	4	4	4	4	3	3	3

Table 5.10 Soil moisture (Scale 1 dry – 10 wet)

Quadrat	1	2	3	4	5	6	7	8	9	10
Oak wood	6.0	7.0	7.0	5.5	8.0	4.8	6.8	8.0	6.0	9.0
Quadrat	1	2	3	4	5	6	7	8	9	10
Douglas fir	1.0	1.5	3.0	2.0	6.0	8.5	2.0	6.0	7.0	2.5

Table 5.11 Infiltration time: minutes and seconds per 5 cm (readings at the litter/mineral soil interface)

Quadrat	1	2	3	4	5	6	7	8	9	10
Oak wood	70:35	0:49	6:26	2:26	10:50	2:26	6:39	23:26	25:54	3:03
Quadrat	1	2	3	4	5	6	7	8	9	10
Douglas fir	30:26	25:48	18:18	14:21	26:58	17:43	14:27	128:0	41:0	17:08

Table 5.12 Interpreting infiltration (*Source:* J. Woodfield)

Preceding weather conditions If the weather has been hot and dry, rainfall will collect on the surface prior to infiltration. Wet weather will leave pores in the soil open and infiltration will continue until the ground reaches saturation point.

Slope position Downslope is likely to become saturated more quickly, and upslope rates will be slower than downslope, since downslope receives throughflow from upslope.

Soil texture and structure Sandy soils have large pore spaces and a crumb structure, so infiltration is usually rapid. Clay has tiny pores and a platey structure, so moisture collects on the surface prior to infiltration.

Slope angles Steep slopes encourage runoff rather than infiltration, and the soil will have a low moisture content. Infiltration should be quick on steep slopes.

Plants Roots can channel the water down quickly through macro-pores.

Rock type Limestone which has been dissolved by acidic rainfall may have macro-pores. Other rocks, such as chalk and sandstone, have porous properties. Some are impermeable.

Soil compaction Trampling and compaction of soil will reduce the pores available for infiltration.

Nature of precipitation Large raindrops may compact the soil and slow down infiltration.

Frozen ground Obviously slow rates may be due to permafrost or temporary freezing of the subsurface.

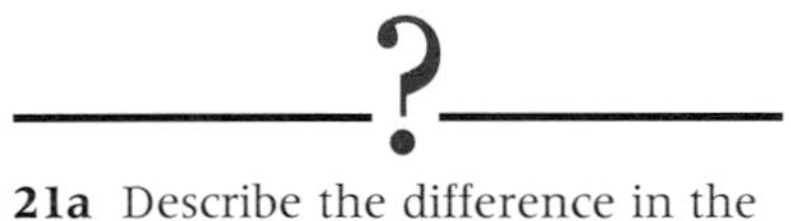

21a Describe the difference in the amount of light recorded at 2m above the ground and at ground level (Table 5.13). Give reasons for this.
b Why are the light levels in Site A consistently higher than in Site B? Use Table 5.15, Figures 5.24–5.25 to explain your answer.

22a Calculate the height of the trees in each site using the formula:
Height of tree $x = y \times \tan. A + 2$
b Which site has the tallest trees?
c From Table 5.14 identify which forest area has the largest range of plant species.
d Suggest reasons to account for the differences in species. Use data from Tables 5.9–5.17 to support your argument.

Woodland structure and diversity

Table 5.13 Light at the ground and at 2 metres above (Scale: 0 = dark, 10 very bright. Conditions: cloudless sky. Time of readings 11.30 a.m. – 12.30 p.m.)

Oak wood										
Quadrat	I	2	3	4	5	6	7	8	9	10
Ground	3	3	2	4	1	4	4	5	5	2
2 metres	4	4	2	7	7	7	6	6	5	7
Douglas fir										
Quadrat	1	2	3	4	5	6	7	8	9	10
Ground	1	0.5	< 0.5	< 0.5	< 0.5	1	1	1	< 0.5	0.5
2 metres	1	0.5	1	1	0.5	2	3	2	< 0.5	0.5

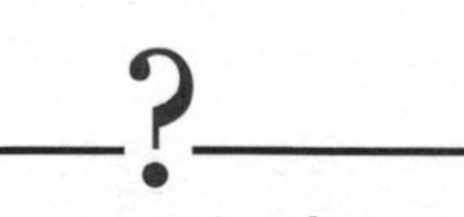

23 Date the trees using the formulas below Table 5.16. Do they match the dates on Figure 5.26?

24a Calculate the amount of wood *V* produced per tree using the formula:
$V = {}^{1}/_{3}\, \pi\, r^2\, h$ (r = radius of base, h = height)
b Now calculate the ratio of age to the amount of timber produced. The highest ratios indicate the greatest rates of timber production.
c Which trees have the highest rates of growth at Nagshead?

25a A forester wants to manage a coniferous forest so that it is more like a broadleaved plantation. How would the results of your investigation (from question 22) be of use to the forester?
b Using your scientific understanding of the factors that affect coniferous and broadleaved ecosystems, give the forester some advice while you wait for the results of your investigation. Justify your advice using data from the tables and your knowledge of plant growth requirements.

26 Essay: Describe the advantages and disadvantages of non-native conifers in upland areas and lowland areas. Refer to the following:
- timber production
- recreation
- wildlife
- water supply
- landscape
- employment.

Table 5.14 Flora: cover scores per metre square quadrat (H = high, M = medium, L = low)

	Oak wood										Douglas fir									
Quadrat	1	2	3	4	5	6	7	8	9	10	1	2	3	4	5	6	7	8	9	10
Flowering plants and ferns	H	H	M	H	H	H	H	H	H	–	–	–	–	L	–	–	–	–	–	–
Bryophytes (mosses)	L	–	L	L	–	M	L	H	H	–	L	L	M	H	H	M	L	–	–	–

Table 5.15 Canopy: percentage cover

	Oak wood										Douglas fir									
Quadrat	1	2	3	4	5	6	7	8	9	10	1	2	3	4	5	6	7	8	9	10
%	50	80	90	0	50	0	50	0	15	10	100	100	50	70	100	5	45	75	100	100

Growth rates

Table 5.16 Tree girth (Height of person = 1.67m)

Tree	Girth at 1.5m	Angle of tree	Distance from tree to person (m)
Oak wood			
A	251	58	20
B	145	48	20
C	261	52	26
D	207	42	31
E	234	55	24
F	197	35	28
G	247	50	22
H	129	37	33
I	186	61	15
J	231	49	24
Douglas fir			
A	105	56	21
B	77	51	17
C	116	60	14
D	89	47	21.5
E	91	67	15
F	91	45	22.5
G	117	53	19
H	103	61	13
I	98	50	27
J	88	51	19

$$\text{Age of oak trees} = \frac{\text{girth} \times 1.5 \times 2}{2.5 \text{ cm}}$$

(based on the assumption that an oak's girth increases by approximateley 2.5 cm per year and that a factor of 2 be applied to trees that grow in woodlands as opposed to in isolation)

$$\text{Age of Douglas fir} = \frac{\text{girth} \times 1.5 \times 2}{6.25 \text{ cm}}$$

(Douglas fir increases in girth by 5–7 cm per year)

Table 5.17 Temperature at ground level and 2 metres above ground (Readings taken between 11.30 a.m. and 12.30 p.m.)

	Oak wood										Douglas fir									
Quadrat	1	2	3	4	5	6	7	8	9	10	1	2	3	4	5	6	7	8	9	10
°C at ground	9	9	9	9	8	9	8	9	9	9	9	8	7	8	8	8	9	8	9	8
2 metres	8	8	8	8	8	8	8	8	8	8	9	9	9	9	9	9	9	9	9	9

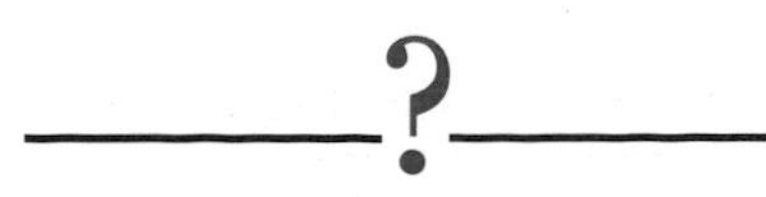

27a Using information contained in Chapter 2 and this chapter, collect data for tropical rainforests and boreal forests which can be compared with the information for temperate deciduous forest in Table 5.18.
b Construct a spider diagram to show the comparison between the three biomes.

Table 5.18 Characteristics of temperate deciduous forests

Climate
Precipitation 600–2250 mm per year. Distributed evenly. Temperature range 29–38˚C. Higher latitude, so less solar radiation input.

Structure and flora
One tree layer, example oak. Shrub (e.g. hawthorn) and herb (e.g. bluebell) layers frequently present. Low species diversity may lead to single-species domination.

Characteristic fauna
Some stratification of fauna. Herbivore density can determine regeneration rate. Example of mammal is a deer.

Mean net primary productivity
1200 g/m^2/year

World net primary production
8.4 billion tonnes per year

Mean biomass of standing crop
30 kg/m^2

Nutrient cycling
Active soil fauna and mixing with mineral soil. Litter layer always present but not thick. Nutrients stored almost equally in soil, biomass and litter.

Soils
Brown earth: leaf litter mixed into soil by earthworms. Humus mildly acidic (mull). Clay deposited at lower layers only. Some leaching as water moves down, creating a moderately to weakly acid soil. Mineral input from bedrock also important.

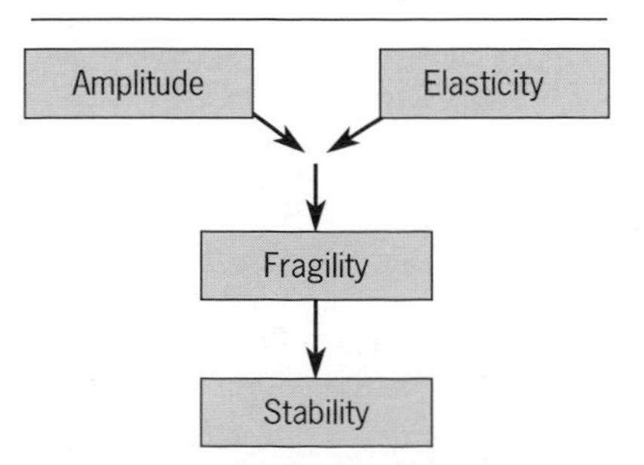

Figure 5.27 Flow diagram of amplitude, elasticity, fragility and stability

5.5 Tropical rainforests, temperate deciduous forests and boreal forests: a comparison

When inputs to world-scale ecosystems vary, this will inevitably form biomes which differ from each other. Unlike the tropical rainforest which has a high input of energy and precipitation, temperate deciduous and boreal biomes have lower inputs and characteristics linked to a high level of seasonality.

Ecologists thought that the greater the number of species that a system had, the greater its ability to withstand changes. These changes could be either within the external environment or in the number of individual living things. The basis of this theory was that large numbers of food chains and intricate webs would provide many **negative feedback** loops where changes could be absorbed and controlled (see Chapter 1.) However, we now know that in a complex ecosystem small environmental changes which cause **positive feedback** can have drastic effects.

Factors influencing stability

Early **seral stages** in succession are far less stable than later ones (see Chapter 6). One of the major characteristics of a **climatic climax community** is its stability. Stability is influenced by fragility. Fragility is the ease with which an ecosystem can be disrupted. This is related to the resilience of the system. This in turn depends on its amplitude, i.e. how far the system can change from its threshold of tolerance before its equilibrium is disrupted completely or replaced with another system. It is also dependent on the elasticity, i.e. the rate at which recovery from disturbance can occur. (Figures 5.28 and 5.29 show these relationships.)

Tropical rainforests

Ecologists consider tropical rainforest to be one of the most fragile of climatic climax communities. Fragility is usually linked with a high proportion of energy and nutrients stored in the above-ground biomass, to youthful unstable soils or to 'islands' which are limited in extent (see Fig. 3.26).

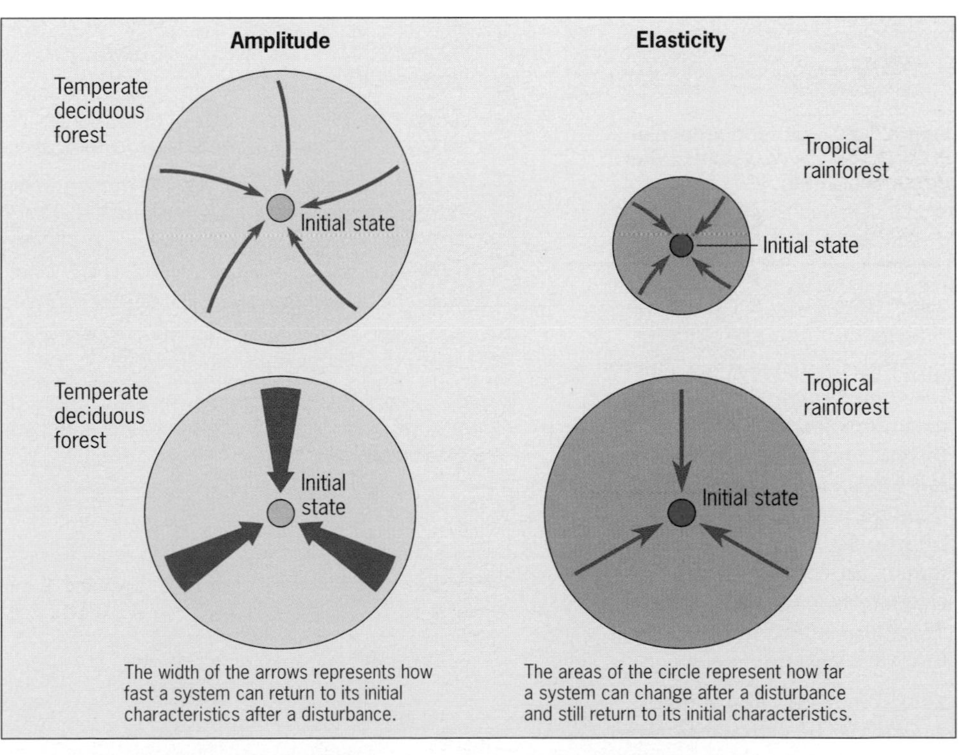

Figure 5.28 Elasticity and amplitude in temperate deciduous forests versus tropical rainforests (after Orians in van Dobben and Lowe McConnell, 1975)

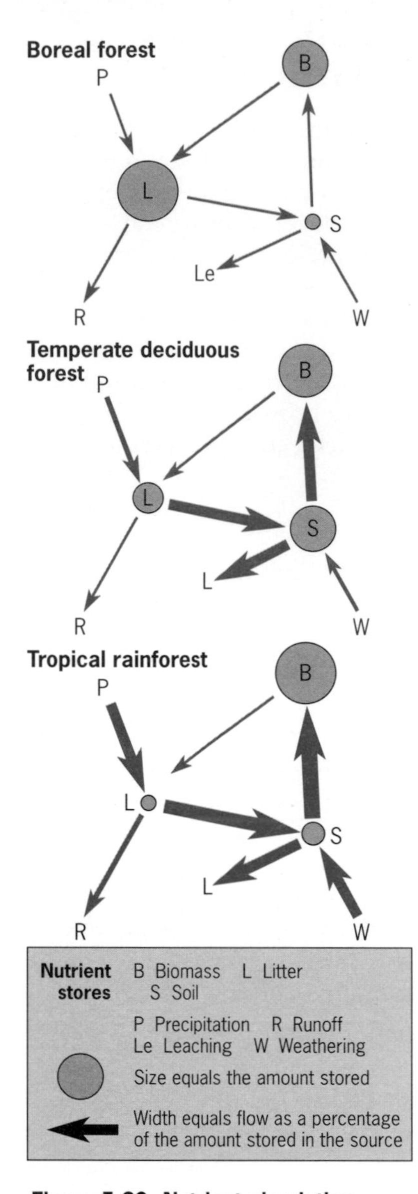

Figure 5.29 Nutrient circulation in three forest ecosystems (after Gersmehl, 1976)

Tropical rainforest fragility is associated with the ecosystem's vulnerability to human intervention. This is due to:
a its above-ground biomass (see Fig. 2.20), where most nutrients are stored,
b high species diversity and a high degree of **specialisation** of many species with narrow **niches**, which means that, if a creature or plant disappears, it is likely to affect others more dramatically than in less specialised systems,
c parts of the ecosystem becoming isolated after destruction, which means that movement of species into the isolated area is restricted.

Temperate deciduous and boreal forests

Temperate deciduous and boreal forests are much less fragile. Their resilience is much greater. This is largely due to fewer species, a more equal distribution of nutrients throughout their **stores**, biomass, litter and soil, plus efficient methods of nutrient movement.

Parts of the temperate deciduous system can be temporarily shut down during the non-growing season by the leaves being lost. Both boreal and temperate deciduous trees have adaptations for survival in low temperatures.

Summary

- Boreal forest and temperate deciduous forest are adapted to survive temperatures below 6°C, but boreal replaces deciduous when the growing season is below six months and has a frost-free period below four months.
- Most trees in boreal forest areas are evergreen conifers with adaptations to combat extreme cold.
- In upland Great Britain, native coniferous species such as Scots pine are associated with a rich flora and fauna. However, uniform plantations of sitka spruce and other non-native species do not have such high wildlife value.
- Forestry can have local economic benefits in some upland areas, providing an income for local communities.
- Much of the temperate deciduous biome has been cleared and replaced by agricultural land but, within the boreal zone, wood is used for timber and paper production.
- Temperate deciduous forests tend to form base-rich humus and brown earth soils, whereas coniferous forests tend to form acidic humus and podsol soils.
- Temperate deciduous forests have more nutrients stored in biomass and soil than boreal forests, which have more in the litter. Acidic conditions and lower temperatures limit the activity of soil fauna in boreal regions, meaning that litter breakdown is slower.
- Boreal species have been introduced further south as part of large-scale afforestation schemes. Higher temperatures ensure rapid growth rates. This higher productivity leads to a faster turnover of money.
- In lowland Great Britain, non-native conifers have been associated with a reduction in ground flora, less diverse fauna and a reduction in the attractiveness of the landscape.

?

28a Using Table 5.18 and the information you gathered in question 27, explain the differences between inputs within temperate deciduous forests and boreal forests, and inputs within tropical rainforests for precipitation and temperature.
b From Figure 5.29 identify what impact these have on the characteristics of the temperate deciduous forest and the boreal forest ecosystems.

29 Essay: How does climate determine the characteristics of world-scale forest biomes.

6 Changing ecosystems, natural succession and management

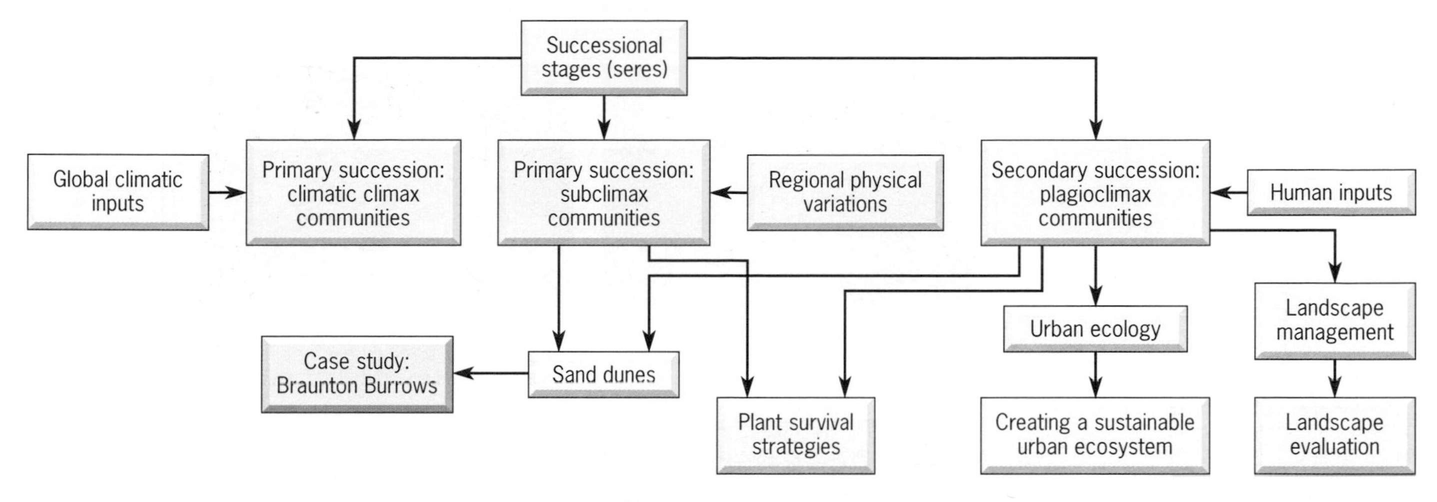

6.1 Introduction

The type of plant community found in any area depends partly on the physical environment, which is made up of rock type, landforms, climate, water supply and soil. In Chapters 2, 4 and 5 we saw that the distribution of plant communities on a global scale is influenced more by climate than by any other factor. On regional and local scales, climatic effects are modified by other variables.

Within Great Britain we would expect the **climatic climax community** to be mainly temperate **deciduous** woodland or **boreal** forest in parts of Scotland. However, in reality we have a large number of different **habitat** types. These reflect a range of physical variables – altitude, differing drainage conditions, a variety of rock and soil types – as well as levels of human and animal activity. These factors affect the **successional** stages or **seres** in the development of a community (Figs. 6.1–6.6).

6.2 How communities change

Primary succession

Although **ecosystems** are largely self-regulating, they are subject to change. Ecologists note that succession always appears to follow a similar pattern. In this orderly sequence of events one community paves the way for the next seral stage by the accumulation of **biomass** which decomposes, enriches the soil and provides increasing numbers of **niches** for **herbivores**, **carnivores** and **detritivores**. This can also be used to explain the process whereby ponds succeed to dry woodland (Fig. 6.7).

Ecologists call this hydroseral succession: a process of gradually advancing vegetation zones, aided by the ever-increasing build-up of silt and **humus**, towards the middle of the pond. In this case people are considered to have no part in the start of this process, which is called primary succession. It is important to note that no one has been able to observe the process of primary succession because of the length of time involved.

Secondary succession

When people are involved, secondary succession takes place. As with primary succession, the various communities that succeed each other are called seral

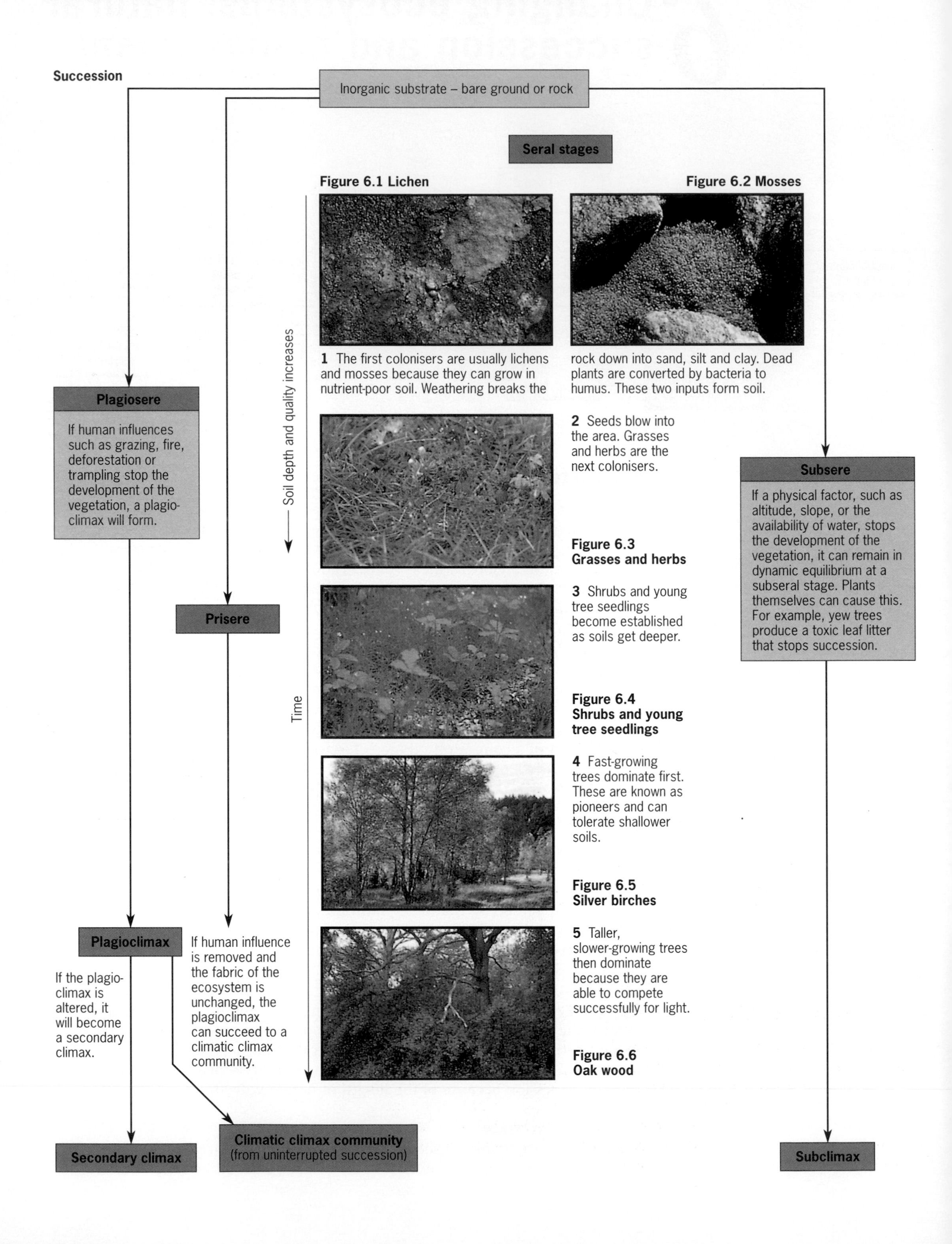

Succession
Inorganic substrate – bare ground or rock
Seral stages
Plagiosere
If human influences such as grazing, fire, deforestation or trampling stop the development of the vegetation, a plagio-climax will form.
Prisere
Subsere
If a physical factor, such as altitude, slope, or the availability of water, stops the development of the vegetation, it can remain in dynamic equilibrium at a subseral stage. Plants themselves can cause this. For example, yew trees produce a toxic leaf litter that stops succession.
Soil depth and quality increases
Time
1 The first colonisers are usually lichens and mosses because they can grow in nutrient-poor soil. Weathering breaks the rock down into sand, silt and clay. Dead plants are converted by bacteria to humus. These two inputs form soil.
2 Seeds blow into the area. Grasses and herbs are the next colonisers.
3 Shrubs and young tree seedlings become established as soils get deeper.
4 Fast-growing trees dominate first. These are known as pioneers and can tolerate shallower soils.
5 Taller, slower-growing trees then dominate because they are able to compete successfully for light.
Plagioclimax
If the plagio-climax is altered, it will become a secondary climax.
If human influence is removed and the fabric of the ecosystem is unchanged, the plagioclimax can succeed to a climatic climax community.
Secondary climax
Climatic climax community (from uninterrupted succession)
Subclimax

Figure 6.1 Lichen

Figure 6.2 Mosses

Figure 6.3 Grasses and herbs

Figure 6.4 Shrubs and young tree seedlings

Figure 6.5 Silver birches

Figure 6.6 Oak wood

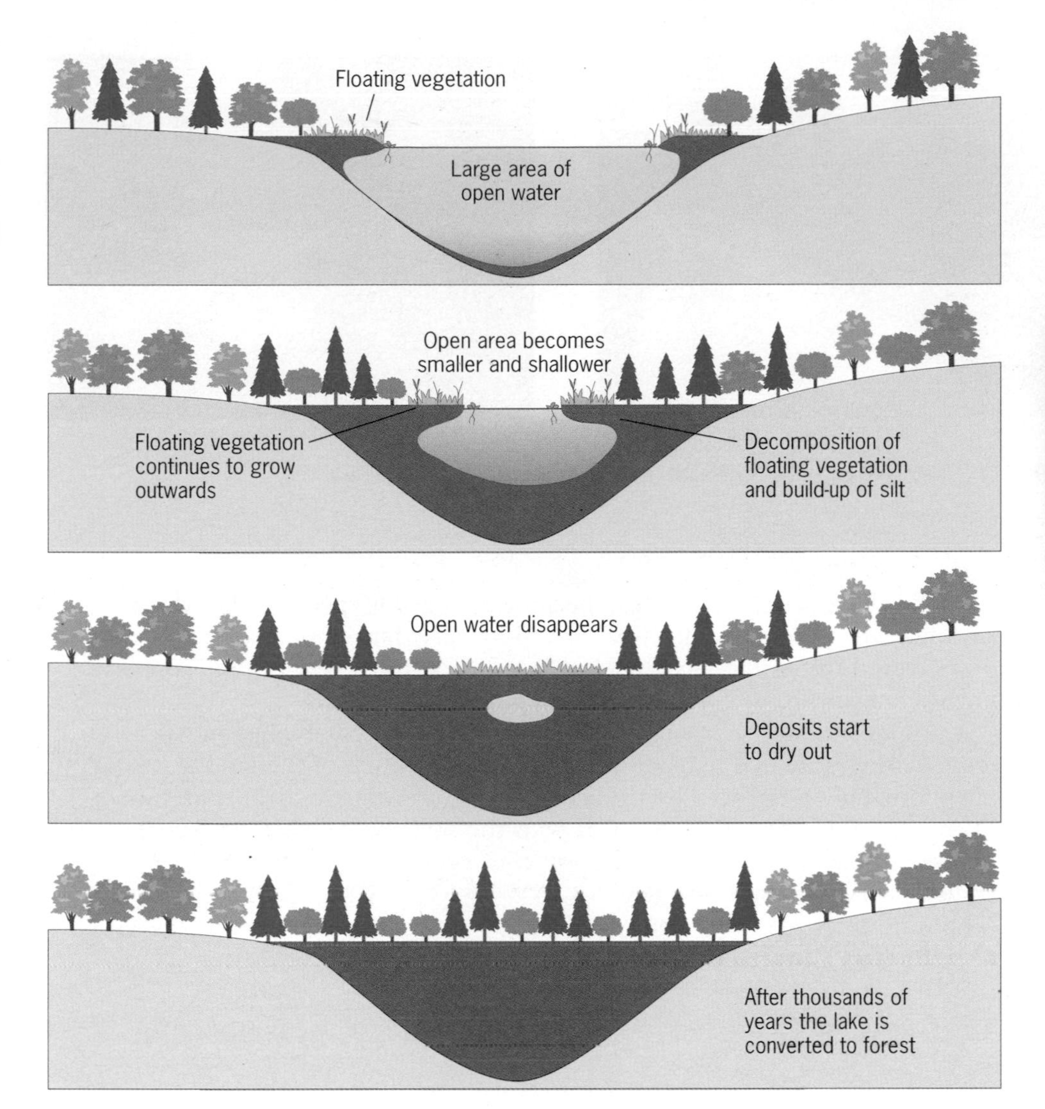
Floating vegetation
Large area of open water
Open area becomes smaller and shallower
Floating vegetation continues to grow outwards
Decomposition of floating vegetation and build-up of silt
Open water disappears
Deposits start to dry out
After thousands of years the lake is converted to forest

Figure 6.7 A bog lake succession

Figure 6.8 Bog lake succession in Poland

stages on the way to the development of a climatic climax community, a community that is apparently self-duplicating and not subject to further change unless there is a change in the climate.

Fields, once created as clearings within woodlands, eventually revert to woodland of a similar composition to the woods that surround it. This process is known to many farmers and was measured at Rothamsted Experimental Station in Hertfordshire, where a wheatfield has been left fallow since 1882. It is now covered in woodland, with trees over 18 metres high. This sort of change occurs to many grassland areas neglected by people. Arable weeds and coarse grasses begin to dominate the community, to be replaced by tall grassland, which in turn gives way to tree **species**, and woodland covers the ground in much the same way as it did before early people began to cultivate the land (Figs 6.9–6.11).

?

1 Which seral stage shown on Figure 6.7 does Figure 6.8 represent?

6.3 Climatic climax communities

Climatic climax communities are not all exactly the same. A mosaic of communities may exist as a result of differing soil conditions: forest fires could have removed organic matter from the soil, or hydroseral succession could contribute organic matter. In each case the climatic climax communities that occur on these soils may permanently differ from the climatic climax communities that surround these areas. Ecologists call these areas **subclimax communities** (see Chapter 7).

Figure 6.9–6.11 Heathland on Hollesley Heath, Suffolk between 1965 and 1981

?

2 Rearrange the photographs (Figs 6.9–6.11) of heathland succession in order.

Causes and consequences of change

Climatic climax communities can become changed by people. The African savanna was once forest but was cleared by early farmers; the soils in these areas were further impoverished by this activity and now can only support grasses and shrubs that are tolerant of drought and fire.

A similar situation exists in the Mediterranean, where evergreen oak woodland was cleared by people for agriculture. Much of the land was later abandoned to be replaced by the scrubby, spiny, aromatic shrubland known as garigue (see Chapter 7). Here, as with the savanna, the soil has become impoverished and is unable to support the regeneration of the original climatic climax community (Fig. 6.12).

Communities that have been significantly altered in this way are known as **deflected climaxes** or **plagioclimaxes** (Chapter 7).

Figure 6.12 Garigue vegetation

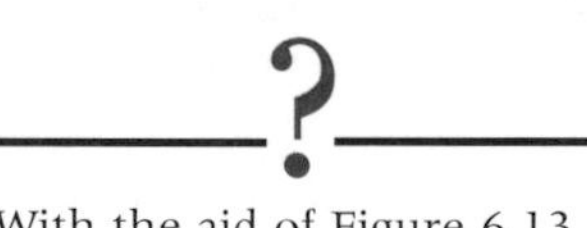

3a With the aid of Figure 6.13, explain what has happened to the landscapes shown in Figures 6.14 and 6.15.

b What do you think the author of this diagram means by 'excessive' and 'reasonable' exploitation?

4 Using Figure 6.16, describe how the changes in plant species are affected by:

a rock type **b** water levels
c soil type **d** organic matter.

e Describe how the ecosystem succeeds from yellow dune to woodland.

f How does the structure of the dune system affect the higher **trophic levels**?

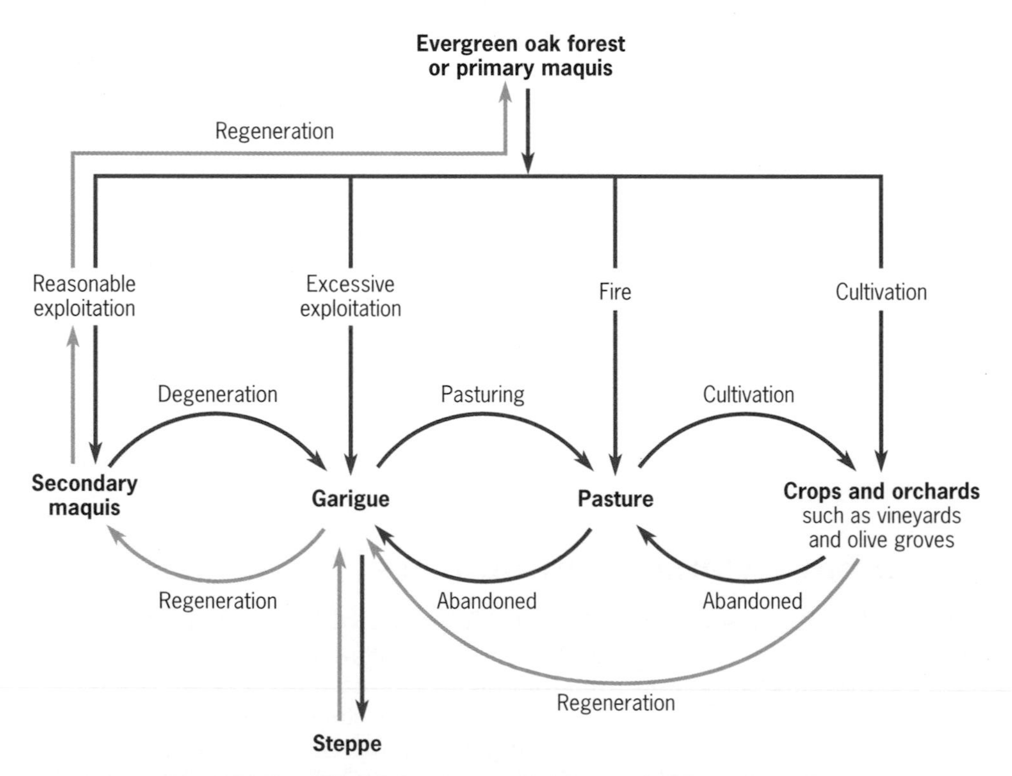

Figure 6.13 Degeneration and regeneration of Mediterranean plant communities due to human activities (after Polunin and Huxley, 1981)

Figure 6.14 Cyprus landscape

Figure 6.15 Edge of Lassithi on the way to Mallia, Crete

Sand dune ecosystems

Sand dunes are found in coastal areas throughout the world. Away from human interference, they are subclimax communities limited by the action of the sand from the sea. As the distance of the dune systems increases away from the shore and the sand stabilises, they begin to succeed towards a climatic climax community (Fig. 6.16).

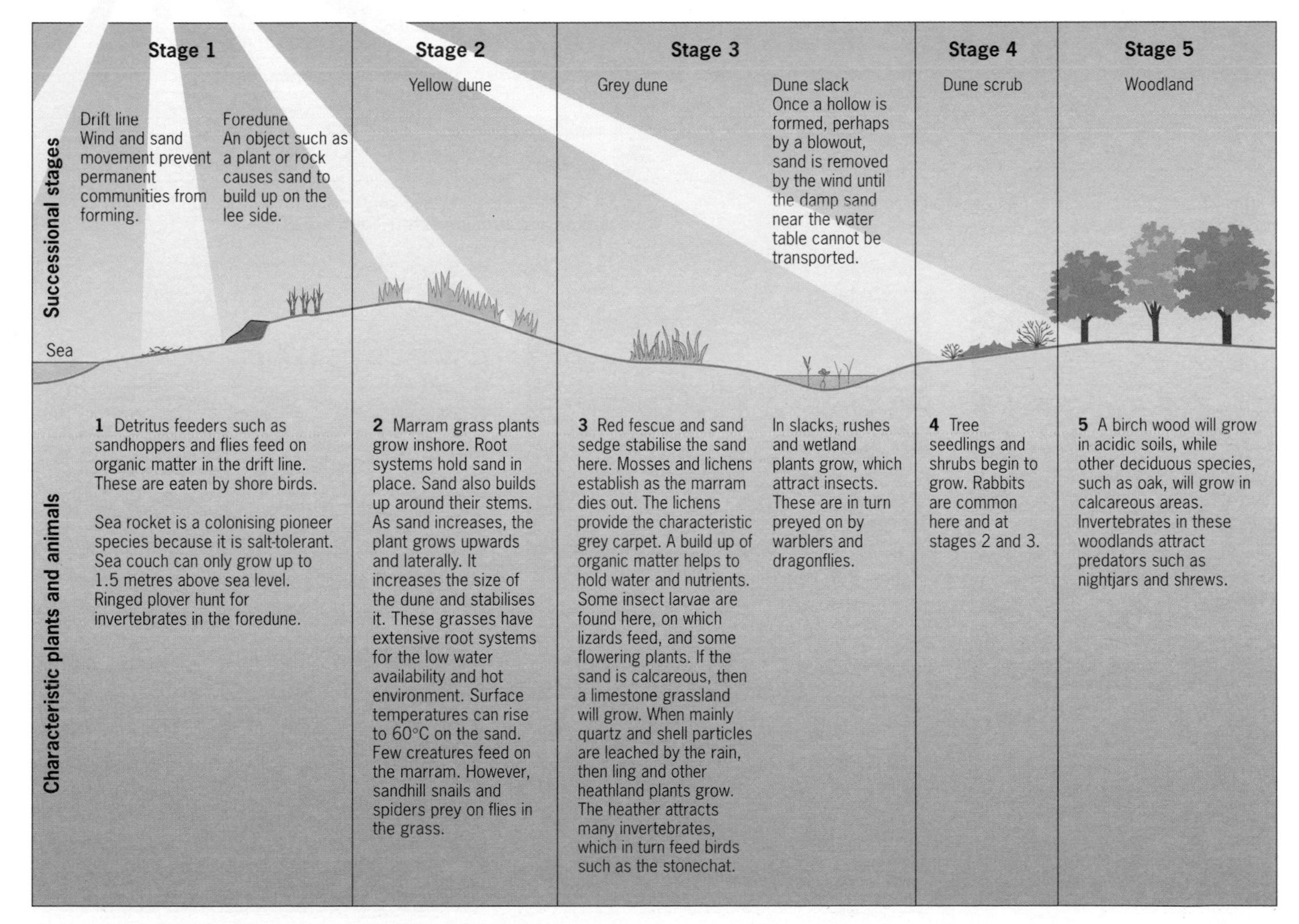

Energy from the Sun
Successional stages
Characteristic plants and animals
Sea
Stage 1
Drift line
Wind and sand movement prevent permanent communities from forming.
Foredune
An object such as a plant or rock causes sand to build up on the lee side.
Stage 2
Yellow dune
Stage 3
Grey dune
Dune slack
Once a hollow is formed, perhaps by a blowout, sand is removed by the wind until the damp sand near the water table cannot be transported.
Stage 4
Dune scrub
Stage 5
Woodland
1 Detritus feeders such as sandhoppers and flies feed on organic matter in the drift line. These are eaten by shore birds.
Sea rocket is a colonising pioneer species because it is salt-tolerant. Sea couch can only grow up to 1.5 metres above sea level. Ringed plover hunt for invertebrates in the foredune.
2 Marram grass plants grow inshore. Root systems hold sand in place. Sand also builds up around their stems. As sand increases, the plant grows upwards and laterally. It increases the size of the dune and stabilises it. These grasses have extensive root systems for the low water availability and hot environment. Surface temperatures can rise to 60°C on the sand. Few creatures feed on the marram. However, sandhill snails and spiders prey on flies in the grass.
3 Red fescue and sand sedge stabilise the sand here. Mosses and lichens establish as the marram dies out. The lichens provide the characteristic grey carpet. A build up of organic matter helps to hold water and nutrients. Some insect larvae are found here, on which lizards feed, and some flowering plants. If the sand is calcareous, then a limestone grassland will grow. When mainly quartz and shell particles are leached by the rain, then ling and other heathland plants grow. The heather attracts many invertebrates, which in turn feed birds such as the stonechat.
In slacks, rushes and wetland plants grow, which attract insects. These are in turn preyed on by warblers and dragonflies.
4 Tree seedlings and shrubs begin to grow. Rabbits are common here and at stages 2 and 3.
5 A birch wood will grow in acidic soils, while other deciduous species, such as oak, will grow in calcareous areas. Invertebrates in these woodlands attract predators such as nightjars and shrews.

Figure 6.16 A generalised sand dune ecosystem and succession

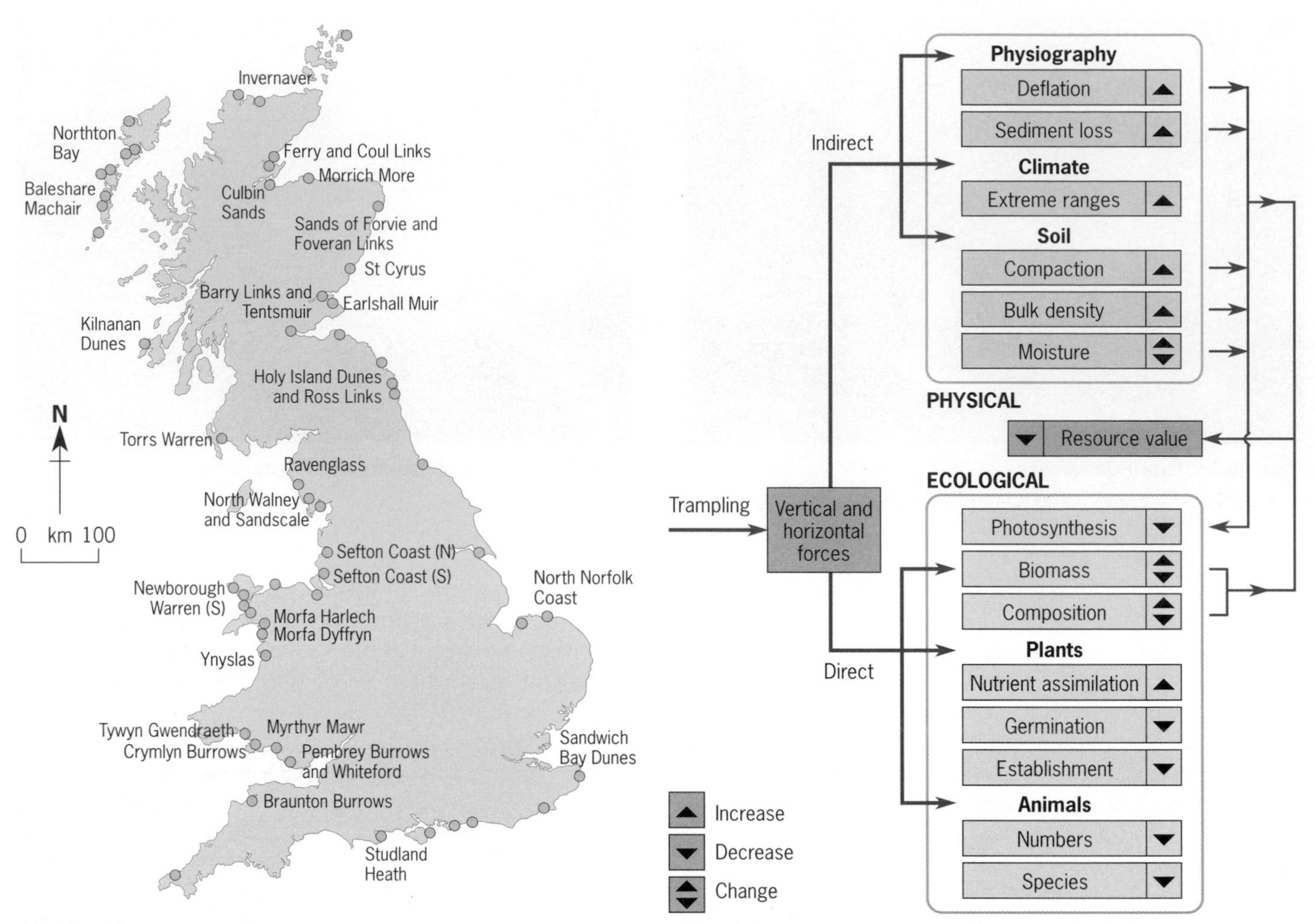

Figure 6.17 Major sand dunes of wildlife importance (after Davidson, 1991)

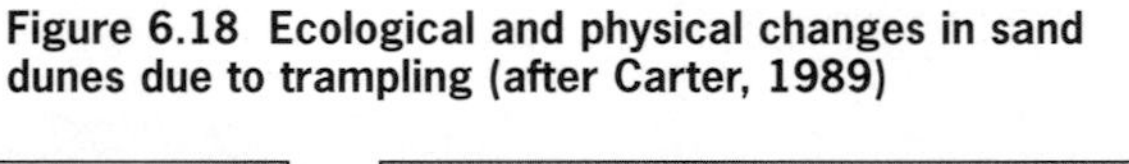

Figure 6.18 Ecological and physical changes in sand dunes due to trampling (after Carter, 1989)

Figure 6.19 Blowout at Sefton, Merseyside

Figure 6.20 Car park area at Sefton

Figure 6.21 Christmas trees planted in dunes at Sefton

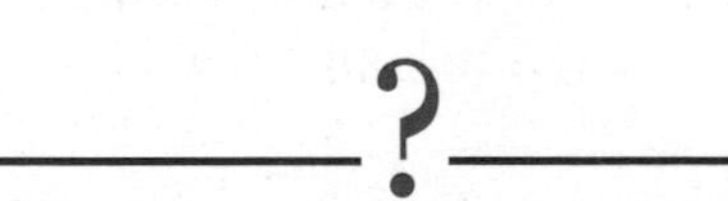

5 Using Figure 6.18, explain how people may affect the succession of a sand dune ecosystem.

6 What type of damage shown in Figures 6.18 and 6.19 is caused by people at Sefton in Merseyside?

Figure 6.22 A cordoned-off area at Sefton

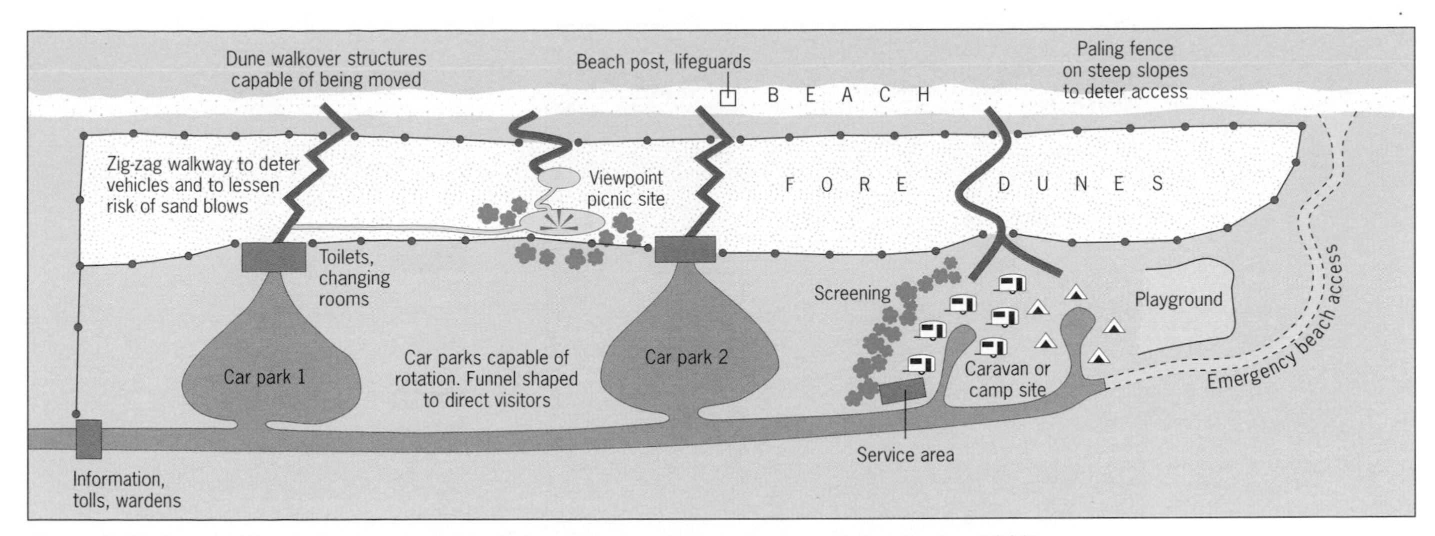

Figure 6.23 Commonly used management techniques to stop damage to dunes (after Carter, 1989)

The total area of sand dunes in Great Britain is about 56 000 ha (Fig. 6.17). Many of these sand dunes are influenced by the action of people. Dune soils are very fragile, and trampling by humans and burrowing by rabbits can break through the soil, releasing fresh wind-blown sand to create blow-outs. This is erosion (Fig. 6.18).

Most Local Authorities recognise the importance of sand dunes as a form of coastal defence, in addition to their importance for wildlife. They are also used for recreation for sites such as golf links and picnic areas. Any damage to the system is therefore of concern to Local Authorities.

?

7 From Figures 6.20–6.22, describe what the rangers have been doing at Sefton to manage the habitat.

8 From Figure 6.23 describe other measures that they could use to reduce damage to the ecosystem.

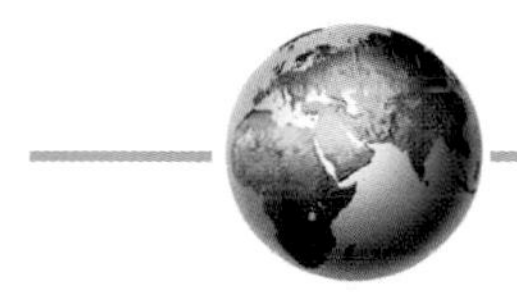

Braunton Burrows

Figure 6.24 Plot A

At 970 ha, Braunton Burrows SSSI is one of England's largest dune systems. It is an international Biosphere Reserve recognised by UNESCO. Its large size and limited access mean that it does not suffer much from the effects of tourists. It is owned by a private landowner, Christie Estate Trustees, but 603 ha to the south is leased to the Ministry of Defence (MOD). They have used the dunes since the Second World War when US soldiers trained there before the Normandy Landings, causing much damage to the dune system.

English Nature appointed a warden to run a National Nature Reserve, sub-leasing it from the Ministry of Defence in 1964. The management plan of the SSSI involved grazing with cattle and sheep to prevent scrubby vegetation taking over and dominating the rare orchids. The landowner would not allow this to happen and English Nature withdrew their management of the site. The MoD now employ the warden but he is unable to stop the scrub from taking over. Even the rabbits cannot help. Their population numbers have declined due to a recent viral epidemic. Although they can graze areas of short grass, any areas that have been taken over by the scrub and aggressive grasses, such as Yorkshire Fog, are now impossible for them to eat.

The dune system does not follow a classic pattern of successional change as shown in Figure 6.16, and its problems are not related to visitor damage but to control of succession by grazing. While there is a pattern of stages 1, 2, 3, 4 and 5 overall, within the dune system there are often areas where stage 4 vegetation occurs in place of stage 3.

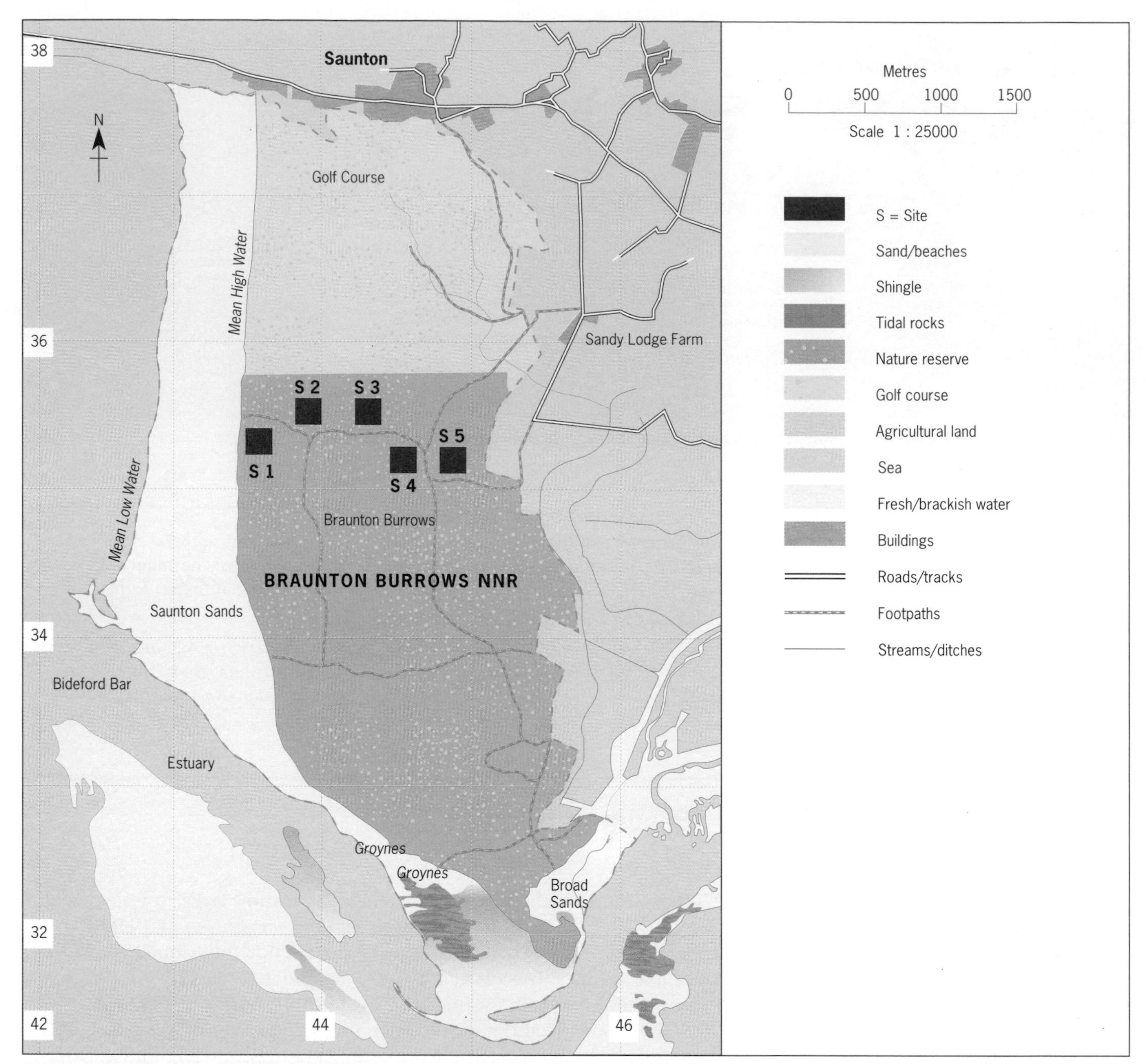

Figure 6.25 Location of Braunton Burrows

Figure 6.26 Plot B

Figure 6.27 Plot C

Table 6.1 Data for each site

Average results	Site 1	Site 2	Site 3	Site 4	Site 5
Soil temperature (°C)	24.5	24.8	17.6	20.6	24.6
Slope angle (degrees)	23	23	14	2	10
pH of soil	6.3	6.5	6.5	5.1	6
Water content (driest = 1; wettest = 10)	1.6	1.2	1.8	10	1.2

Site 1

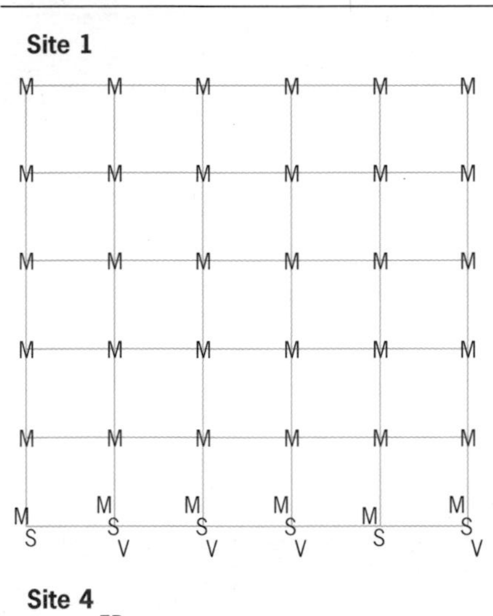

Site 2

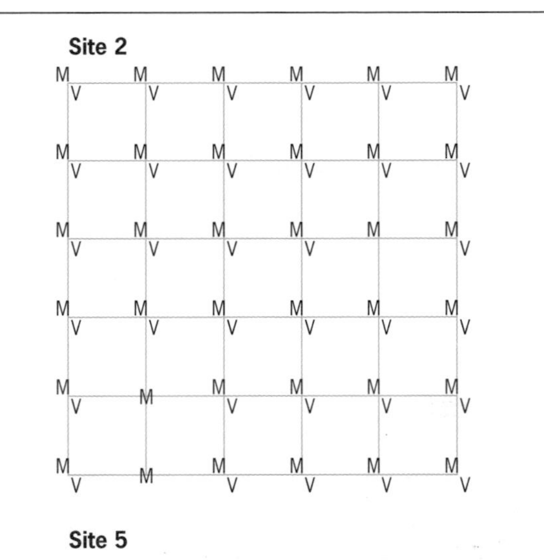

Site 3

RH YF | RH YF | RH YF | RH YF | RH YF | RH YF
RH | RH YF | RH YF | RH YF | RH YF | RH YF
RH YF | M RH YR | RH YF | M RH YF | RH YF | RH YF
M RH | M RH YF | M RH | M RH YF | M RH YF | M RH YF
M RH | RH YF YR | M RH | M RH YF | M RH YF | M RH
M RH YR | M RH | M RH | M RH | M RH | M RH

Site 4

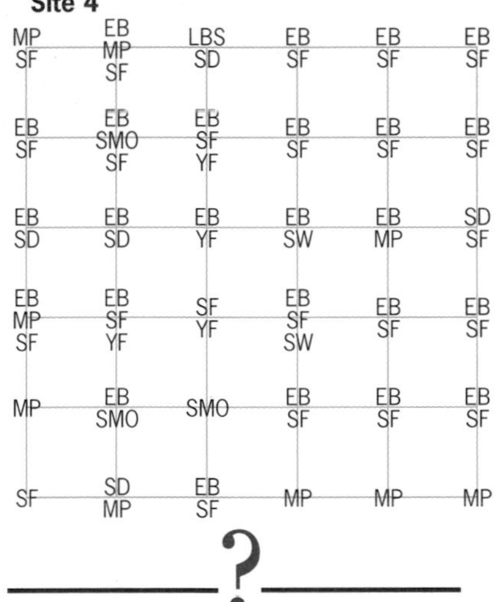

Site 5

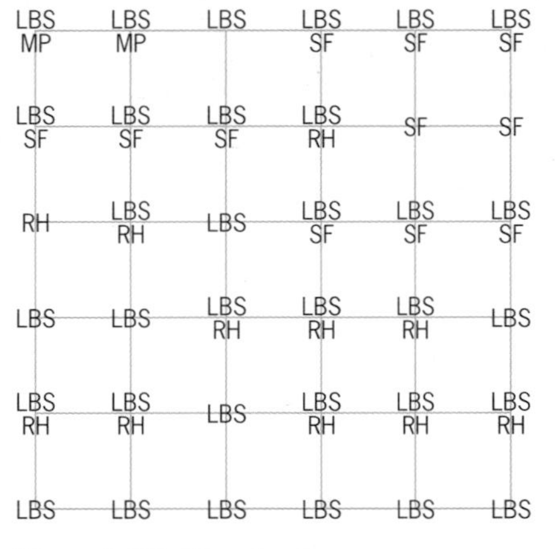

EB	eyebright
LBS	lady's bedstraw
M	marram grass
MP	marsh pimpernel
RH	restharrow
S	sand
SD	sedge weed
SF	sheep's fescue
SMO	southern marsh orchid
SW	silverweed
V	sea bindweed
YF	Yorkshire fog
YR	yellow rattle

Figure 6.28 Sites 1 – 5

?

9 Look at the data table and match each photo to one of the sites. Explain your choice.

10 Calculate the percentage cover for each species. Describe the changes you observe from site 1 to site 5. How does this relate to the data in the table?

11 Match each photo to one of the quadrats and explain your choice.

12 Look at each photo and study the data. Describe any management problems that you observe.

Figure 6.29 Plot D

6.4 Regional variations

In the British Isles the landscape has been formed by the interplay of human and physical variables. Most habitats have been managed in some way. Often it is difficult to see where physical factors dominate over human ones, and vice versa. Upland vegetation is limited by the variables shown in Figure 6.34. Some lowlands are governed by flooding, others by forced drainage. Rock and soil type also play a role in determining the nature of some communities (Figs. 6.30–6.33).

Ecosystems constantly respond to change. They do this naturally as they follow successional routes until climate or other physical factors cause them to reach a **dynamic equilibrium**. As ecosystems are deflected into plagioclimaxes by people's actions, their wildlife diversity may increase.

The main climatic climax communities within the British Isles are woodlands. However, as we learnt in Chapter 4, these have all been affected by people in some way and are more likely to be secondary climaxes.

Figure 6.30 Alpines have a small amount of biomass to enable them to survive in stressful environments.

Figure 6.31 Chalk is porous and water drains away quickly. Plants are adapted to the physiological drought conditions in the soil. Human activity has prevented it from succeeding to shrubland.

Figure 6.32 Infertile soils are one factor which causes a heath community to form. Human activity prevents it from succeeding to woodland.

Figure 6.33 Impermeable rock or the presence of a soil iron pan cause water to build up. Peat communities form in these water-logged conditions.

Figure 6.34 An upland landscape

Figure 6.35 Annual meadow grass is the only annual grass in Britain. It produces many seeds and colonises bare ground.

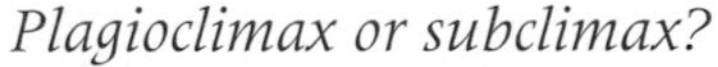

Plagioclimax or subclimax?

Separating plagioclimaxes from naturally-induced subclimax communities is extremely difficult. One way of attempting to do this is to divide plants up according to their survival strategies.

Plant strategies

Ecologists agree that there are three basic plant strategies which are a response to the effects of **competition**, **stress** and **disturbance**. Competition is when plants compete for limited resources (see Chapter 1). Stress is caused by external factors such as unfavourable temperature, poor soils, or exposure and reduces the growth rate and biomass of the plant. Disturbance consists of mechanisms which also limit the plant biomass, but this is by partial or total destruction.

Plants respond in one of three ways:

STRESS TOLERATORS

These are able to endure very unproductive habitats which are very hot, dry, cold, wet or have very poor soils. They reduce the amount of biomass they produce in order to survive. Many of the Alpine plants such as eidelweiss possess such characteristics (Fig. 6.30).

RUDERALS

These are short-lived species which are highly fertile. They are found in productive areas where there has been some disturbance. Examples of ruderals include annual meadow grass, which is the only annual grass in Great Britain. It produces many seeds and **colonises** bare ground rapidly (Fig. 6.35).

COMPETITORS

These have a highly competitive ability to capture nutrients, sunlight and water. They occur in relatively undisturbed conditions. Many of the aggressive grasses such as false oat grass are competitors (Fig. 6.36).

Figure 6.36 This tall species of false oat grass prefers fertile soils. It is intolerant of grazing, mowing or disturbance. As it grows, litter builds up. Grass litter takes a long time to break down; hence light is reduced and the microclimate altered.

?

13 Figure 6.37 shows a triangular graph on which the relative position of plants can be plotted according to their strategy for survival. Refer to Figure 6.38 and answer the following questions by using Figure 6.37.

a Which type of plant is associated with most disturbance?

b Which type of plant is associated with the most stressful conditions?

c Which type of plant appears affected by all three equally?

14a Make a list of the habitats shown on Figure 6.38. Then, using Figure 6.37, write alongside them the strategy of the plants associated with them.

b Using Figure 6.38, say what type of plants shown in Figure 6.39 are likely to be found in each habitat.

c How does management appear to affect plant communities?

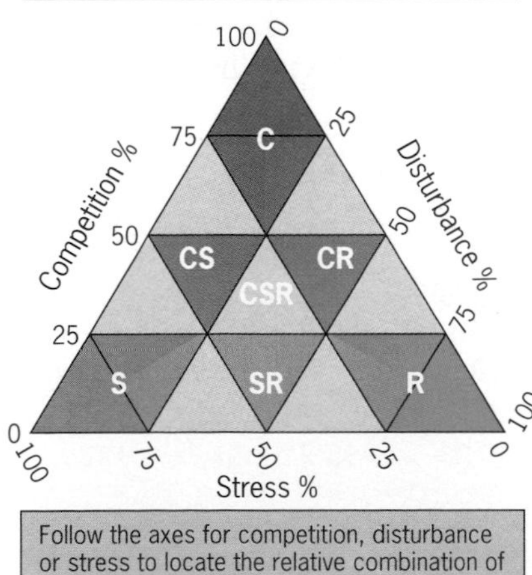

Follow the axes for competition, disturbance or stress to locate the relative combination of plant strategies exhibited in a sample. Figures 6.38 and 6.39 contain such samples. For example, CS represents a combination of competitor and stress-tolerator strategies.

Figure 6.37 The relative importance of competition, disturbance and stress to plant strategy categories (after Grime, 1977)

Figure 6.38 The types of strategies used by six groups of vegetation (after Grime, 1977)

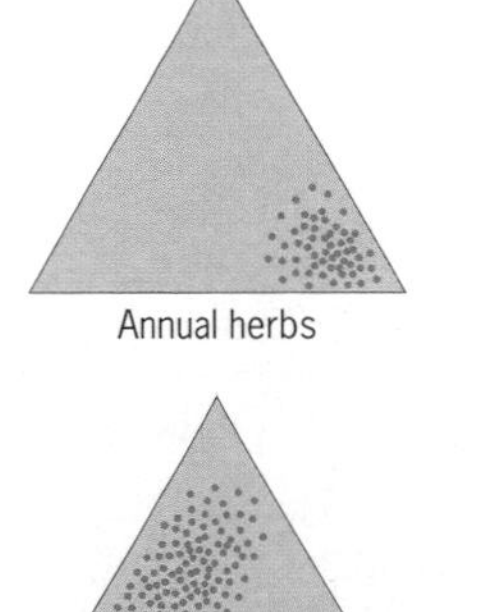

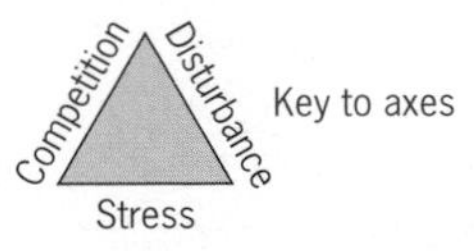

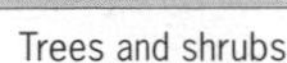

The clusters represent the relative importance of competition, disturbance and stress on each of these types of plants. The clusters also represent the range of strategies exhibited by each of these six types of plants.

?

15a Grazing is needed to manage Braunton Burrows dune system. Use Figure 6.39 to suggest two types of land management that may be used to create the same type of vegetation mix at stage 3 of the dune succession.
b How could trampling benefit the system at Braunton?

16a What are the three most common forms of vegetation shown in Figure 6.40?
b Using Figure 6.40, explain why there are many of these vegetation types.
c To what extent do the climatic factors shown in Table 6.2 seem to affect the distribution of some vegetation types in these cities?

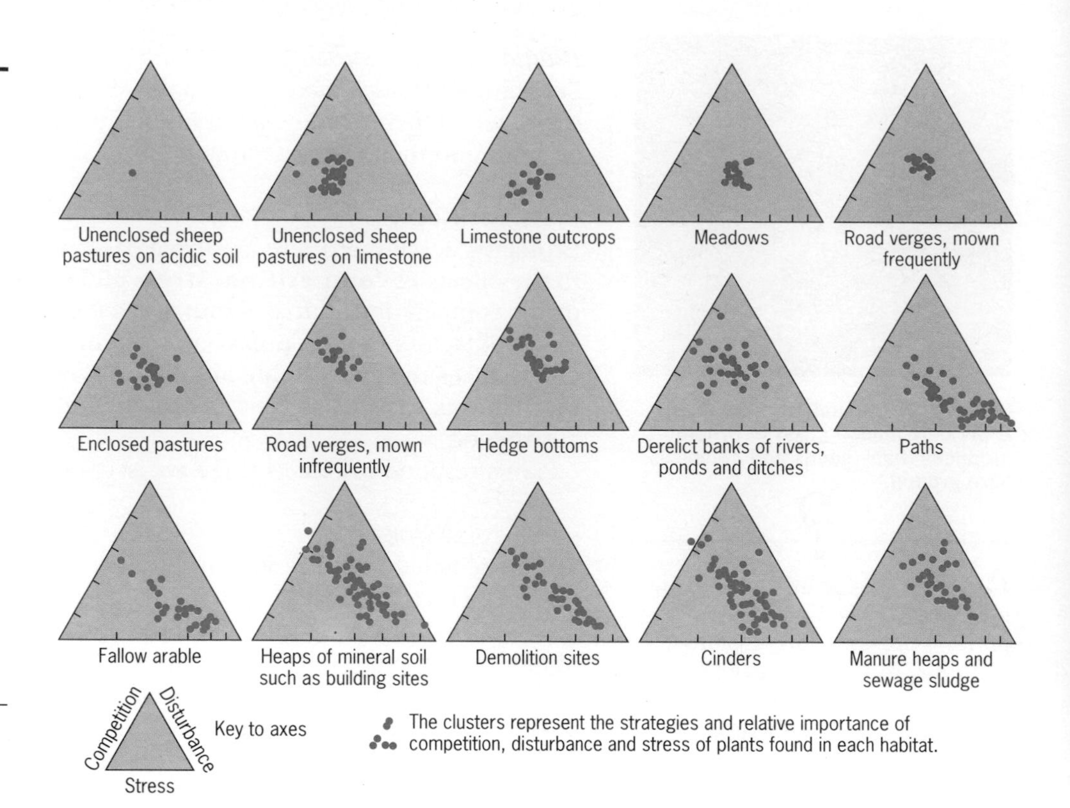

Figure 6.39 Clusters of vegetation samples taken from 15 types of habitats and their relative response to competition, disturbance and stress (after Grime, 1974)

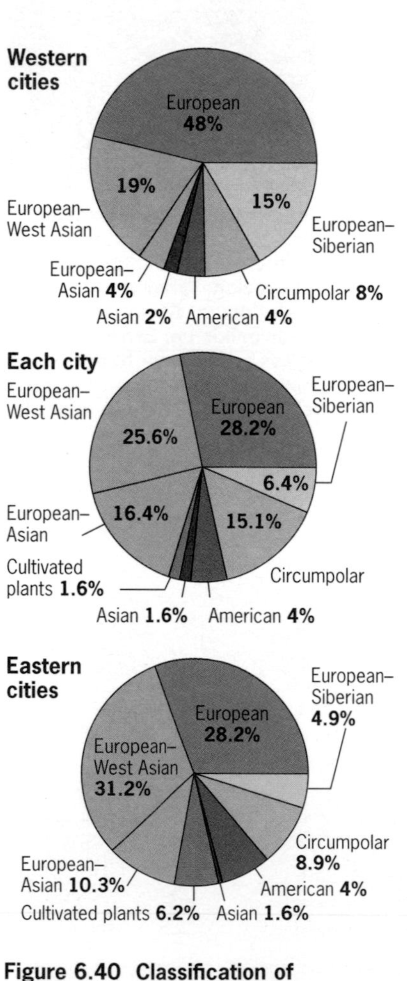

Figure 6.40 Classification of vegetation in European cities by geographical origin (after Kunick, 1982)

6.5 Urban ecology

Urban landscapes are the most managed landscapes and therefore have a **flora** and **fauna** which will always form a plagioclimax community.

Within towns and cities, regional climatic influences can often be changed by the effect of the urban heat island (where temperatures are higher than in surrounding areas). Inputs into urban ecosystems are therefore different from those of surrounding rural areas. The physical structure of the urban area also creates opportunities for precipitation **inputs** to be reduced, since groundwater is extracted and runoff is fast over impermeable surfaces.

These impermeable surfaces can either be human-made surfaces or compacted soils. Much of the vegetation is artificially controlled, many species are non-native (Fig. 6.40) or planted rather than arriving by means of colonisation.

Table 6.2 Climatic data from seven European cities (*Source:* Wolfram Kunick, 1982)

	Amsterdam	Brussels	Duisburg	Hanover	Berlin	Poznan	Lodz	Warsaw
Altitude above sea level (m)	2	108	20-50	57	35	58	97	121
Average temperature (°C)	10.0	9.7	9-10	9.0	8.4	8.3	7.9	9.5
Average annual rainfall (mm)	648	835	750-800	644	577	513	567	545
Area (km²)	314	160	1280	225	480	115	214	445

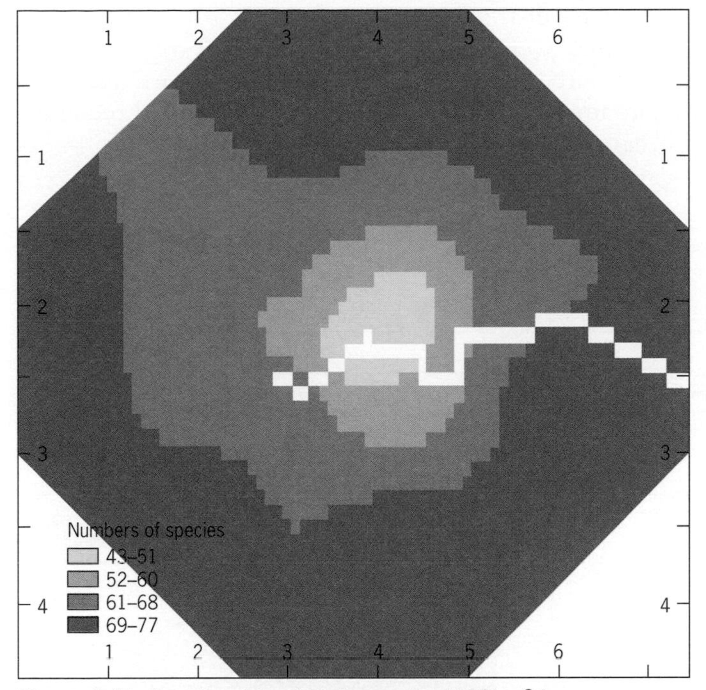

Figure 6.41 Density of land birds species/100km^2 in Greater London (after Cousins, 1982)

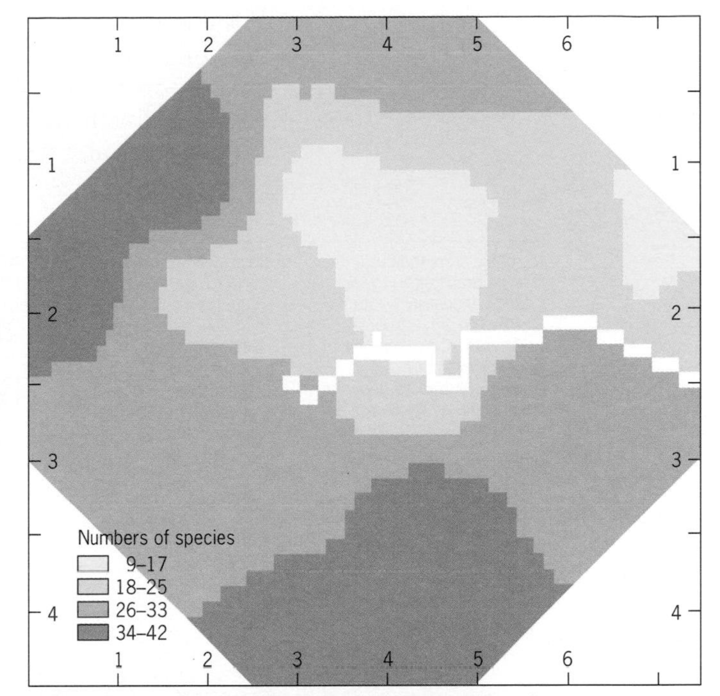

Figure 6.42 Density of land snail species/100km^2 in Greater London (after Cousins, 1982)

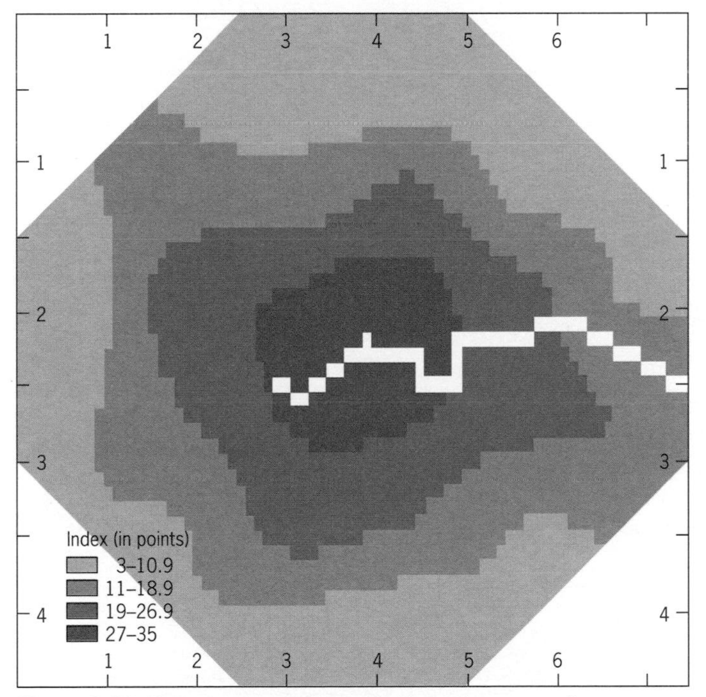

Figure 6.43 Pattern of an index of built environments in Greater London (after Cousins, 1982)

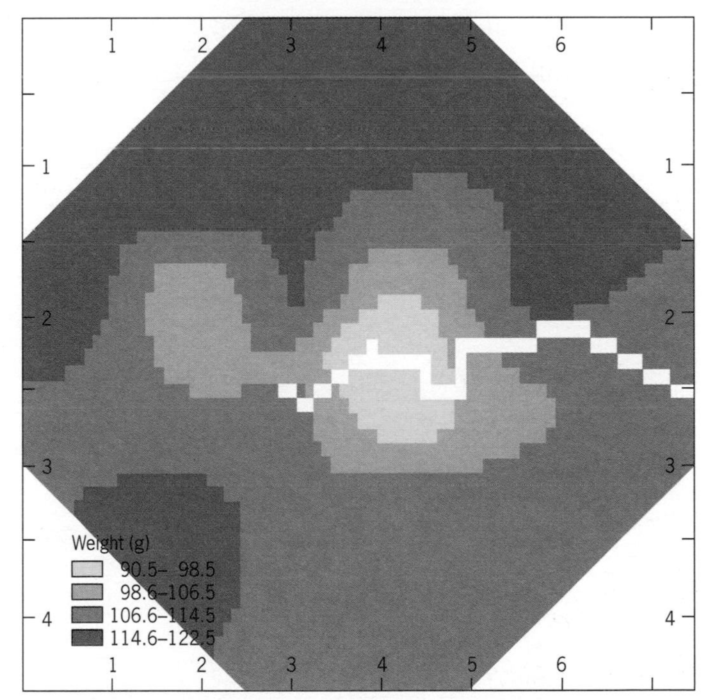

Figure 6.44 Average species weight of land birds/100km^2 in Greater London (after Cousins, 1982)

?

17a Describe the distribution of birds and snails in Greater London from Figures 6.41 and 6.42.
b How does this reflect the pattern of the built environment shown in Figure 6.43?
c Why do you think birds and snails have different distributions in relation to the built environment?
d Explain the distribution of these creatures with reference to Figure 6.45. Your explanation should include references to all ecosystem inputs as well as the flora.

18 From Figure 6.44, describe the pattern of bird weight in relation to the built environment.

The more northerly and westerly distribution of heavy birds around Greater London shown in Figure 6.44 may well reflect the distribution of some of the areas of open water around the urban area. Many former gravel workings have been filled with water and are important areas for roosting gulls, and breeding waterfowl. Water within urban areas can attract a wealth of wildlife.

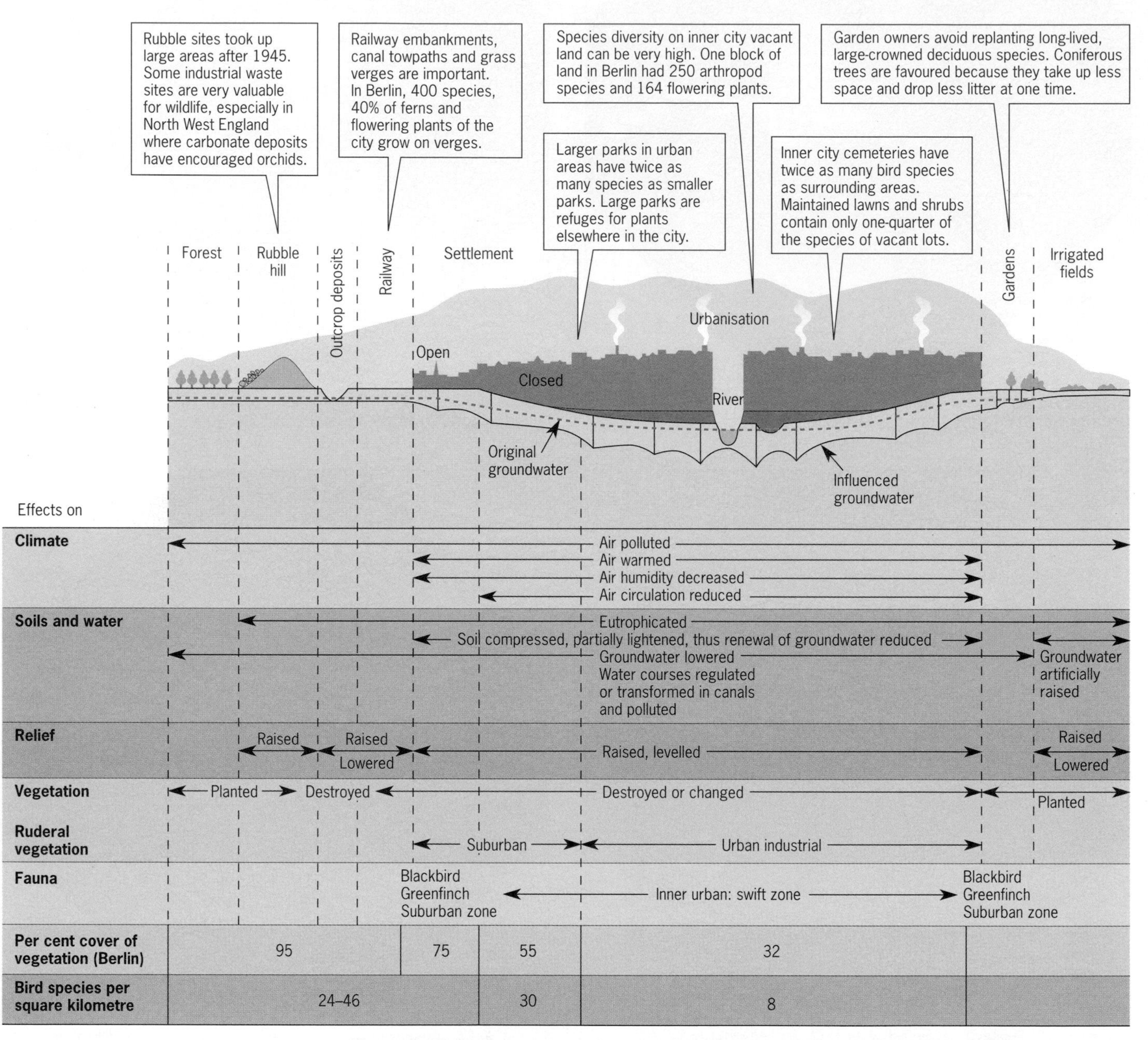

Figure 6.45 A transect through an urban environment (after Horbert et al., 1982)

6.6 Landscape management

People manage landscapes for different aims and objectives (Fig. 3.2). Sandwell Valley Nature Reserve (Fig. 6.46) has been managed to create a range of habitats which favours diversity of bird species and provides amenity value for the visitors to the reserve. This is considered to be an important aim within an urban area. If the reserve had been managed for woodland species alone, its management plan would have encouraged succession.

Table 6.3 Linton's landscape evaluation categories

Urbanised and industrialised	-5
Continuous forest	-2
Treeless farmland	+1
Moorland	+3
Varied forest and moorland	+4
Richly varied farmland	+5
Wild landscapes	+6
Lowland	0
Low uplands	+2
Plateau uplands	+3
Hill country	+5
Bold hills	+6
Mountains	+8

Figure 6.46 The island at Sandwell Valley Nature Reserve, 1992

Landscape evaluation

Designing a management plan for a natural area is never going to be an objective process, since people like different species of wildlife and different landscapes. Also many environments do not have the general appeal that some do. Linton developed landscape evaluation categories, with points according to the landscape value of particular environments (Table 6.3).

?

19a Give your own score, from 1 to 10, to the landscapes in these photographs: Figures 3.18, 4.31, 5.14, 6.33, 7.1, 8.17.
b Compare your list with another student's.
c Evaluate the landscapes using Table 6.3.
d Comment on the relevance of your findings for landscape management.

20 Essay: Evaluate the significance of natural succession in the creation and management of a landscape for a specific purpose.

6.7 Creating a sustainable urban ecosystem

Over half of the world's population live in cities. The challenge is to improve environmental conditions and reduce the demands on the earth's resources. The 1992 Rio Earth Summit prompted many urban communities to measure the state of their environments and then to apply strategies to improve urban ecosystems. Local agenda 21a have been developed which involved communities in environmental planning. Social and environmental issues developed simultaneously, in keeping with the theme of development and environment of the Rio Earth Summit.

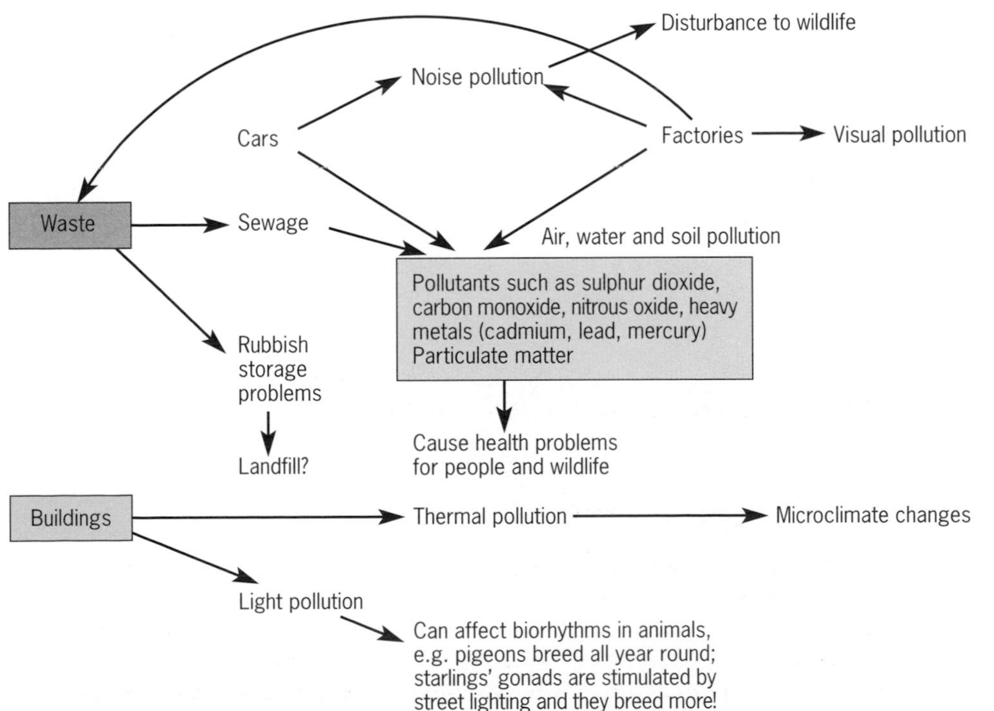

Figure 6.47 Problems in an urban area

The project

- Improve the area around the station and integrate paths, cycle routes and community bus routes. Introduce traffic calming.
- Make sure that all buildings follow green building guidelines, such as sustainable resource use and energy efficiency.
- Begin environment programmes such as urban gardening, planting of native trees, habitat restoration, energy conservation and environmental education involving partnerships with local schools.
- Fund raise to build the project and use financing mechanisms such as energy-efficient mortgages where homebuyers gain credit for the money they save from reduced transportation or energy costs.
- There should be a mix of land uses and diversity of housing types.
- The economic status of the residents should be improved through environmentally orientated business, skill building and training.

Conditions in some inner city areas have forced people to move out into the suburbs, extending the need for more building land and reducing habitats for wildlife on the edge of cities. In Cleveland, Ohio, in the USA, they were desperate to attract people back to the centre, particularly to an area in the city around a dilapidated station – the W65 Street Rapid station. This area is now being redeveloped into an eco-village. Robert Gilam defines an eco-village as 'a small-scale, full-featured settlement in which human activities are harmlessly integrated into the natural world in a way that is supportive of healthy human development and can be successfully continued into the future'.

An eco-village recycles materials and replicates a natural ecosystem where inputs balance outputs. It is a settlement where people live, work, spend leisure time and carry out business activities in balanced amounts. The health of individuals and the principle of sustainability are at its heart.

In 1997, 60 residents and experienced designers discussed their ideas for the community. They won funding for the project and are beginning to develop their plans.

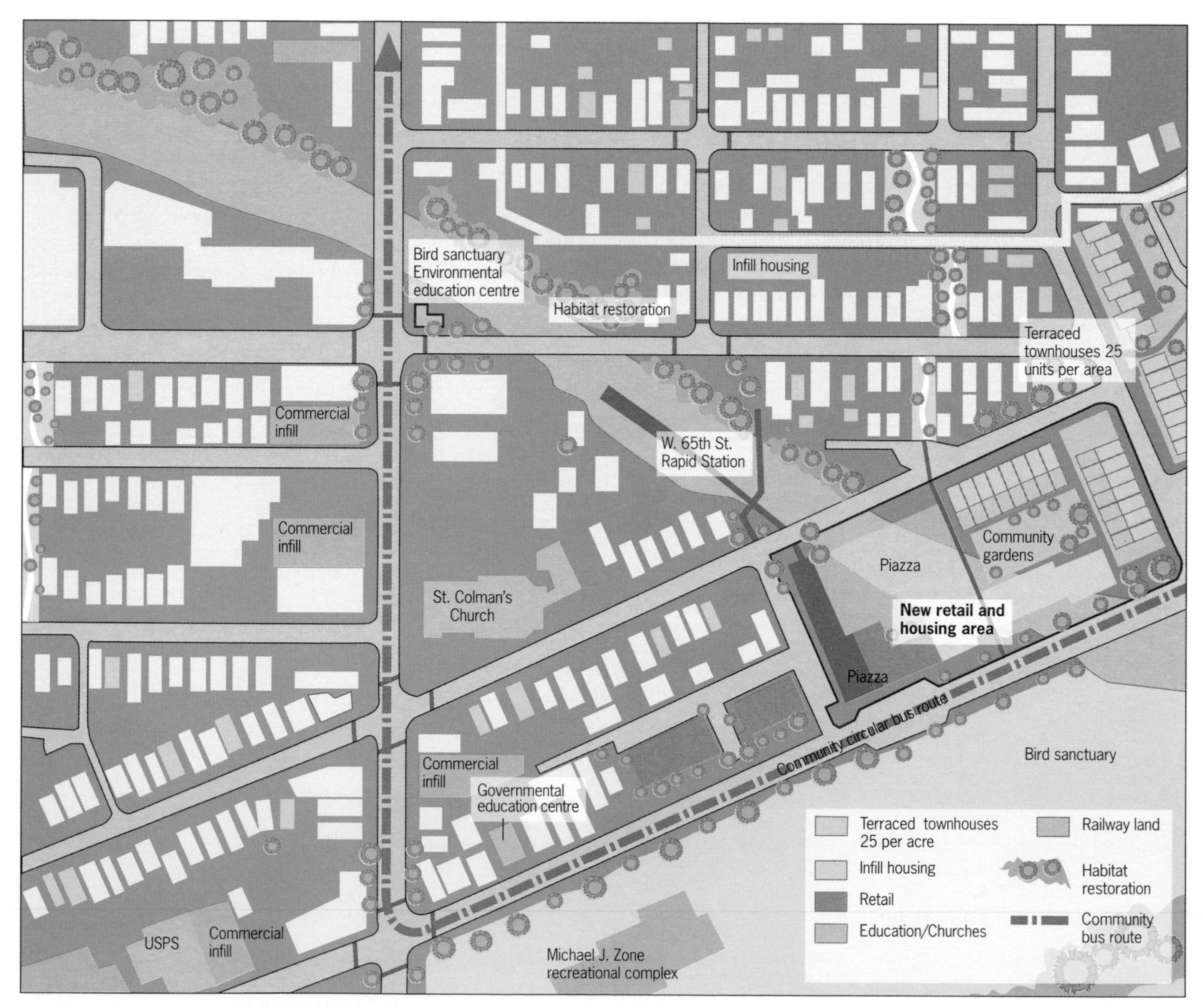

Figure 6.48 Layout of the Cleveland eco-village

A day in the life of the Cleveland eco-village, 1 June 2010

Julie woke early with sunlight in her face. She loved waking up in her new bedroom. It was so light and airy that she often felt like she was floating. Although it wasn't a large bedroom, it felt open because of the abundant windows and skylights.

Windows and overhangs were precisely orientated to let in sunlight in the mornings and evenings, and also during the winter months when the sun was low in the sky. At noon on a hot summer day, her place stayed cool and shady. The whole development of town houses and apartments had been designed to take advantage of the angles of the sun throughout the year in Northeast Ohio. The result was energy efficient, comfortable and delightful.

Julie's mornings weren't as hectic now. She used to live in an apartment out in the suburbs and, as a single parent, it has always been a struggle to get downtown to work on time. First there had been a two-mile drive one way to drop her son, Josh, at school. Then she had to endure a long and stressful commute on I-71. If the weather was good and there were no unusual traffic jams, she could make it in 50 minutes.

Now it took less than 20 minutes. Josh's new school was a couple of blocks away in a renovated church building. From there she walked another block to the W65 Street rapid Transit Station and took the train two stops to downtown. The law firm where she worked had its offices in the Terminal Tower, above the downtown Rapid Station.

Josh's school was different too. The Kirtland Ecology School was a charter school within the Cleveland Public Schools and had a customised Montessori curriculum based on environmental stewardship and neighbourhood development. The students planted organic gardens, studied maths and measurement on construction sites, and learned geography by studying the migration routes of songbirds. Although Kirtland school was just three years old, it was already gaining a reputation for motivating kids to achieve. Scores of state proficiency tests were as high as scores at many suburban schools.

The eco-village had become a living laboratory for all kinds of projects and research activities. High school students developed a recycling and composting program that halved the neighbourhood's waste stream. Urban planning students from Cleveland's State university helped plan a bikeway and pedestrian routes. NASA scientists were testing hydrogen fuelled cells and photovoltaic panels as power supplies for public buildings. Business students were helping to recruit companies to the village's eco-industrial park in which waste by-products from one plant became the valuable feedstocks of another.

The combination of the townhouse's smart solar design, super insulation and high efficiency appliances made Julie's utility bills so low that at first she thought there must be some mistake; she kept part of the money in her pocket as extra disposable income and another part for her mortgage. In fact, a big reason that she was able to afford a new home was the innovative financing package offered in the eco-village. Not only did a portion of her home's energy savings help qualify her for a bigger mortgage, but so did part of her transportation savings. By living in a neighbourhood with such good transit service, she didn't need a car.

If she couldn't find what she needed in the shopping court built over the Rapid station, she could hop on a community circulator bus and to other shopping areas.

If she needed to carry a lot of groceries or wanted to go somewhere that was hard to get to by bus, she could always rent a car. There were so many car-free people in the eco-village that a co-op rental business had developed. Anything from a small compact to a mini van was available, including electric vehicles that reduced pollution. It was a lot cheaper to rent a car for the few times she really needed one than to own one all the time and she was glad to give up the worries of car ownership and maintenance.

With the transit, retail services and housing all mixed together, the plaza was a centre of activity all day long. Old-time residents told Julie that the old W65th Street Rapid Stop had once been a decrepit and dangerous place that people avoided; now it was a lively fun place and it was attracting more users.

Figure 6.49 Life in the Cleveland eco-village (David Beach, Director of EcoCity, Cleveland, from the *EcoCity Cleveland Journal*)

In Nottingham a green commuter plan has been put in place to help improve the health of people and the environment. Green commuter plans occur where employers improve the environmental impact of their business by developing actions that encourage employees to use public transport, cycle, walk and use pool cars instead of commuting to work. Thirty-five companies in the city are following these plans.

?

21 Explain how the eco-village at Cleveland will help to solve the problems identified in Figure 6.47.

22a Draw a plan for your local area to develop it into an eco-village.
b How would you persuade people that this is a good idea?

23 Choose three of the green commuter plan actions that you think would persuade employees to reduce the use of their cars. Why do you think these would be the most effective? Which three do you think would be difficult to implement? Give your reasons.

Table 6.4

Types of measures to encourage staff

- Staff discount travel cards for public transport systems and park-and-ride travel cards
- Information is listed at the workplace of public transport routes and times. The firm may also negotiate with travel companies to make extra stops made near to the offices or factories involved
- Improve footpaths
- Introduce loans for bike purchase or mileage allowances for cycles
- Ensure safe cycle storage and changing facilities
- Introduce teleworking, allowing employees to work from home for a number of days each month
- Allow flexible working hours to avoid congestion
- Have contracts with taxi firms and electric pool cars so that company cars are not needed
- Encourage car sharing using a workforce data base.
- Reduce workplace car parking and allowances to discourage car ownership

Benefits for business	Benefits for Nottingham
• Increases staff motivation and morale	• Cleaner city with less traffic congestion
• Reduces the need for car parking places and allows the land to be used more profitably	• Reliable and improved public transport
• Can result in savings	• Better health due to air quality, exercise and fewer accidents
• Usually benefits more employees and leads to equal treatment	• A quality city
• Reduces staff stress due to driving	• Releases road space, allowing more cycle ways and bus lanes in the long term
• Helps recruitment and retention of staff	• Improves the image of the company

Summary

- Ecosystems usually change by following a sequence of seral stages in which one community succeeds another. They ultimately form a climatic climax community which does not change naturally unless the climate does.
- Not all of the British Isles are covered by woodland, which was originally the climatic climax vegetation. Other variables have led to the formation of a number of different habitats.
- Where regional variations exist, subclimax communities may form which never attain a climatic climax. Similarly, human activity may deflect succession to form a plagioclimax community.
- If human action is stopped, a plagioclimax may follow a successional route to reach a climatic climax, unless the ecosystem has been changed so that it can only reach a secondary climax.
- Once sand is stabilised away from the sea, the sand dune ecosystem succeeds to form a climax community. This could be a climatic climax if human influence was absent.
- The main ways of creating a plagioclimax are grazing, fire, extraction of soil and rock, drainage, disturbance, fertilisation, deforestation and afforestation. Many habitats are affected by these to varying degrees.
- Plants can have one of three strategies: a response to stress, disturbance and competition. They are known as competitors, stress tolerators and ruderals. Different proportions of these plants help to identify the extent to which an ecosystem has been managed.
- Urban areas are the most managed of all habitats and have high numbers of non-native species.
- In urban areas there is a positive correlation between distance from the centre and species numbers. This reflects changes to soil, climate, vegetation and water, which are higher in the most built-up areas.
- Landscape management is influenced by people's perception of the environment.

7 Management of shrubland communities

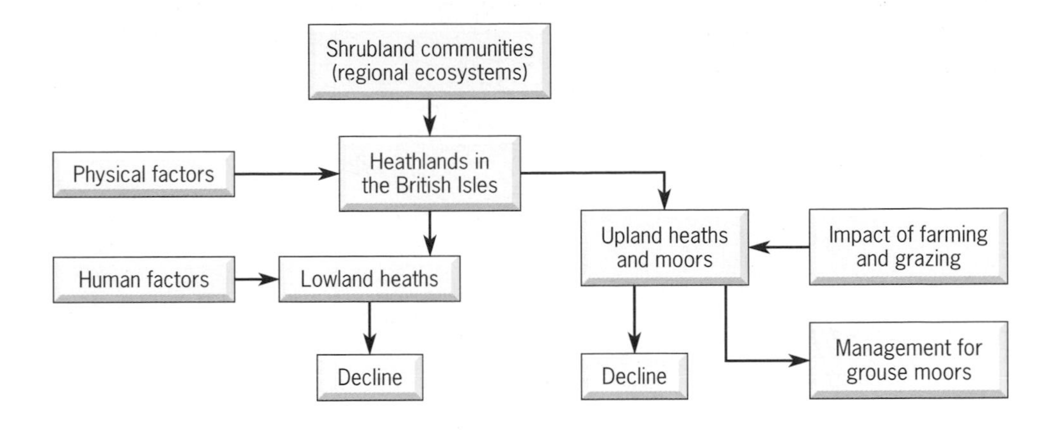

7.1 Introduction

Shrubland occurs for two main reasons: first, a **seral** stage on the way to developing into a climatic climax or secondary **climax forest** community, either as a plagiosere (Fig. 7.1) or prisere or, second, as a **subclimax community** where tree growth is prevented – for example, on unstable slopes or in exposed **habitats** (Fig. 7.2). Important shrubland communities include the garigue or maquis of the Mediterranean and the chaparral of the Americas.

?

1 Refer to the figures for woodland and shrubland on Table 1.4 and explain how and why the **biomass** is different from that produced by forest **biomes**.

Figure 7.1 Chalk shrubland, Sussex

Figure 7.2 Upland heath, Wales

7.2 Heathlands: a shrubland community

All heathlands are made up of dwarf evergreen shrubs with woody branching stems and very small leaves, adapted to reduce water loss (Fig. 7.3). Despite the ability of heathland shrubs to withstand exposed conditions (heather plants in the Alps can withstand temperatures of -24°C), they are unable to tolerate prolonged dry periods. Within Europe the ideal conditions for heathland are:

1 Soils with low nutrient availability
2 Soils with a pH of 3.5–6.7
3 Small seasonal fluctuations in air and soil humidity
4 Protection from low temperatures by snow cover
5 Absence of shade.

?

2a Use an atlas to find out climatic data for potential heathland sites (Fig. 7.4) in Denmark, the Netherlands and the British Isles.
b Describe and explain the distribution of heathlands in Figure 7.4.

3 Would heathland plants be mainly **stress** tolerators, competitors or ruderals (see Section 6.4)? Give reasons for your answer.

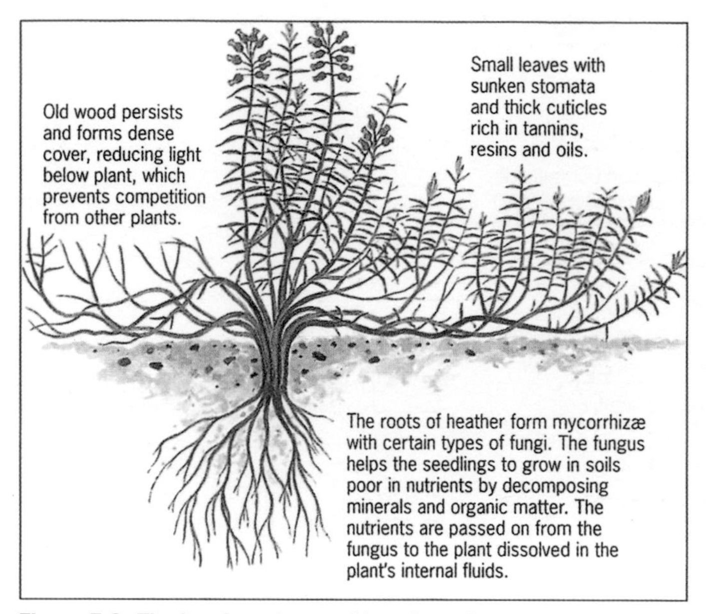

Old wood persists and forms dense cover, reducing light below plant, which prevents competition from other plants.
Small leaves with sunken stomata and thick cuticles rich in tannins, resins and oils.
The roots of heather form mycorrhizæ with certain types of fungi. The fungus helps the seedlings to grow in soils poor in nutrients by decomposing minerals and organic matter. The nutrients are passed on from the fungus to the plant dissolved in the plant's internal fluids.

Figure 7.3 The heather plant and its adaptations

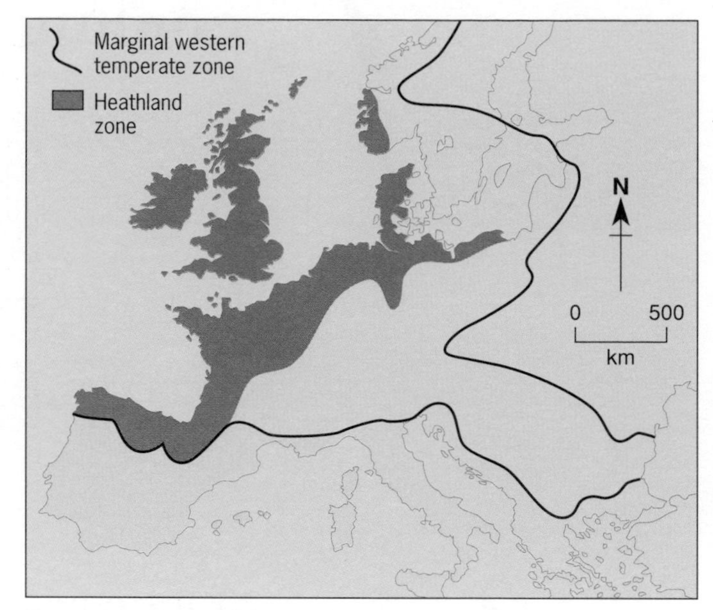

Figure 7.4 The distribution of heathlands in Europe (after Webb, 1989)

?

4a In the British Isles, how might the climatic inputs into a heathland **ecosystem** (Fig. 7.5) differ in upland and lowland areas?
b Suggest the effect that these climatic differences might have on the upland and lowland heathland ecosystems. Consider the effects on the soil, plants and animals.

Lowland versus upland heaths

Heathland vegetation in north-west Europe can be found in both lowlands and uplands. Lowland heaths occur below 250–300 metres, to be replaced by upland heaths or moors at higher altitudes.

Soils

Heathlands occur on free-draining, sandy soils that are generally **podsolised** (see Chapter 5). Evidence from pollen analysis suggests that heathland expansion coincided with climatic change. Increased rainfall during the Atlantic period (5500–2500 BC) may have contributed to the process of podsolisation and this created an important condition for the spread of heath.

Figure 7.5 A heathland ecosystem

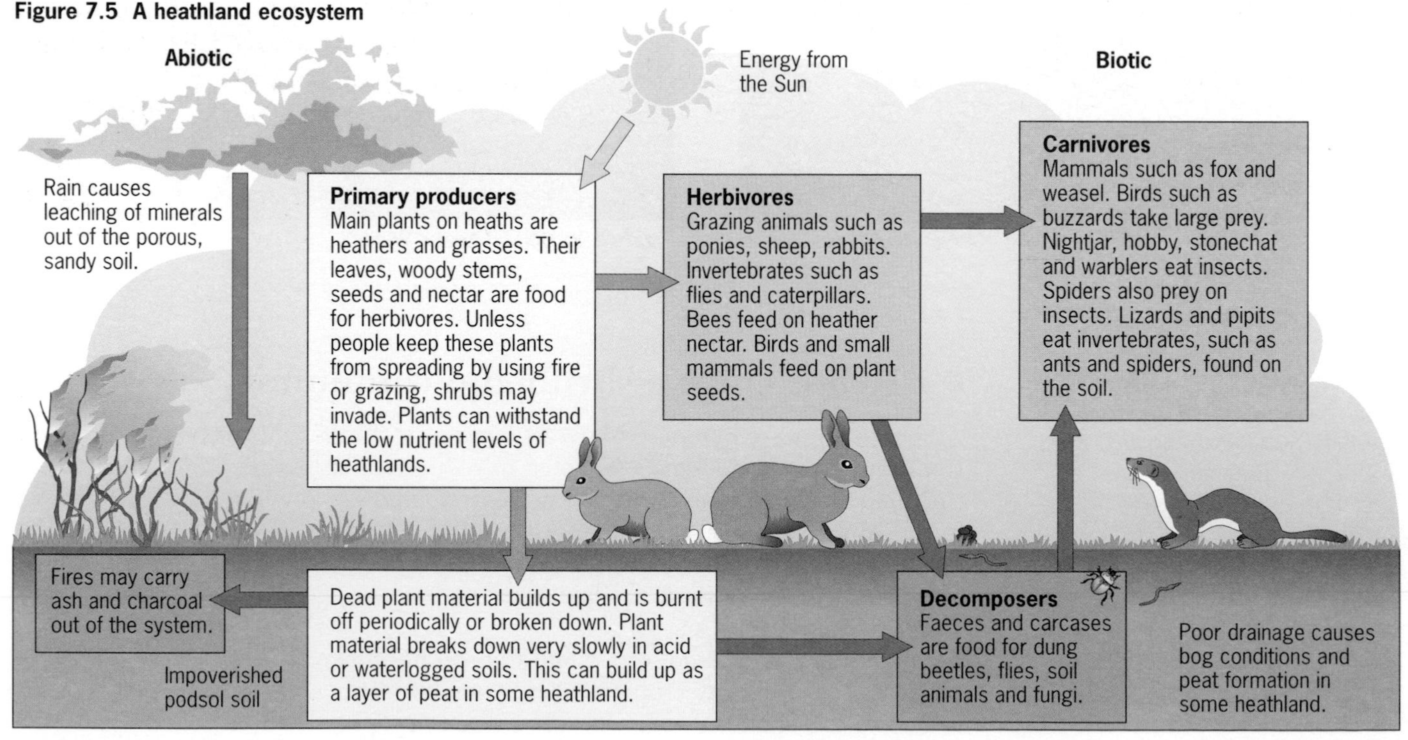

Abiotic
Energy from the Sun
Biotic
Rain causes leaching of minerals out of the porous, sandy soil.
Primary producers
Main plants on heaths are heathers and grasses. Their leaves, woody stems, seeds and nectar are food for herbivores. Unless people keep these plants from spreading by using fire or grazing, shrubs may invade. Plants can withstand the low nutrient levels of heathlands.
Herbivores
Grazing animals such as ponies, sheep, rabbits. Invertebrates such as flies and caterpillars. Bees feed on heather nectar. Birds and small mammals feed on plant seeds.
Carnivores
Mammals such as fox and weasel. Birds such as buzzards take large prey. Nightjar, hobby, stonechat and warblers eat insects. Spiders also prey on insects. Lizards and pipits eat invertebrates, such as ants and spiders, found on the soil.
Fires may carry ash and charcoal out of the system.
Impoverished podsol soil
Dead plant material builds up and is burnt off periodically or broken down. Plant material breaks down very slowly in acid or waterlogged soils. This can build up as a layer of peat in some heathland.
Decomposers
Faeces and carcases are food for dung beetles, flies, soil animals and fungi.
Poor drainage causes bog conditions and peat formation in some heathland.

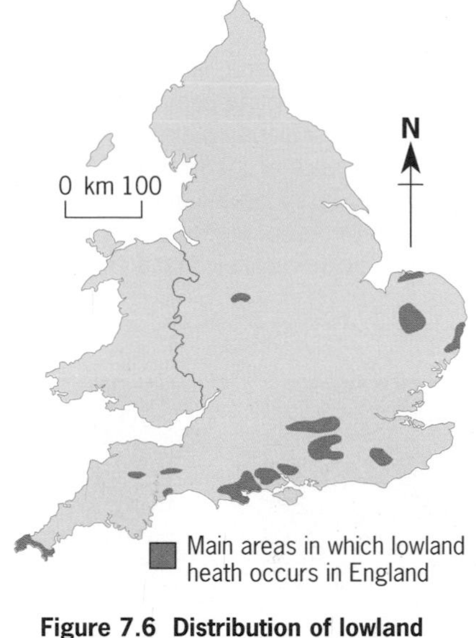

Figure 7.6 Distribution of lowland heath in England (after Tubbs, 1991)

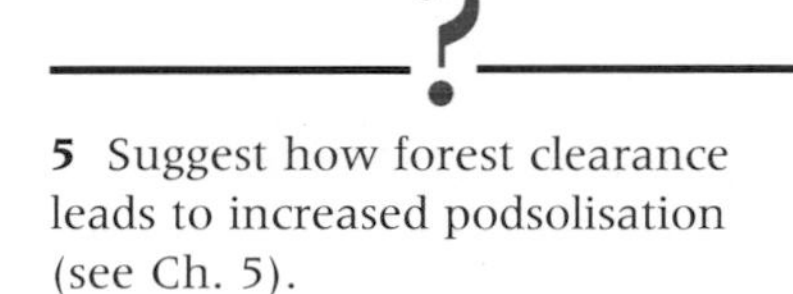

5 Suggest how forest clearance leads to increased podsolisation (see Ch. 5).

6 Use Figure 7.8 to find out when heathland actually expanded in the British Isles.

7 According to Figure 7.4 all of the British Isles fall within the zone suitable for heathland formation, yet heathland does not cover the British Isles. Give two reasons to explain this.

8 What **inputs** were reduced to ensure the survival of the heathland habitat? Draw a simple systems diagram to show these (see Fig. 1.4).

9a How might people's use of heathland (Table 6.2) favour **species** such as the Dartford warbler (Fig. 7.7)?
b Which practices might be harmful to this species?

Figure 7.7 Dartford warbler: requires mature heather with gorse for nest sites and bushy gorse clumps for invertebrate prey and winter shelter

The spread of heathlands

The increase in the spread of heathland may not have been entirely due to climatic factors. Pollen analysis also reveals that the spread of heathland coincided with the activities of early farmers, who removed trees and prevented their regeneration by the grazing of livestock. Forest clearance leads to increased acidity and podsolisation on poor soils, which favours the growth of heathlands.

It is difficult to work out whether the spread of heathland followed or preceded the start of podsolisation. It is now accepted, however, that heathlands expanded into areas that were formerly forest.

Heathland as a subclimax community can only develop where there are factors which prevent the **succession** to woodland, such as grazing by native **herbivores**, or exposure. Where the activities of people prevent the succession to woodland, heathland represents a **plagioclimax community**.

7.3 Lowland heathlands

The areas of lowland heathland that exist today in Great Britain largely evolved as a result of human activities (Fig. 7.6). Heathlands provided rough grazing for livestock, which prevented the regeneration of woodland, as did the practice of burning to encourage the growth of young heather, which is more palatable to sheep. Burning also reduced soil fertility, which ensured the continued survival of the heathland habitat.

The cutting of turves for fuel also kept the nutrient content of heathland soils low. Clay and sands were excavated for building materials, and a wide range of uses was found for various heathland plants (Table 7.1). In the Midlands the heaths are much younger than those in the South-East and South-West. These heaths were largely formed from medieval times to the eighteenth century, when trees were removed for charcoal-burning. All of these activities created a number of differing habitats for a wide range of organisms that were dependent upon heathland for their survival.

The decline of lowland heathland

Seventy-two per cent of south and eastern heathland has been lost from England since the 1820s (Fig. 7.8). However, Great Britain has more heathland than any other country in Western Europe: its 56 000 ha represent 20 per cent of the total remaining.

Many heathland areas within Great Britain are continuing to decline, despite a variety of laws to protect them.

Table 7.1 Some heathland plants and their uses

Heather species	
Thatching	
Turves cut for domestic fuel (turbary) and also to make charcoal for the tin-smelting industry	
Rope-making (Hebrides)	
Trackway foundations	
Gorse species	
Animal fodder (when wilted)	
Boundary hedges	
Thatching ('rafters')	
Fuel for bread ovens and pottery kilns	
Land drains (gorse was used to fill in trenches that were then covered with earth)	
Bracken	
Animal bedding	Potash for glass and soap making and bleaching
Domestic fuel	Thatching
Fuel for brick kilns	

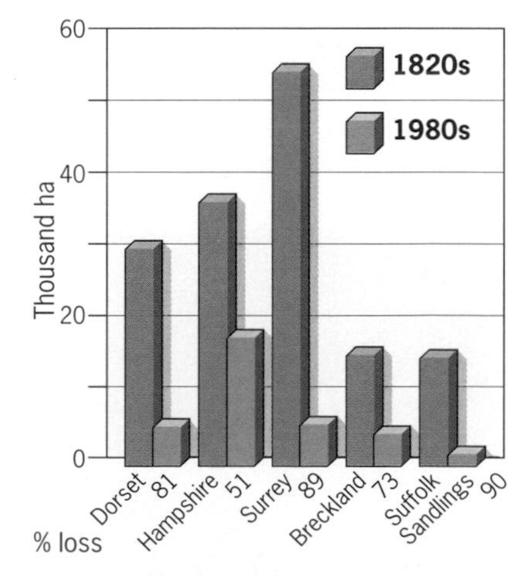

Figure 7.8 Changes in the extent of the main areas of southern and eastern heathland in England since the 1820s (after RSPB, 1989)

?

10a Using Figures 7.9–7.14, draw a spider diagram to show the factors that damage SSSIs on heaths.
b Which of these inputs may have been responsible for the decline in heathland shown in Figure 7.8?
c What in your view, could be done to reverse this decline? Give reasons for your answer.

Figure 7.9 Agriculture

Modern fertilisers and irrigation methods have 'improved' heathland soil for pasture or arable crops. Grazing is less attractive to farmers, and areas have succeeded to woodland. Heathland is now closer to arable land. Eutrophication (nutrient enrichment) caused by spray drift and runoff of agricultural fertilisers has encouraged arable weeds and aggressive grasses.

Figure 7.10 Fire

Fire can be used to manage some heathlands. It encourages the growth of young heather and prevents succession to woodland. In excess, it is harmful. The dry summers of 1975 and 1976 caused 30 per cent of Surrey heaths and 11 per cent of Dorset heaths to be damaged. Deliberate arson also occurs on heathland sites near urban areas.

Large areas, for example 20 per cent of original heath in east Dorset and Hampshire (west of the Avon), were planted by private landowners and the Forestry Commission after World War 2. Although planting has stopped, self-sown pines from plantations threaten further areas.

Figure 7.11 Forestry (New Forest)

Figure 7.12 Urban expansion (Canford Heath)

Development of urban areas and roads has been more of a threat to heathlands than agriculture or forestry. Seventy-five per cent of Dorset's heathland SSSIs have adjoining urban areas. Of the nine public inquiries held between 1977 and 1978, four were for roads, four for residential areas and one for motorbike scrambling. Tourism and recreation pressure are problems in areas like the New Forest.

Figure 7.13 Myxomatosis

The introduction of myxomatosis in 1953 into Britain to control rabbits killed over 90 per cent of them. The subsequent lack of grazing pressure on heathlands encouraged succession.

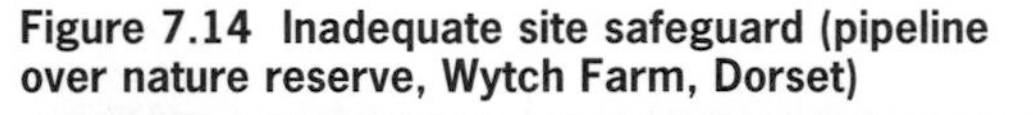

Figure 7.14 Inadequate site safeguard (pipeline over nature reserve, Wytch Farm, Dorset)

Forty-two per cent of damage on heathland SSSIs during 1985 and 1986 was due to consents under Town and Country Planning legislation, 31 per cent due to oil and gas pipelines, 21 per cent due to agriculture and forestry and 6 per cent due to fire.

?

11a From Figure 7.15 identify the factors which prevent temperate deciduous woodland from forming as the climatic climax vegetation in upland moors.
b Do you think the upland moor and heath ecosystems are largely subclimax shrubland communities or plagioclimax shrubland? Give reasons for your answer.

7.4 Upland heaths and moors

Most of British upland vegetation can be described loosely as 'moorland' where it implies open wild country. The heather moorland or upland heath is a shrub community, but upland moorland also contains grassland and peat bogs.

Chapters 4 and 5 highlighted the former extent of forest within the British Isles. Forest was more extensive than at present in upland areas.

Grouse moors

Much of the heather moorland in the British Isles has disappeared in recent years, sometimes through burning and subsequent agricultural improvement, or through overgrazing. In some areas, the continued existence of moorland is due to **management** for red grouse (a game bird).

Figure 7.15 An upland ecosystem

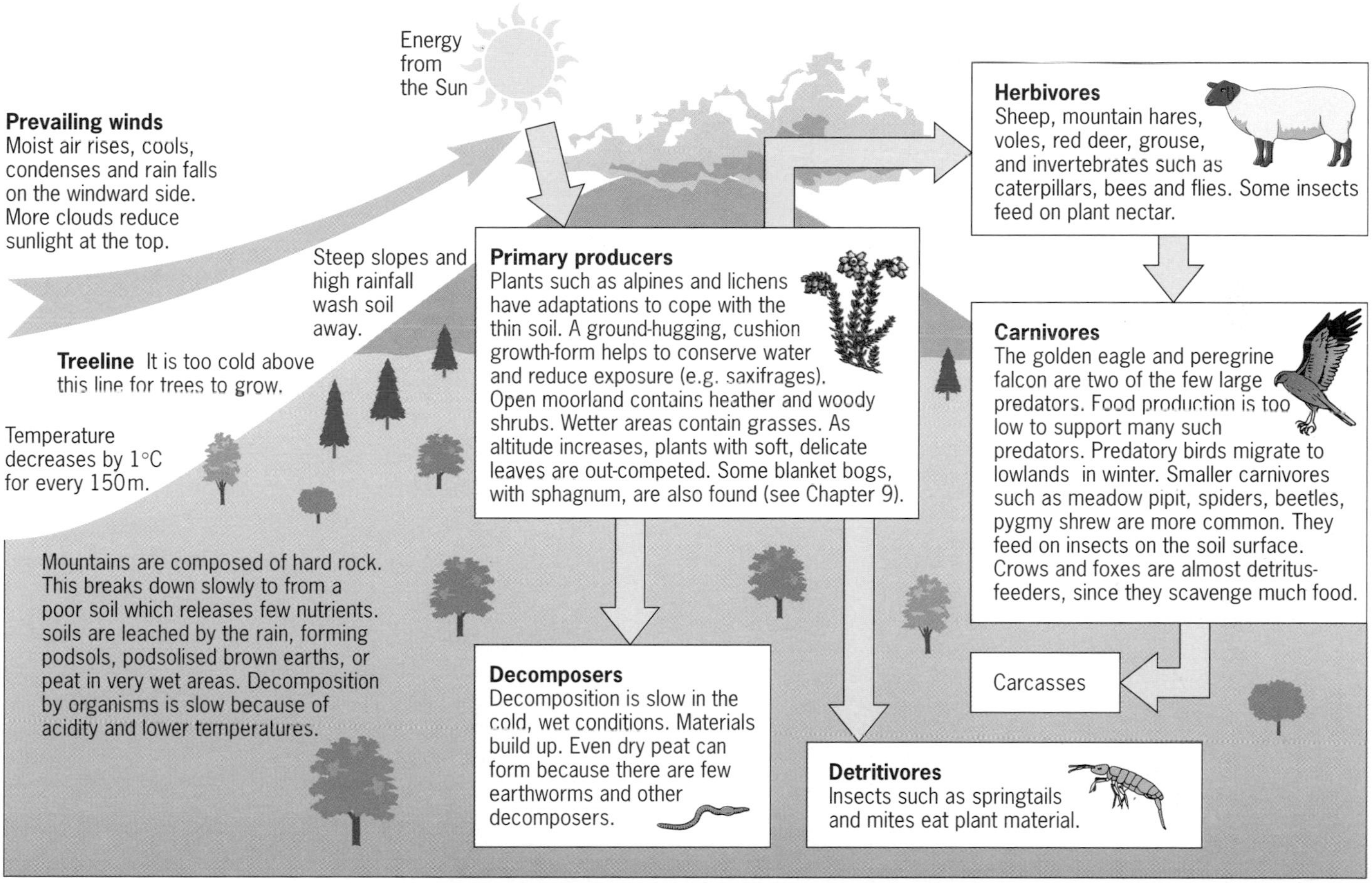

Abiotic

Biotic

Energy from the Sun

Prevailing winds Moist air rises, cools, condenses and rain falls on the windward side. More clouds reduce sunlight at the top.

Steep slopes and high rainfall wash soil away.

Treeline It is too cold above this line for trees to grow.

Temperature decreases by 1°C for every 150m.

Mountains are composed of hard rock. This breaks down slowly to from a poor soil which releases few nutrients. soils are leached by the rain, forming podsols, podsolised brown earths, or peat in very wet areas. Decomposition by organisms is slow because of acidity and lower temperatures.

Primary producers Plants such as alpines and lichens have adaptations to cope with the thin soil. A ground-hugging, cushion growth-form helps to conserve water and reduce exposure (e.g. saxifrages). Open moorland contains heather and woody shrubs. Wetter areas contain grasses. As altitude increases, plants with soft, delicate leaves are out-competed. Some blanket bogs, with sphagnum, are also found (see Chapter 9).

Herbivores Sheep, mountain hares, voles, red deer, grouse, and invertebrates such as caterpillars, bees and flies. Some insects feed on plant nectar.

Carnivores The golden eagle and peregrine falcon are two of the few large predators. Food production is too low to support many such predators. Predatory birds migrate to lowlands in winter. Smaller carnivores such as meadow pipit, spiders, beetles, pygmy shrew are more common. They feed on insects on the soil surface. Crows and foxes are almost detritus-feeders, since they scavenge much food.

Carcasses

Decomposers Decomposition is slow in the cold, wet conditions. Materials build up. Even dry peat can form because there are few earthworms and other decomposers.

Detritivores Insects such as springtails and mites eat plant material.

Summary

- Shrubland communities can be found as subclimax or plagioclimax communities or as a seral stage on the way to a climatic climax community.
- Shrublands are less productive than woodland commuities and have a lower biomass. This is generally due to lower nutrient inputs caused by a range of physical and human factors.
- Technological changes have increased the degree of influence that humans have on shrubland environments. Low levels of interference resulted in some benefits for wildlife. However, more intensive use of the land (prompted by changes in EU subsidies) has resulted in unsustainable systems.

?

12 What further impacts might grazing have on upland ecosystems? Use Table 4.10 to help.

8 Agriculture and trampling in temperate grasslands

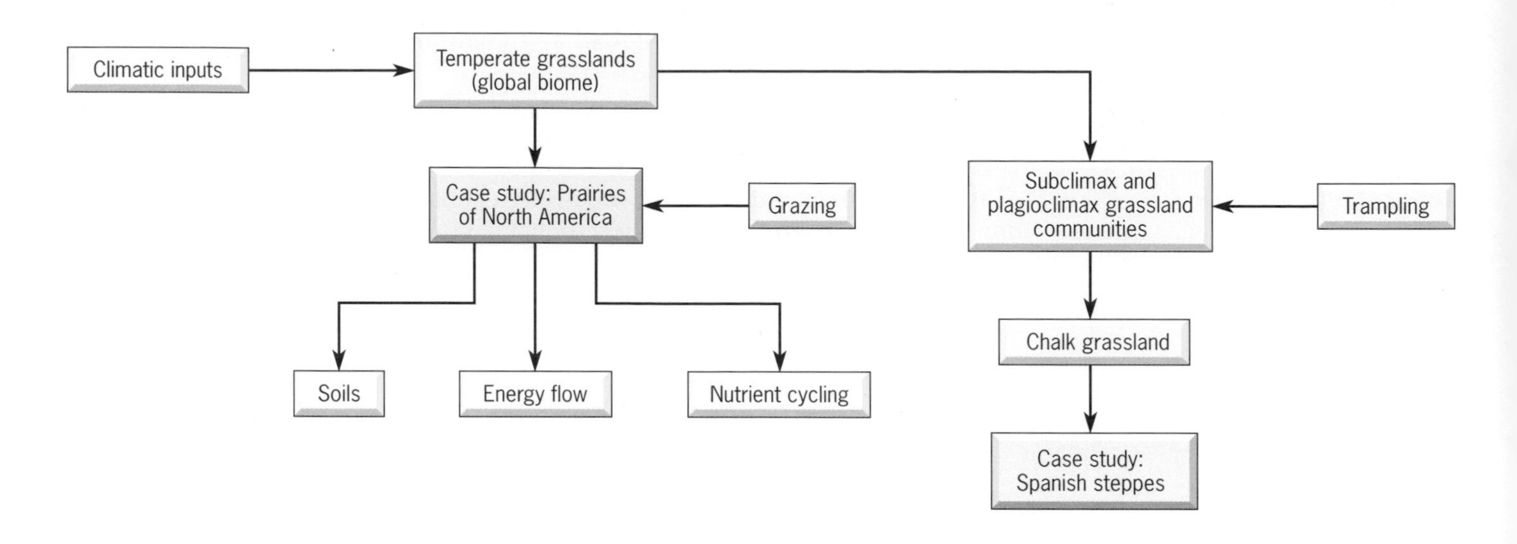

8.1 Grassland ecosystems

Temperate grasslands occur as **biomes** in both cool and warm temperate climatic zones (Fig. 8.1). In cool temperate regions, grassland replaces forest when the mean annual rainfall is approximately 500mm and in warm temperate areas where mean annual rainfall is below 750mm (Fig. 8.2).

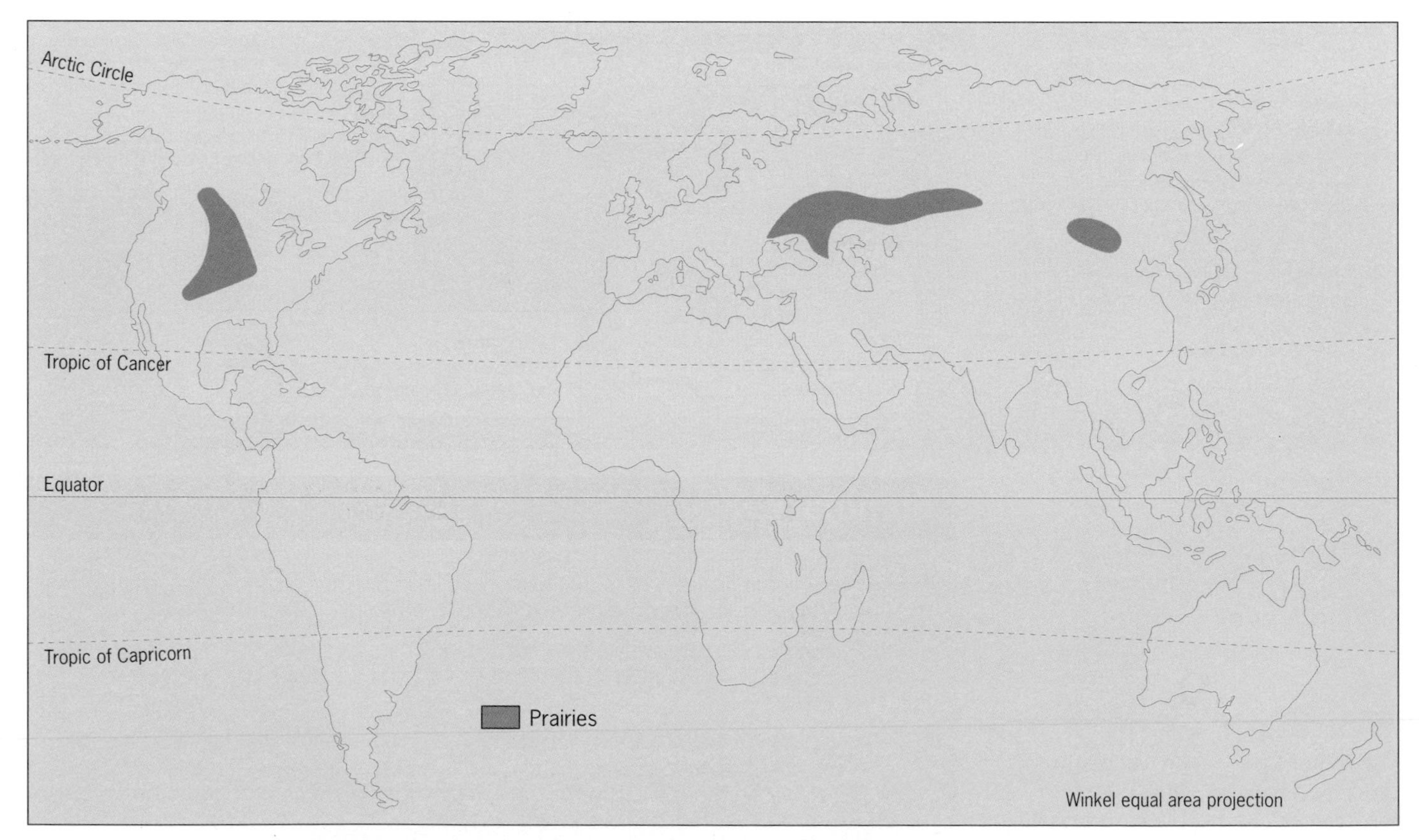

Figure 8.1 Distribution of temperate grasslands which are climatic climax communitites

?

1a Refer to Table 1.1. How does the net **primary productivity** per unit area of temperate grasslands compare with woodland and shrubland?
b Refer to Figure 8.2 and describe the differences in climate **inputs** between shrubland and temperate grassland.
c Explain the variations in productivity between shrubland and temperate grasslands.
d Explain how the climate contributes to the formation of a **chernozem** soil (Fig. 8.3).

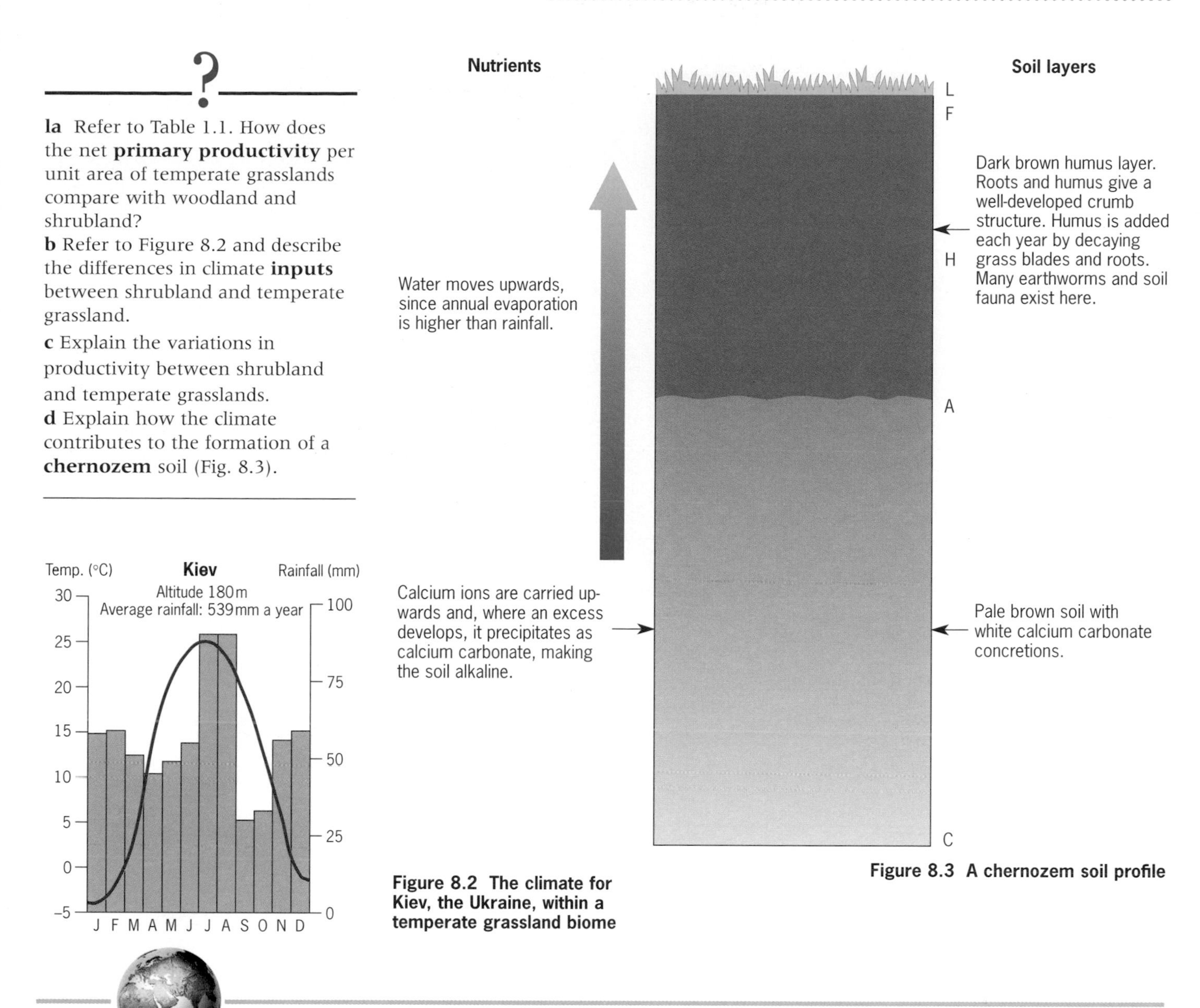
Temp. (°C)
Kiev
Rainfall (mm)
Altitude 180m
Average rainfall: 539mm a year
30
25
20
15
10
5
0
–5
100
75
50
25
0
J F M A M J J A S O N D
Nutrients
Water moves upwards, since annual evaporation is higher than rainfall.
Calcium ions are carried upwards and, where an excess develops, it precipitates as calcium carbonate, making the soil alkaline.
Soil layers
L
F
H
A
C
Dark brown humus layer. Roots and humus give a well-developed crumb structure. Humus is added each year by decaying grass blades and roots. Many earthworms and soil fauna exist here.
Pale brown soil with white calcium carbonate concretions.

Figure 8.2 The climate for Kiev, the Ukraine, within a temperate grassland biome

Figure 8.3 A chernozem soil profile

The prairies of North America

The prairies of North America form a huge area of temperate grassland. Within this area there are regional differences between the height of grasses and also the soil that is below them. However, all of the prairie grasslands consist of a continuous layer of mat-forming and tussocky grasses (Fig. 8.4).

Climate

Rainfall decreases from east to west across the prairies, and this is reflected in a range of different grassland communities.

Soils

Soils are affected by the nature of the vegetation itself and also the amount of water. Short-grass prairie occurs on chernozem soils, while wetter climate and tall-grass prairie result in the formation of prairie soils. Prairie soils have properties halfway between brown earths (see Chapter 4) and chernozems.

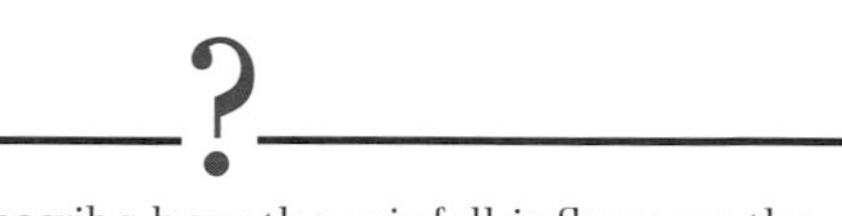

2 From Figure 8.5 describe how the rainfall influences the height of grassland vegetation (above-ground herbage) in Oklahoma, Texas and Colorado.

3 Using Figure 8.5, describe the difference between short and long-grass prairie, noting:
a depth of roots
b above-ground herbage
c above-ground litter.

Figure 8.4 Prairie grassland in Alberta, Canada

Figure 8.5 Vegetation production and standing crop of different grassland areas of the US prairies (after Reymeyer and Van Dyne, 1980)

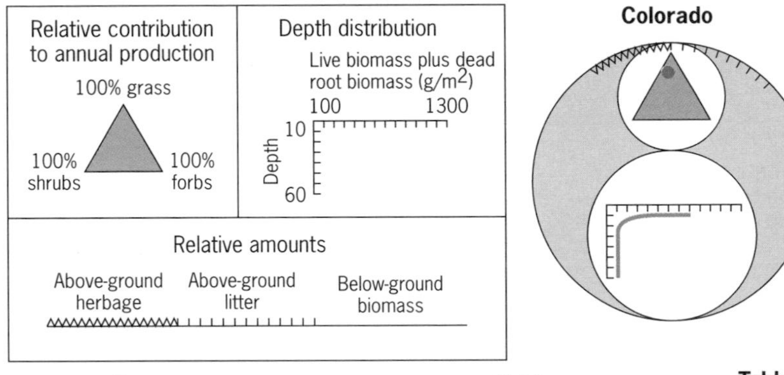

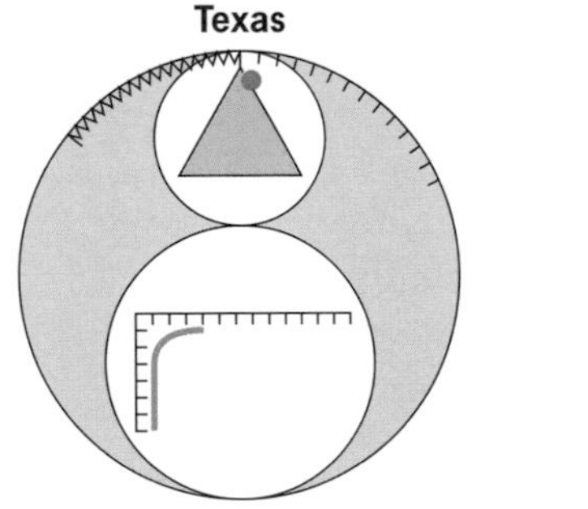

Oklahoma

Table 8.1 Energy flow in short-grass prairie (short-grass prairie net primary productivity: 100 grams per metre square per year)

Percentage of NPP energy in parts of the ecosystem:	
Above-ground primary producers	
Below-ground (crown and roots)	
Percentage of NPP stored in other parts of the ecosystem:	
Above-ground herbivores, cattle, carnivores	0.2
Below-ground herbivores, carnivores, decomposers	21.0

Energy flow

As in all temperate regions, there are distinct seasons. The net primary productivity of prairie grassland is limited during the winter by low inputs of sunlight. However, the productivity of prairie grasses is not limited by total energy inputs but by the **evapotranspiration** caused by increased solar **radiation** during the summer.

?

4 Study Table 8.1.

a Where is most of the energy stored in prairie grasslands?

b Where are most of the **fauna**?

c Why are above-ground **herbivores** likely to be found at lower densities in short-grass areas than long-grass areas?

Nutrient cycling

Humus is added to the soil each year by decaying blades and roots of grasses. Most of the nutrient cycling is carried out below ground, where most of the **biomass** is concentrated.

Grazing

In areas of short-grass prairie, the main way that people have used the **habitat** is by using grazing animals to produce dairy products, leather and meat.

Grazing animals and the plants which they eat have evolved together and are **adapted** to each other. Large herbivores also have **symbiotic** bacteria and protozoa in their digestive tract, which helps to break down plant material. Large mammals do not migrate as far as birds and so need to be adapted to survive the differences between hot summers and cold winters. Large herbivores can store fat and are able to survive on low-quality roughage during the non-growing season.

Niches

Within any **ecosystem** there is **resource partitioning** (see Section 1.7). In prairie ecosystems, there are such divisions between some of the large herbivores.

Cultivation and grazing

The tall-grass prairies have been mainly used for arable cultivation rather than grazing.

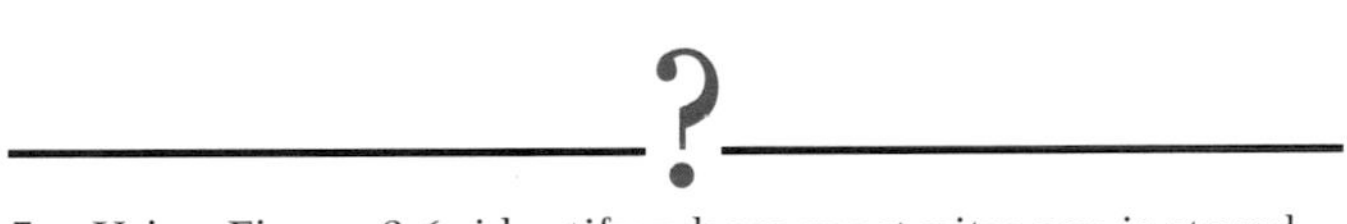

5a Using Figure 8.6, identify where most nitrogen is stored in short-grass prairie.
b Which part of the plant contributes most nitrogen?
c How is the nitrogen transferred from the vertebrates and **invertebrates** to the soil?

6 Study Figure 8.7.
a Describe the relationship between grazing, resilience and stability.
b What is the relationship between net primary productivity, nutrient turnover and grazing intensity?
c Describe the relationship between the amount of herbage after grazing and the number of animal gains per head.
d Explain why the relationships you have identified exist.

7 Study Figure 8.8.
a Which animals do you think would be able to live together without causing too much **competition** for resources?
b Which animals may compete with each other?
c What could happen if two animals exploit the same niche (Section 1.7)?

8a From Table 4.10 make a list of the problems that could be caused by over-grazing on short-grass prairie.
b From the ecosystems details you have studied, list the reasons why people have preferred to cultivate these areas rather than short-grass prairies.

Damage to the prairies

During the 1930s, the prairie lands of south-east Colorado, south-west Kansas and parts of Oklahoma, Texas and northeast New Mexico turned into a dust bowl. This was linked to: poor land management, including overgrazing; a severe drought that lasted several years; cultivation on former grassland areas. These three factors meant that the protective grassland mat and network of water-retaining roots were largely removed. The topsoil of the region was exposed and carried away by strong spring winds. The sun's rays were blocked by dust storms, and dirt piled high in drifts. Some dust storms reached the east coast of the USA. Thousands of families left the area during the peak of the American Depression in the 1930s.

Government aid was made available to stop the wind erosion. Wind breaks were planted, and areas of arable land were restored to grassland. The area recovered by the 1940s.

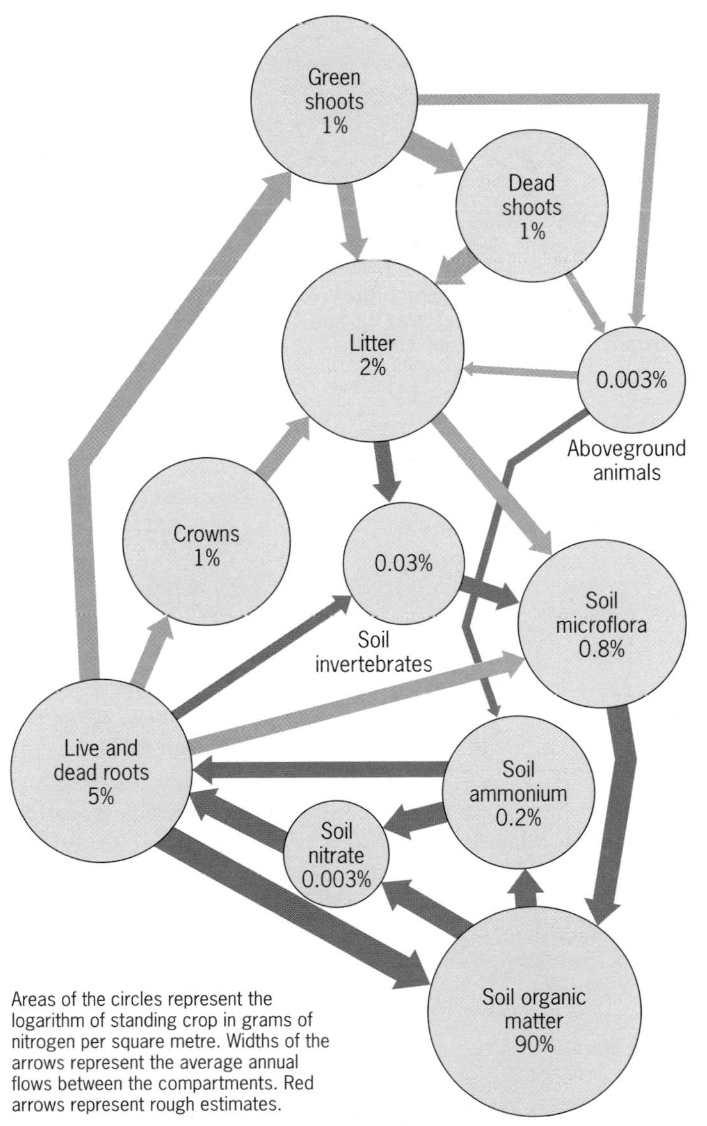

Figure 8.6 Nitrogen cycling in a shortgrass prairie grassland (after Reymeyer and Van Dyne, 1980)

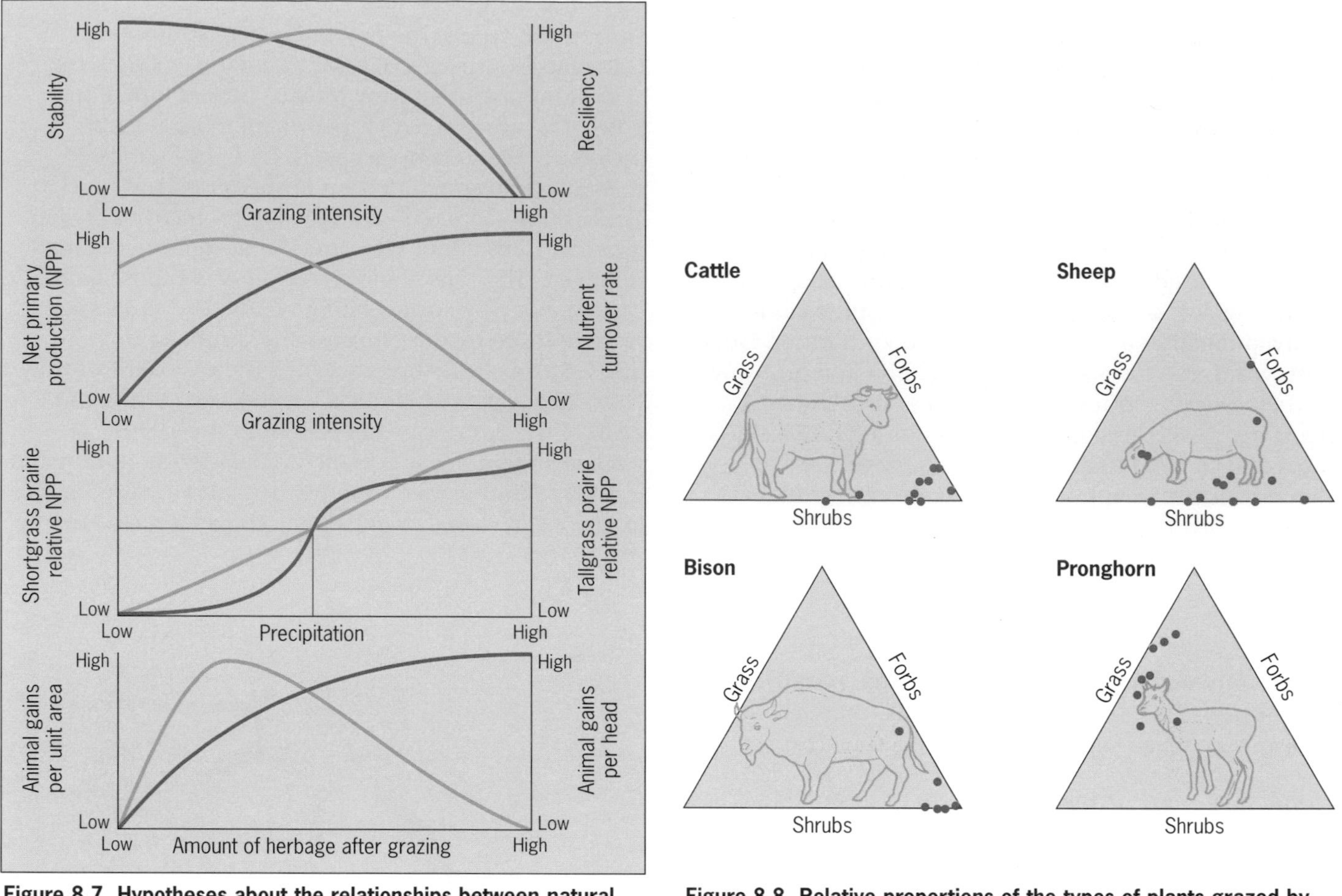

Figure 8.7 Hypotheses about the relationships between natural semi-arid grassland and grazing and precipitation (after Reymeyer and Van Dyne, 1980)

Figure 8.8 Relative proportions of the types of plants grazed by large grassland herbivores (after Reymeyer and Van Dyne 1980)

8.2 Subclimax and plagioclimax grassland communities

On a regional scale, grasslands form **subclimax** or **plagioclimax communities**. Subclimax wetland grasslands form where soil conditions are too wet for tree growth (see Chapter 9). Subclimax upland grasslands form where soils are leached above the tree line.

Lowland grasslands away from the continental interiors shown on Figure 7.1, and much of the upland grasslands, are plagioclimaxes. People sometimes manage them to stop them from moving into the next **seral** stage which is shrubland. As long as the processes which formed grasslands continue, they are balanced semi-natural communities. The individual characteristics of certain types of grassland are determined partly by the soil and rock type but also by their **management**.

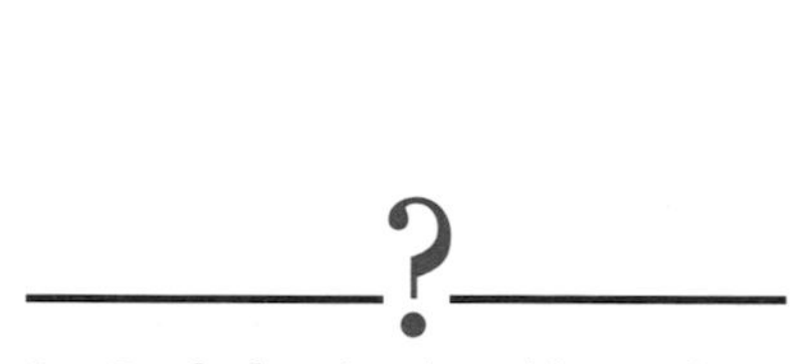

9a Rank the sites in Table 8.2 from high to low, according to the number of **species** they have.
b Which of the following variables could account for differences in species numbers?:
- rock type
- mowing
- altitude
- pH

c Give reasons for your answer to **b**.

Trampling

The effects of seasonal trampling were determined by Harrison (1981) for some sites near London (Table 8.2). The summer treatment consisted of 400 passages once a week for five weeks between July and August 1978, yielding a total of 2000 passages. The winter treatment was 100 passages, once a week for four weeks in February and March 1979 (Figs 8.9 and 8.10). All sites were flat, at the same altitude and had not been trampled previously.

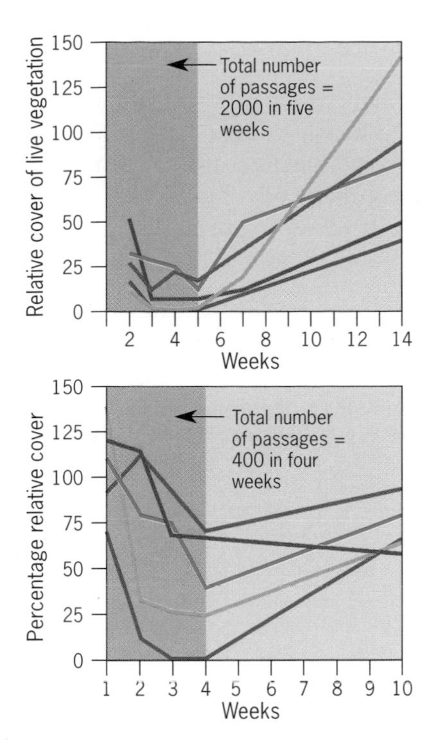

Figure 8.9, 8.10 Relative live cover of vegetation after trampling on summer (above) and winter (below) treatment plots (after Harrison, 1981)

Table 8.2 The effects of trampling on selected ecosystems (*Source:* Harrison, 1980–1)

Site	Species	Percentage cover: At the end of treatment – Control	At the end of treatment – Path	At the end of recovery – Control	At the end of recovery – Path
A: Keston Common (heathland on pebble gravels; podsolic soil; pH 4.5)	Wavy hair grass	2	2	28	16
	Heather	43	1	31	–
	Moss	–	2	–	–
B: Redhill Common (grass heath on Lower Greensands; podsolic soil; pH 4.5)	Yorkshire fog	40	4	42	25
	Wavy hair grass	4	7	9	19
	Bent	3	1	–	–
	Sheeps sorrel	4	1	–	–
	Bracken	48	–	49	4
C: Farthing Downs (annually mown grassland on rendzina soil over chalk; pH6.5)	Yorkshire fog	16	–	14	–
	Cocksfoot	–	1	7	23
	Fescue	15	7	5	5
	False oat grass	23	2	16	10
	Bent	2	–	–	I
	Knalaweed	2	–	–	1
	Thistle	3	–	–	1
D: Happy Valley (annually-mown chalk grassland on clay with flints over chalk; pH 7)	False oat grass	19	20	19	29
	Cocksfoot	15	12	35	20
	Yorkshire fog	14	–	15	–
	Bent	23	7	18	17
	Buttercup	11	–	–	–
	Fescue	6	5	5	–
	White clover	4	–	2	–
	Dock	2	–	–	1
	Dandelion	–	6	–	14
E: Epsom Common (gleyed, silty clay; pH 6)	Bent	17	26	18	16
	Yorkshire fog	21	–	13	12
	False oat grass	46	3	8	26
	Bedstraw	–	2	–	–
	Birch	5	–	–	–
	Potentil	I	–	–	–

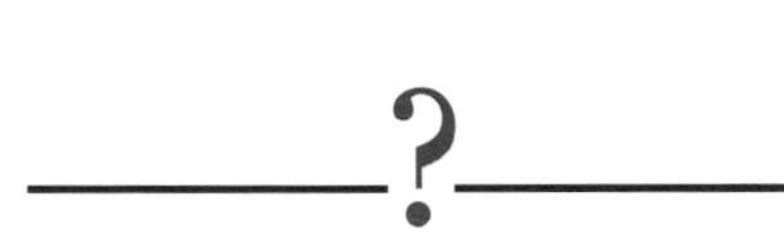

10a From Figures 8.9 and 8.10, identify two plant communities which had least relative cover of live vegetation after summer and winter trampling.
b Using Table 8.2, Figures 6.37 and 6.38, identify the type of plants contained in each of these two sites (competitors, **stress** tolerators or ruderals).
c Why are there fewer competitors on the path?
d In Figures 8.9 and 8.10, what happens to the stress tolerators following trampling? Explain why they may be more able to survive trampling.
e What competitive advantage do grasses seem to have over shrubs when trampling occurs?
f Which two sites seem to have the most resilient vegetation in summer and winter?
g What factors might explain this?

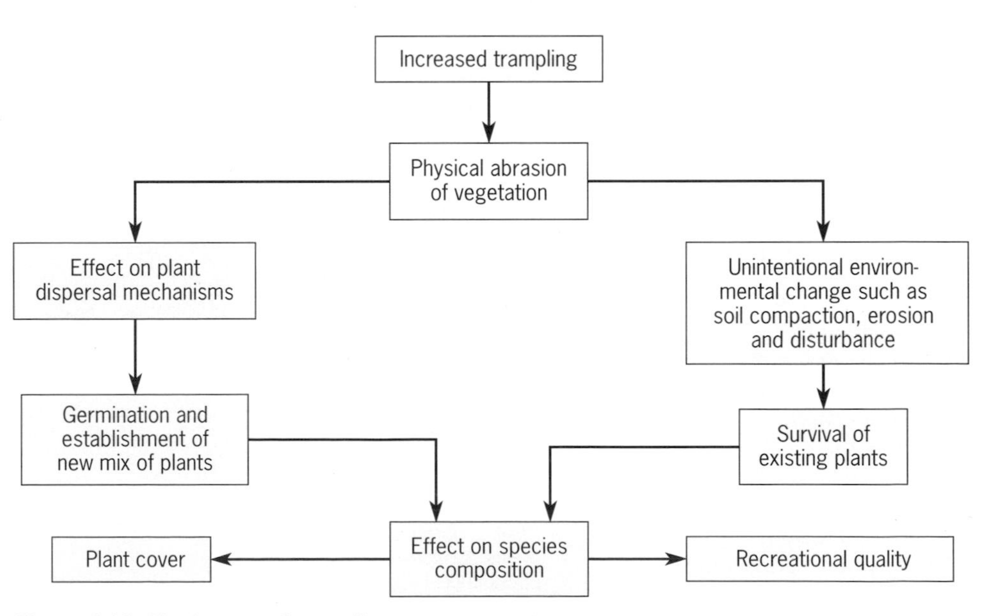

Figure 8.11 The impact of trampling on an ecosystem

?

11a Using Figure 8.11 explain how trampling degrades an ecosystem.
b What do you predict will happen to the vegetation of the paths studied in Table 8.1 after a few years of trampling? Give reasons for your answer.

8.3 Chalk grassland

Figure 8.12 shows a chalk grassland ecosystem. Chalk grassland is partly a plagioclimax caused by sheep grazing, and partly a **biotic** climax caused by the action of rabbits. Such grazing prevents the re-establishment of trees and shrubs, since the animals eat the seed-leaves. The grazing also prevents competitive grasses from dominating and allows a diverse turf of flowers, herbs and grasses to thrive. This rich **flora** provides a suitable habitat for many insects, especially butterflies.

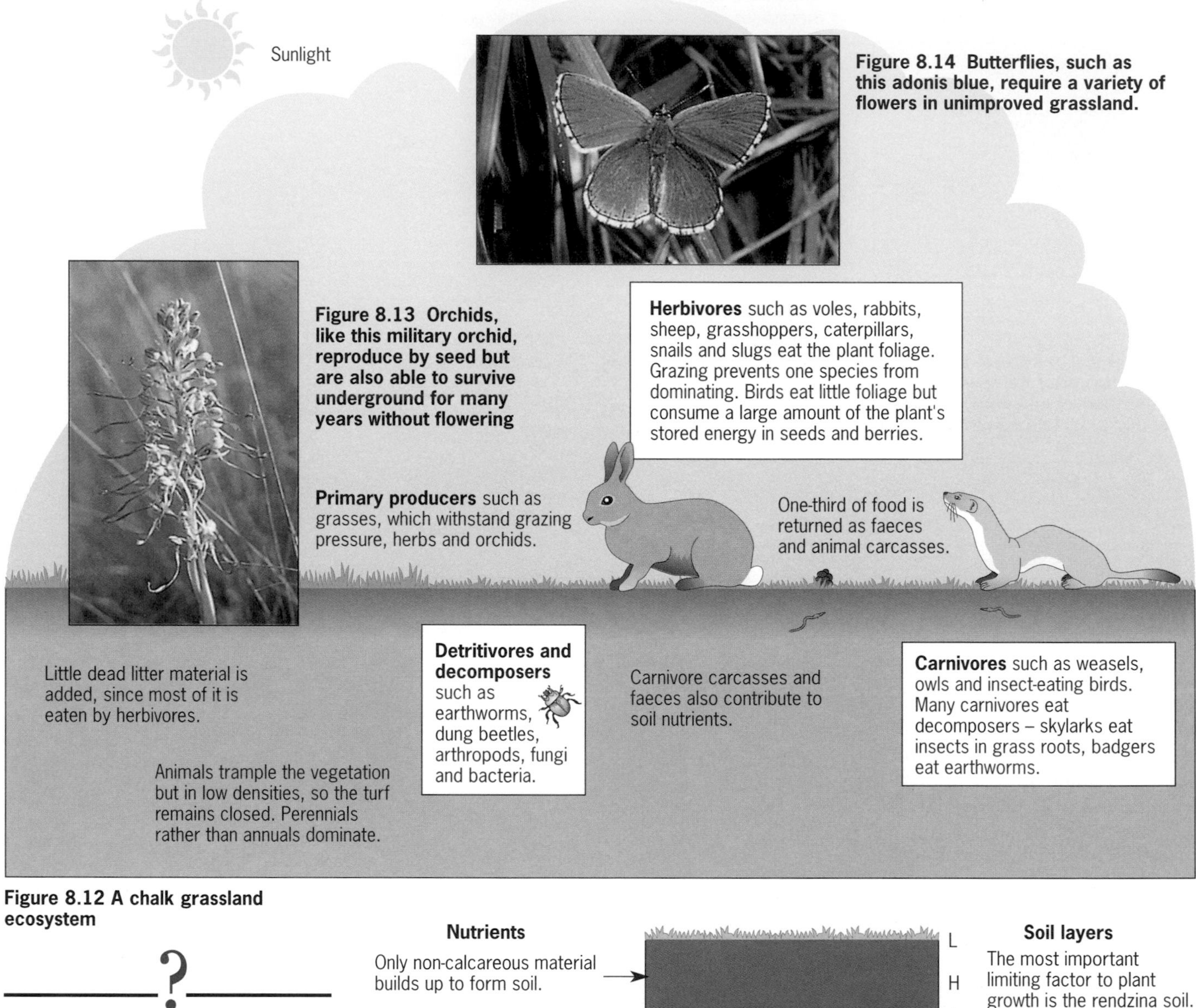

Sunlight
Herbivores such as voles, rabbits, sheep, grasshoppers, caterpillars, snails and slugs eat the plant foliage. Grazing prevents one species from dominating. Birds eat little foliage but consume a large amount of the plant's stored energy in seeds and berries.
Primary producers such as grasses, which withstand grazing pressure, herbs and orchids.
One-third of food is returned as faeces and animal carcasses.
Little dead litter material is added, since most of it is eaten by herbivores.
Detritivores and decomposers such as earthworms, dung beetles, arthropods, fungi and bacteria.
Carnivore carcasses and faeces also contribute to soil nutrients.
Carnivores such as weasels, owls and insect-eating birds. Many carnivores eat decomposers – skylarks eat insects in grass roots, badgers eat earthworms.
Animals trample the vegetation but in low densities, so the turf remains closed. Perennials rather than annuals dominate.

Figure 8.14 Butterflies, such as this adonis blue, require a variety of flowers in unimproved grassland.

Figure 8.13 Orchids, like this military orchid, reproduce by seed but are also able to survive underground for many years without flowering

Figure 8.12 A chalk grassland ecosystem

?

12a Using Figures 8.15 and 8.3, describe the difference between rendzina and chernozem soils.
b How does the nature of the rock affect the chalk grassland ecosystem (Fig. 8.12)?
c How do you think a chalk grassland community would differ from the prairie grassland shown in Figure 8.4?

Nutrients

Only non-calcareous material builds up to form soil.

The calcium carbonate in limestone combines with rainwater that contains dissolved carbon dioxide in the form of carbonic acid. Calcium carbonate is formed and carried away in solution.

Chalk is porous, so water drains downward, causing a physiological drought.

L
H
A
C

Soil layers

The most important limiting factor to plant growth is the rendzina soil.

Well-mixed, dark brown to black mull humus, 25–30 cm deep.

Limestone

Water table

Chalk

Figure 8.15 A rendzina soil profile

Table 8.3 Agricultural use of chalkland in Great Britain since 3000 BC

3000–1000 BC	Farmers cleared woodland for growing crops and grazing livestock.
AD 50	Most of the woodland was cleared and arable farming was thriving. Forest soils had been eroded, leaving immature rendzinas.
1300	Britain had an international reputation for woollen cloth. Sheep grazed on the downs. Rabbit warrens were kept for a supply of meat and fur. The rabbits grazed the grassland. Most of the chalk was covered by calcareous grassland,
1500	Peak of profitability of sheep.
1700	Overproduction of sheep. Enclosure of land took place for rotational agriculture. Some downland was used for sheep. The sheep restored the nutrients to the arable fields.
1800	Napoleonic Wars meant that areas were ploughed to provide grain for horses used in battle. Newly-improved breed of sheep, the Southdown, allowed farmers to increase wool and meat output. More sheep pasture was released for arable crops.
1846	Repeal of the Corn Laws reduced the price of grain, and areas were again used for sheep.
1900	Sheep farming became less profitable as cheap imports of Australian and New Zealand wool and mutton appeared.
1916	World War I caused agricultural intensification. Fertilisers replaced sheep as a means of restoring nutrients.
1930	Rabbits began to compete with livestock for food.
1945	Government grants for ploughing and cereal support payments encouraged arable farming. Downland was now only left on steep slopes and Ministry of Defence ranges such as Porton Down on Salisbury Plain. Pastures were improved by adding fertilisers, which encouraged competitive species, reducing plant diversity. Grass leys became more common than improved pasture. These were seeded with perennial grasses.
1950	Arable farming was more profitable. Areas not ploughed were abandoned and grazed by rabbits. In 1953 myxomatosis was introduced and rabbit grazing was removed.
1960	Shrubs were established and took over downland.
1970	Overall decline of butterfly and bird species, along with flora. Silver-spotted skippers and adonis blue butterflies are very rare. Large blue butterfly is extinct.
1985	Less intensive farming has allowed some areas of downland to return.

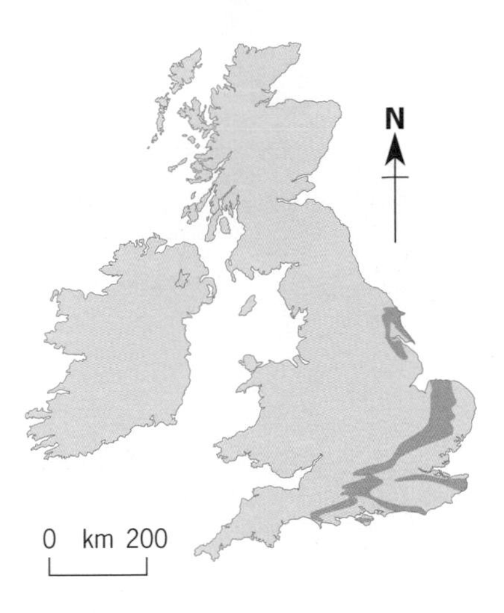

Figure 8.16 Distribution of chalk in Britain

Table 8.4 Management techniques for chalk grassland

1. Apply fertiliser to the grass.
2. Spray herbicide on thistles and docks.
3. Cut vegetation, as two-thirds of downland plants have buds at soil level and 10 per cent have buds in the soil. Cut once a year in May before aggressive plants take advantage of their carbohydrate stores. Undertake a second cut during September after seeding.
4. Put some sheep on the land, as they graze between flower stalks, allowing aggressive grasses to be controlled and flowers to produce seed. Take care over stocking density to ensure that they only have to graze their preferred food which is grass.
5. Put cattle on the land as a short-term measure, since they can clear areas of dead plant litter. Their weight can cut open the turf and remove soil.
6. Remove grass cuttings to remove mulching effect.
7. Divide a large site into sections and graze them at different times. This way you will encourage a mosaic of different flowers.
8. Weed out unpalatable, invasive and shade-giving plants such as yew, thistles, hawthorn and brambles.
9. Introduce myxomatosis to the area.

?

13a Using Table 8.3, explain what has happened to the chalk grassland in Figure 6.31.
b What do you think originally grew on the grasslands in Figure 8.16? Give reasons for your answer.

14 Using Table 8.3, draw a spider diagram to show agricultural influence on chalk grasslands up until the 1980s.

15a Choose the role of either a country park ranger who is managing a site for recreation or a farmer who is within a chalk **ESA** (see Chapter 10).
b Identify your value positions and priorities.
c With these values in mind, choose management techniques from Table 8.4 and rank them in priority order from high to low.
d Which management techniques did you avoid? Why?

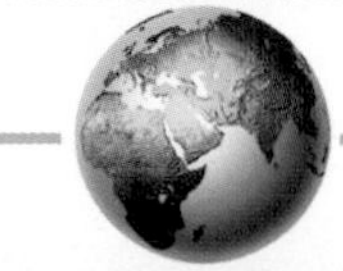

The Spanish steppes

Figure 8.17 The Spanish Steppes

Figure 8.18 The Spanish steppes and ESAs within them

The Spanish steppes (Figs 8.17 and 8.18) represent the last significant area of steppe in the EU. They are a mixture of low perennial shrubs under one metre tall, grassland and a high proportion of bare ground. Although the steppes are fairly uniform, the climate, soils and vegetation vary widely. They are a collection of diverse habitats. In very dry, arid areas, dry shrub steppes are used for livestock grazing. Within less arid areas, grassland is used as pasture and dry cereals are cultivated in rotation with fallows. Agricultural **intensification** has caused this type of habitat to disappear from the few steppe areas which existed in western Europe.

Wildlife

A wide variety of wildlife has evolved to depend on the steppe habitat. Many of the species are now either globally threatened or seriously declining. In particular, there are three globally threatened birds that find a refuge on the steppes. The great bustard is one of Europe's most spectacular birds (Fig. 8.19). The Spanish steppes are a last hope for the species, and hold approximately 75 per cent of the world population. The little bustard (Fig. 8.20) and the lesser kestrel are also globally threatened bird species, for which the Spanish steppes are a crucial habitat.

Vegetation and land use

The Spanish steppes resulted from the clearance of woodland and dense evergreen shrub vegetation by human activity over several thousand years (Fig. 8.21). They are characterised by a flat or rolling open landscape, often cut by deep gorges, with a dry climate (Fig. 8.22). The vegetation is a mixture of uncultivated scrub and rough pasture, and unirrigated arable fields.

Figure 8.20 Little bustard

Figure 8.19 Flock of great bustards

Figure 8.21 These wood pasture systems, called *dehesas*, are rich wildlife habitats. They are maintained by low-intensity cattle and sheep grazing

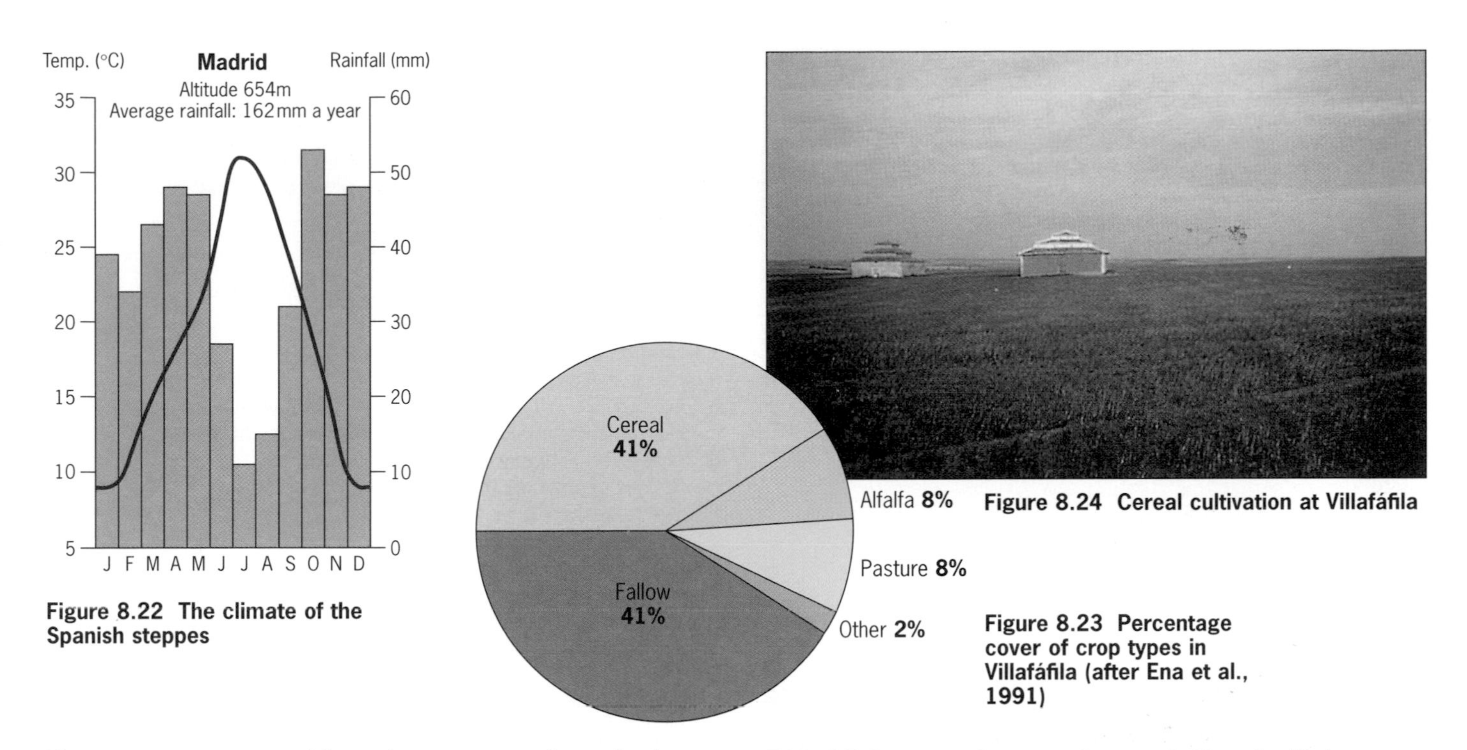

Figure 8.22 The climate of the Spanish steppes

Figure 8.23 Percentage cover of crop types in Villafáfila (after Ena et al., 1991)

Figure 8.24 Cereal cultivation at Villafáfila

The most common cultivated crops are: wheat, barley, oats, and alfalfa. Chickpea, beans, sunflower, sugar beet, rye and lentils are also grown. Traditional land use is a mixture of livestock (sheep) and dry cereal farming. Generally, the steppes are a mixture of uncultivated scrub, cereal field, legume fields, grass/pasture, stubble, ploughed earth and fallow, with sparse trees and occasional oak woodland.

Figure 8.23 shows the percentage cover of different crop types in an area called Villafáfila (Fig. 8.24). Non-irrigated cereal comprises 82 per cent of the crops grown. These are cultivated with a fallow period every other year. Depending on the time of year, these lands are therefore either planted with cereals, left as stubble after harvesting or left fallow. The main cereal crops grown are barley and wheat, with small amounts of oats and rye.

Table 8.5 Important features of steppe habitats for birds

- Open undivided landscape, important for large and shy species such as the great bustard.
- A mosaic of arable fields, stubble, ploughed earth and fallow. Provides an essential variety of habitats for displaying, nesting and feeding.
- Very low level of human activity, and hence little disturbance.
- Timing of activities such as planting and harvesting to minimise the risk of nest destruction or chick mortality.

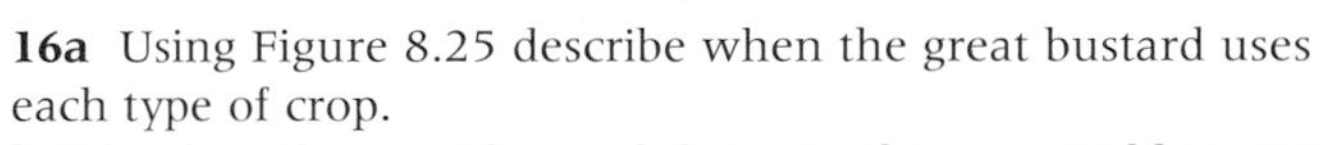

16a Using Figure 8.25 describe when the great bustard uses each type of crop.
b Why does the great bustard thrive in this area (Table 8.5)?

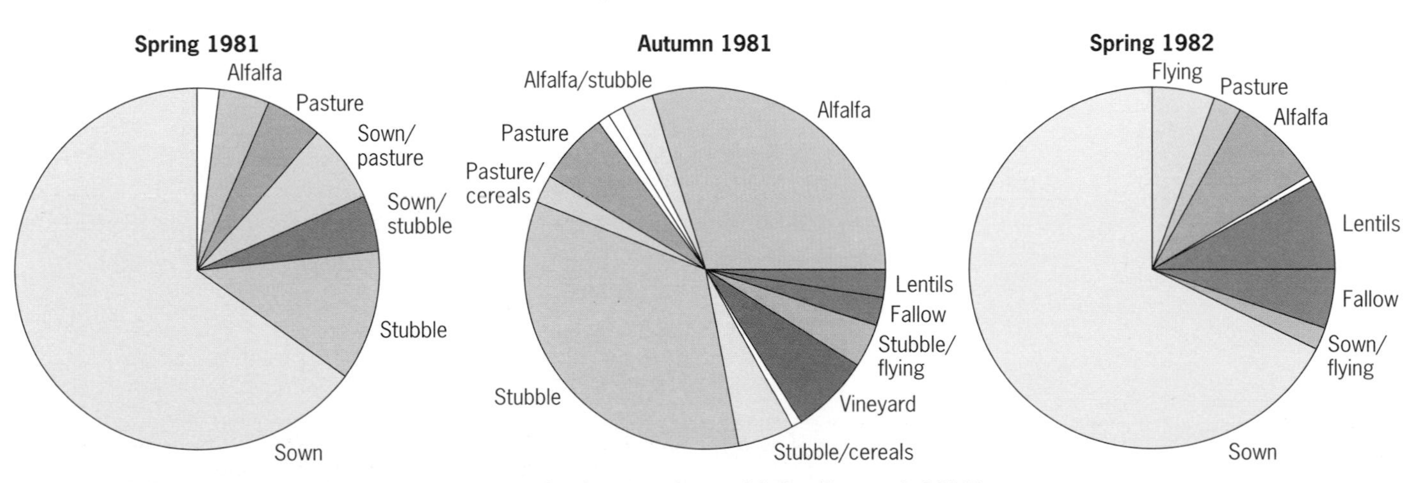

Figure 8.25 Percentage use of different crop types by the great bustard (after Ena et al, 1991)

Figure 8.26 Irrigated steppes at Tierra de Campos

Figure 8.27 Irrigated crops at Tierra de Campos

Figure 8.28 Cultivated *dehesa* in the Tietar Valley

?

17a Using the information in Table 8.6, rank the threats to the Spanish steppes from high to low.
b Suggest ways in which the threats you have identified could be solved to protect the people and the wildlife of the area.

Table 8.6 Threats to the Spanish steppes

1 Irrigation and intensification (Figs 8.26 and 8.27)
a Loss of semi-natural grasslands.
b Loss of crop rotation and fallow, so mosaic of vegetation is lost.
c Annual cropping with fast-growing dense varieties of cereals.
d Higher levels of agrochemicals.
e Increased disturbance from machinery.
f Earlier harvesting.
g Loss of arable habitats to melons, strawberries and irrigated orchards.
h Power lines for irrigation schemes cause deaths of low-flying bustards.
i Abandonment and/or intensification of *dehesas* (woodland pasture systems) causes loss of trees (Fig. 8.28). Trees encouraged water retention, stability of the water table and enrichment of the thin poor soils.
j The EU paid for half of the new irrigation systems and improvements.

2 Rural depopulation and land abandonment
Because of the difficulty in making a living, many farmers have been forced to move away. In turn, the diversity of vegetation and habitat is lost and soil erodes where there was once terracing.

3 Afforestation
In some areas authorities encourage planting of pine and poplar on abandoned land to stop erosion.

4 Eucalyptus
Some farmers sold their farms to pulp companies who bought the land to grow eucalyptus for paper. This is more of a problem in Portugal.

5 'Concentration'
This is a government policy of change from small scattered landholdings into larger single units. This destroys open steppe habitat through fencing for untended livestock and reducing crop diversity.

?

18 Essay: Compare and contrast the impact of agricultural practices on grassland ecosystems.

Summary

- Temperate grasslands occur as a climatic climax community in continental interiors where evaporation exceeds rainfall totals. They form a soil called chernozem.
- On a regional and local scale, grasslands form as subclimax and plagioclimax communities.
- Trampling affects plant communities by changing the relative numbers of competitors, ruderals and stress tolerators.
- The more competitive grasses there are in a grassland community, the fewer herbs.
- The addition of fertiliser increases the success of competitive plants and reduces species diversity in chalk grassland.
- Chalk grasslands are affected by human influence. They are plagioclimaxes because of grazing by rabbits, sheep and mowing. Most of the deeper woodland soils built up on the chalk by succession to woodlands were eroded following deforestation.
- Immature rendzina soils on chalk grassland create stressful conditions of physiological drought and some plants have features adapted to very dry conditions.

9 Wetlands: agriculture and peat extraction

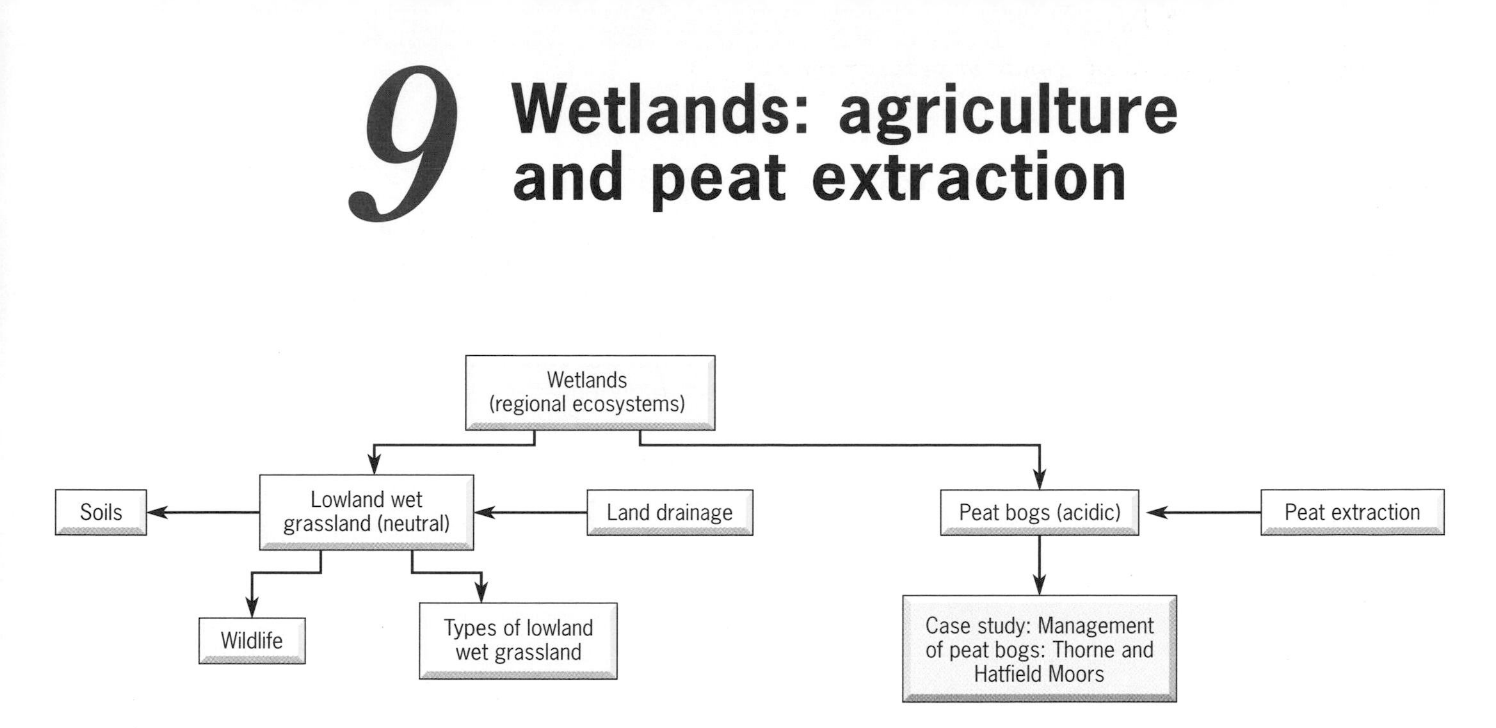

9.1 Introduction

Natural wetland **ecosystems** (such as swamps, marshes, bogs and fens) are found throughout the world in places which are permanently or seasonally flooded (Fig. 9.1). Soils are waterlogged, and plants growing in such areas need to have hydrophytic **adaptations**. Many plants have stems made of cells with thin walls that have large air spaces between them. This allows them to float and to carry air down towards the roots.

Wetlands have presented many opportunities for people. Swamps and marshes are some of the most productive ecosystems (see Table 1.1), since they have huge nutrient **inputs** from flood waters. Many crops and animals are fed on the nutrient-rich lands of fertile floodplains. Once these are drained, they offer very fertile land. The rice that feeds half of the world's population, and two-thirds of the total harvest of fish and shellfish, come from wetlands.

Wetlands are also used to prevent flooding, both alongside rivers and on the coast. They store large quantities of water and release it slowly. On the coast they reduce the destructive effect of waves and water on the landscape. Plants bind sediments and stabilise cliffs, which saves money on sea defences. Wetlands slow down the movement of water, and this allows silts and sediments to settle out. Wetland plants, and the bacteria associated with them, absorb and decompose nitrates, phosphates and organic wastes. For this reason, reedbeds are being used as biological filters to purify water. At Highgrove in Gloucestershire, Prince Charles has been using a system of reedbeds to filter domestic sewage. Research is currently being carried out to work out whether reedbeds could be used by large water companies.

Figure 9.1 Types of natural wetland ecosystems

?

1 Why do you think the areas of grazing marsh in Figure 9.2 were drained?

9.2 Lowland wet grassland

The Association of Drainage Authorities has classified 1200 000 ha of land in England and Wales as 'Areas of Special Drainage Need'. Most of these are under intensive agricultural use, but a tiny proportion still contain wetlands which have not been drained, including one type of wetland known as lowland wet grassland. Lowland wet grasslands occur at low altitudes (below 200 metres) and are subject to periodic flooding. They consist of a variety of semi-natural, **managed** grasslands. These range from the East Anglian washlands to reclaimed freshwater marshes and fens, flood meadows and coastal grazing marshes behind the sea wall.

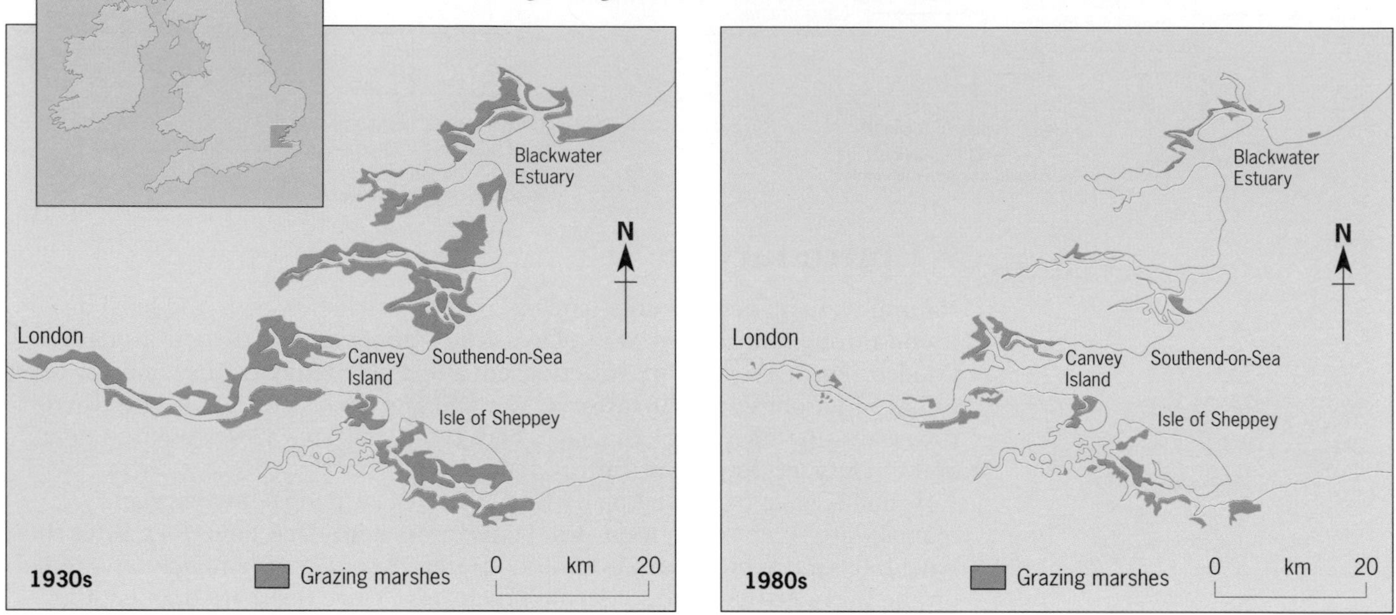

Figure 9.2 Changes in extent of grazing marshes in the greater Thames area (RSPB, 1991)

Wildlife

Lowland wet grasslands, with their associated ditches and banks, are home to plants and animals that are associated with dry land, open water and the transitional area (**ecotone**) between. Some of these lowland wet grasslands, especially ancient water meadows, are botanically rich. One third of the UK's 1750 flowering plant **species** are associated with wetland **habitats**. Some of these plants may also be rare. Lowland wet grassland may contain up to 40 per cent of the nationally scarce species and 20 per cent of the nationally rare species.

Lowland wet grasslands also support nationally rare **invertebrates**, such as the scarce emerald dragonfly, the marsh carpet moth, the little whirlpool ramshorn snail and the lesser marsh grasshopper. Otters also occur in the ditch systems of some sites.

In addition to supporting many plants and animals that are rare or unknown in other habitats, lowland wet grasslands are also major sites for more common wildlife such as reed bunting and sedge warbler, the common frog, common newt and grass snake.

?

2a Using an atlas, list the landforms in the British Isles that are likely to be associated with lowland wet grassland.
b Are there any regional differences in the distribution of these landforms?

Soils

Soils that lie beneath lowland wet grasslands are frequently mottled with yellow-brown streaks and blotches (gleyed). This colour variation is due to the action of micro-organisms in the soil, which in the absence of oxygen convert ferrous salts (blue-grey) to ferric oxides (yellow-red). As many of these soils have formed from river sediments, they are neither strongly alkaline nor very acid. For this reason, many lowland wet grasslands are also described as neutral grasslands, although there are exceptions to this rule.

?

3a Using Figure 9.3 describe the stages you think the natural succession would go through.
b How would grazing prevent the succession from taking place?
c Why would soils under this habitat be particularly good for crops if the land was drained?

Types of lowland wet grassland

Freshwater grazing marsh

In its natural state, freshwater marsh and fenland vegetation displays a **succession** from wet reedswamp to a type of woodland known as fen carr. In most areas this process was stopped by farmers who dug drainage ditches and developed ploughing techniques to drain surface water away, thus lowering the water table.

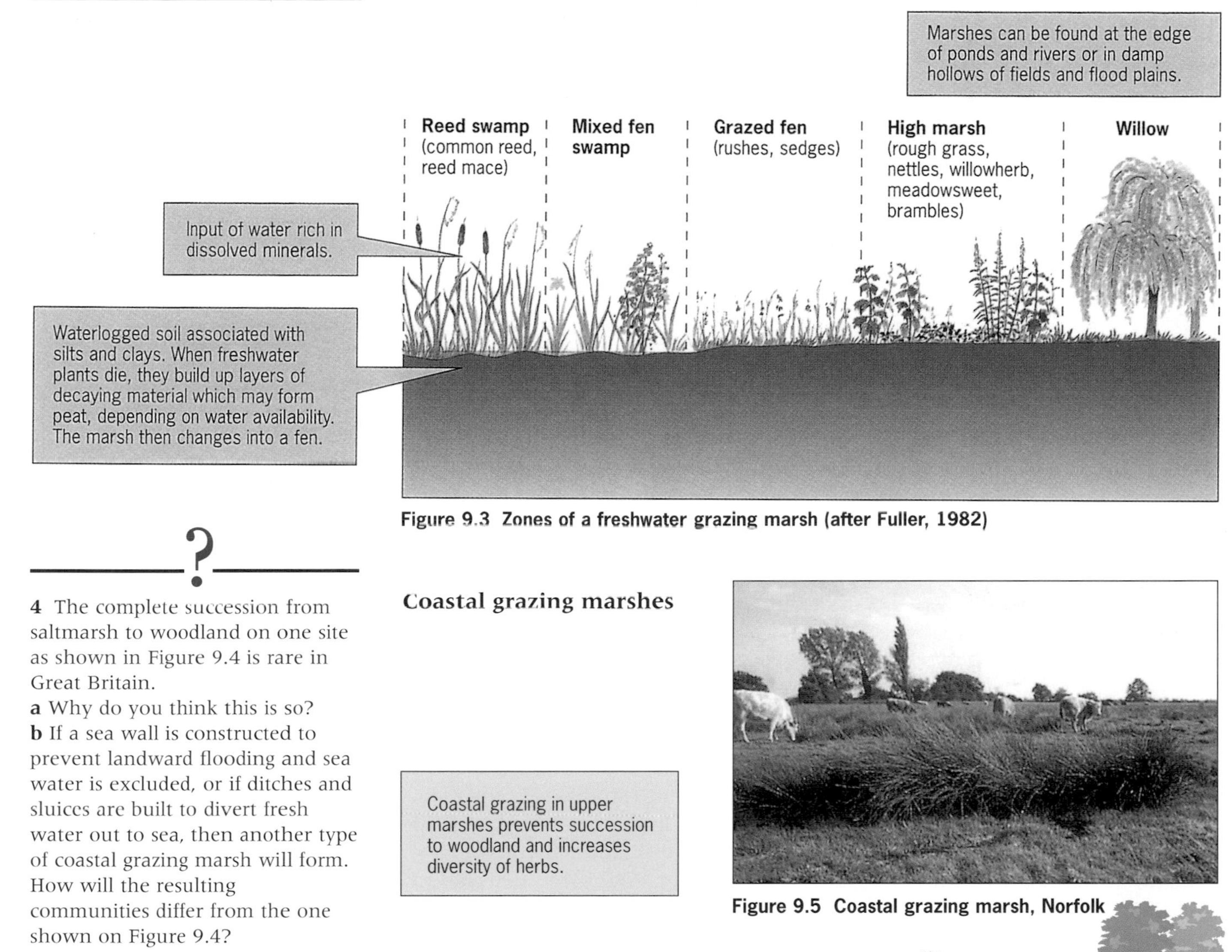

Figure 9.3 Zones of a freshwater grazing marsh (after Fuller, 1982)

?

4 The complete succession from saltmarsh to woodland on one site as shown in Figure 9.4 is rare in Great Britain.
a Why do you think this is so?
b If a sea wall is constructed to prevent landward flooding and sea water is excluded, or if ditches and sluices are built to divert fresh water out to sea, then another type of coastal grazing marsh will form. How will the resulting communities differ from the one shown on Figure 9.4?

Coastal grazing marshes

Coastal grazing in upper marshes prevents succession to woodland and increases diversity of herbs.

Figure 9.5 Coastal grazing marsh, Norfolk

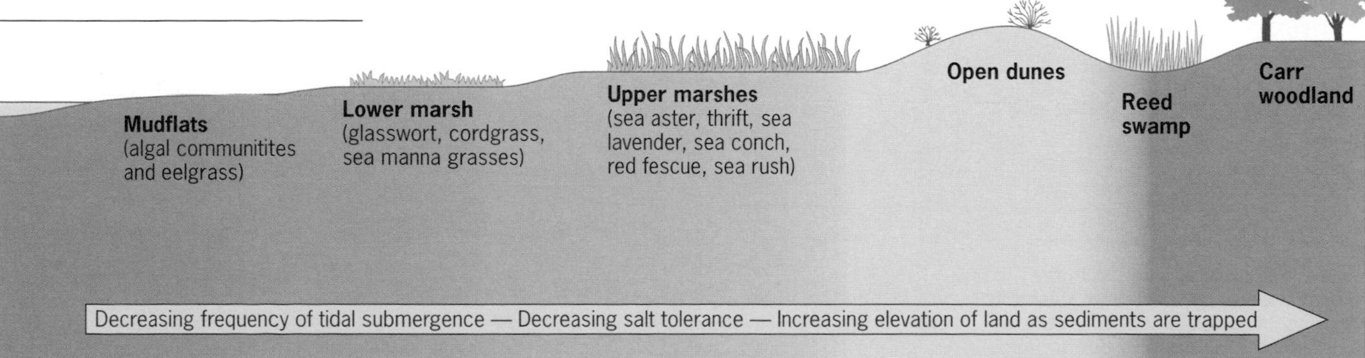

Figure 9.4 Plant zones and succession of a salt marsh

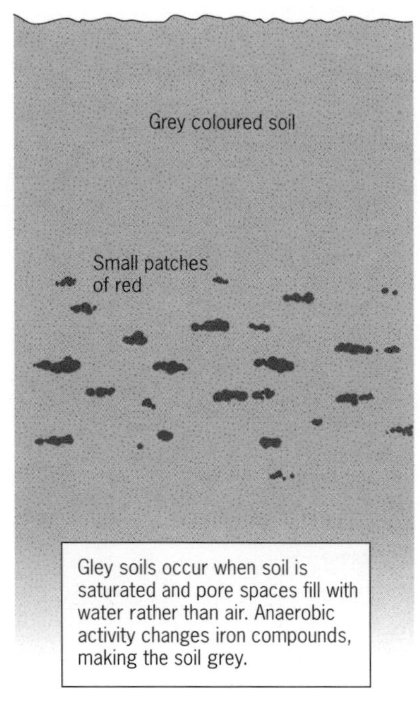

Grey coloured soil
Small patches of red
Gley soils occur when soil is saturated and pore spaces fill with water rather than air. Anaerobic activity changes iron compounds, making the soil grey.

Figure 9.7 A gley soil

Figure 9.6 Flood meadows, Sussex

Flood meadows

Low-lying land adjacent to rivers is often flooded in winter. The floodwaters deposit nutrient-rich sediments. Early farmers converted these wet areas into pastures called flood or water meadows (Fig. 9.6). They did this by digging ditches that were deep enough to drain away surface water, yet retain enough water to act as wet fences to control livestock. The fertile soils provided good grazing, and surplus grass was cut for hay. The farmers used the fertilising effect of river sediments by periodically flooding the meadows. This was achieved by opening sluice gates when river levels were high, thus reversing the flow of water through the ditches. The water poured on to the fields depositing a rich layer of silt.

Flood meadows were considered a valuable resource and required intensive skilled management throughout the year. Ditches and sluices were constantly dredged and maintained to ensure maximum efficiency of water control which, together with the grazing of livestock, prevented the flood meadows from succeeding to fen woodland (Fig. 9.4).

Land drainage

Farmers may obtain substantial benefits (Table 9.1) by draining lowland wet grassland. They can intensify the management as a result of drainage.

Advantages of draining have led to the development of a variety of sub-surface drainage techniques involving clay tile drains and wind- and steam-driven pumps (Fig. 9.8). Large areas of lowland wet grassland were converted to more productive arable land after 1945.

Table 9.1 Benefits of drainage for farmers

- Soil aeration is increased, thereby improving soil structure. This reduces the damage done by heavy machinery and livestock trampling. It also improves seed germination and plant response to fertilisers.
- Drier soils warm more quickly in spring, which encourages earlier plant growth, thereby extending the grazing season.
- Crops form deeper-rooting systems, making them more resistant to drought
- Crop and stock disease is reduced.
- Grassland management can be intensified. Pastures can be reseeded with more productive grass species which respond to increased fertiliser use.
- Grassland can be replaced with arable crops.
- Stocking rates can be increased as a result.
- Grass can be cut for silage several times each season instead of once or twice a year for hay.

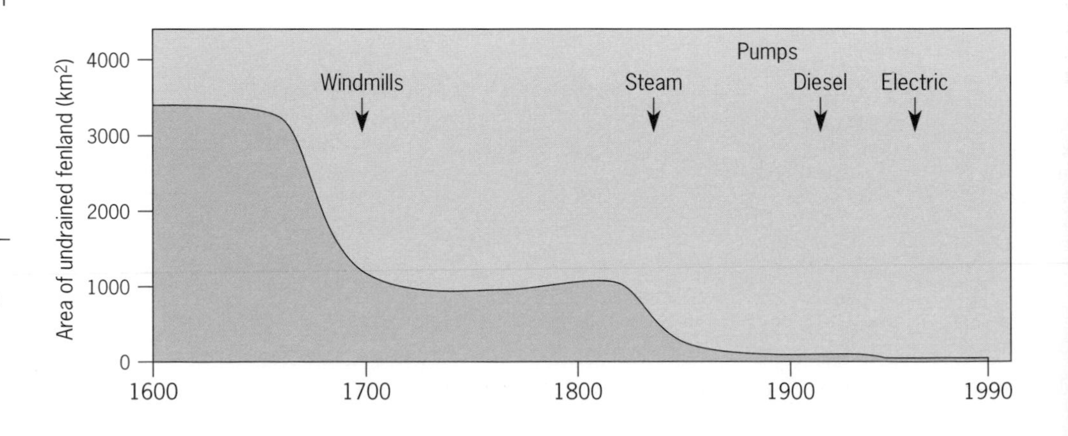

Figure 9.8 Loss of wetland in the fens of Cambridgeshire and adjacent Lincolnshire and Norfolk since the mid-seventeenth century due to land drainage by various methods (after RSPB, 1987)

?

5a It is now more difficult to obtain grants for the draining of lowland wet grasslands. If you were a farmer with a lowland wetland site, how would you try to persuade decision-makers that land drainage was a positive practice? Write a letter to *Farmers' Weekly* with your argument.
b Write a reply to the letter from the conservation officer of a County Naturalist Trust.

6a Using Figure 9.9 explain why peat bogs are less productive than other wetland ecosystems.
b Why do peat bogs have a relatively simple **food chain**?

Effects on wildlife

The results of land drainage and subsequent agricultural improvements that followed were good for the farmers and their customers but proved catastrophic for the wildlife of lowland wet grasslands. Of the 19 plant species that have become extinct in Great Britain in the last 200 years, five were wetland species. According to English Nature, 95 per cent of all lowland neutral grasslands (including dry hay meadow) now lack any significant wildlife interest and only 3 per cent have been left undamaged by agricultural **intensification**. Much of this is lowland wet grassland, because drainage difficulties made further improvement difficult.

As in the uplands, increasing use of herbicides, fertilisers and the ploughing and reseeding with non-tussock-forming species has removed important plants from lowland wet grasslands. It has also deprived many breeding birds of their nest sites. Increased stocking levels (the number of sheep and cattle per unit area) resulted in increased trampling of nests and young. As a consequence, populations have declined and ranges have contracted among the breeding wading birds of lowland wet grassland, notably redshank, snipe and lapwing in eastern and southern Great Britain. Almost one-third of the breeding waders recorded during a national survey of wet meadows in 1982 were concentrated

9.3 Peat bogs

Unlike swamps and marshes, peat bogs are ecosystems with relatively low productivity levels. They do not have large inputs of nutrients from rivers or sea water. Their wetness is caused largely by excess rainfall or trapping of water within a drainage basin.

Most peat bogs are formed due to a build-up of organic matter, usually on waterlogged land and in a basin where continued flooding encourages sedge and fen vegetation (Figs 9.9–9.11).

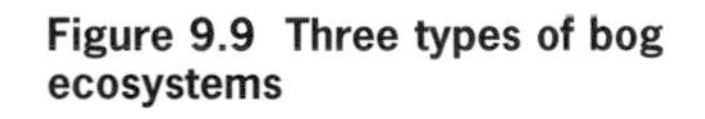
Figure 9.9 Three types of bog ecosystems

Nutrient cycling and the peat bog food web

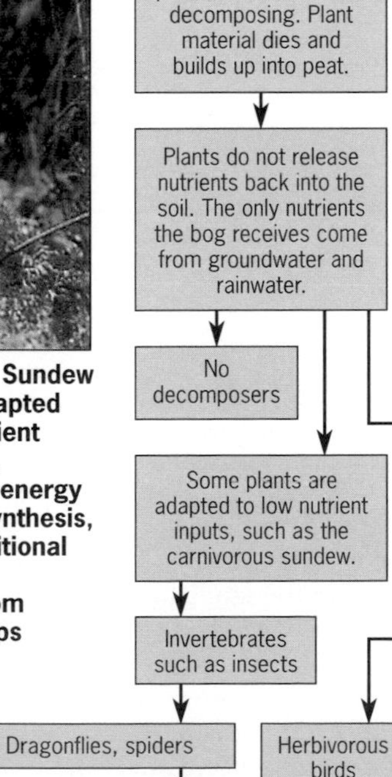

Waterlogged soil prevents material from decomposing. Plant material dies and builds up into peat.
Plants do not release nutrients back into the soil. The only nutrients the bog receives come from groundwater and rainwater.
No decomposers
Some plants are adapted to low nutrient inputs, such as the carnivorous sundew.
Invertebrates such as insects
Dragonflies, spiders
Herbivorous birds
Insectivorous birds (nightjar)

Figure 9.10 Sundew is a plant adapted to a low nutrient environment. It obtains its energy from photosynthesis, but gets additional nourishment (nitrogen) from insects it traps and digests.

Fen
Fens occur in calcium-rich waterlogged soils.

Raised bog (low rainfall: 1010mm per year)
Some can be up to 10 metres higher than the surrounding land.

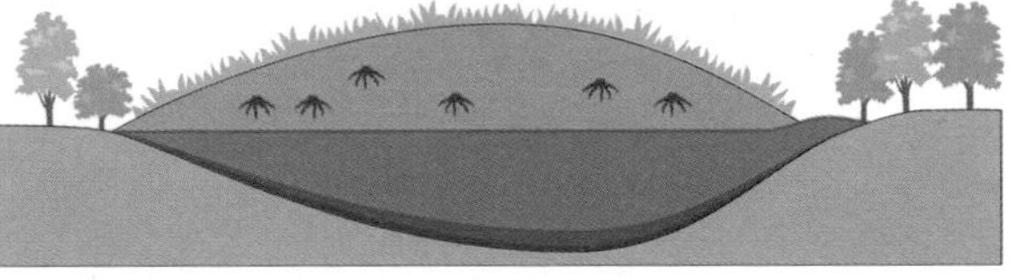

Blanket bog (high rainfall: over 1390mm per year)
Bog spills out of the basin and blankets the surrounding land.

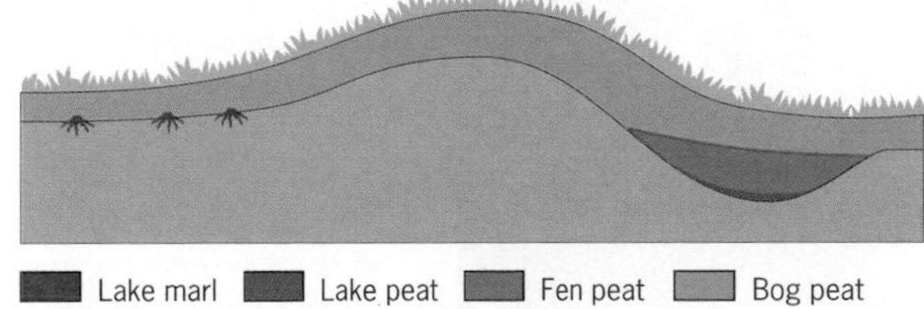

The role of sphagnum moss

In acidic areas, sphagnum moss becomes the most successful bog plant. It holds up to 20 times its own weight in water. It absorbs positive ions in rainwater and releases hydrogen ions, thereby increasing the acidity of the soil water. It then becomes an increasingly unsuitable habitat for decomposers, thus speeding up the accumulation of plant remains. The sponge-like qualities of sphagnum help to increase the water table as sphagnum grows, preventing contact with mineral layers below.

Figure 9.11 Sphagnum mosses are responsible for the growth of the peat bogs.

Management of peat bogs: Thorne and Hatfield Moors

Peat bogs contain few natural nutrients and so have been managed to increase their productivity for agriculture. Thorne and Hatfield Moors (Fig. 9.12) are the largest lowland raised bogs in Great Britain. During the seventeenth century Hatfield Chase was the site of the first marshland reclamation in Great Britain for farming. A method was devised to increase the soil fertility of the former bog areas. River and estuarine deposits are extremely fertile, since they contain many materials washed from the land. Areas of former bog were artificially flooded and covered with silt (wet warping) during the first half of the nineteenth century.

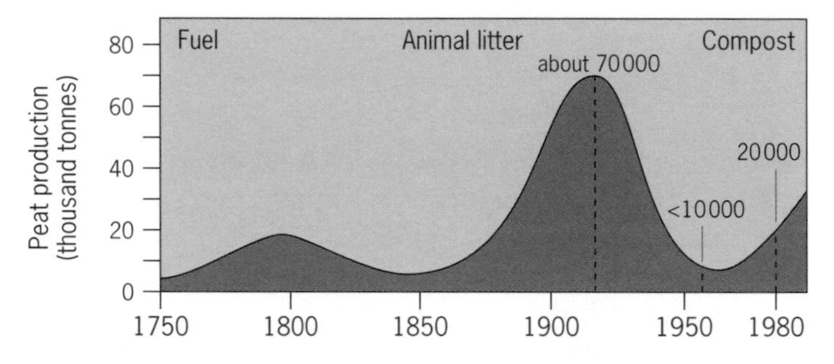

Figure 9.13 Peat production on the Thorne and Hatfield Moors, 1750 to 1980 (after Bain and Eversham, 1991)

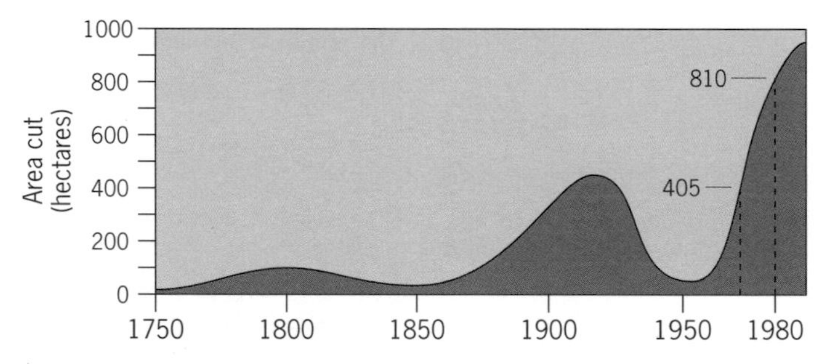

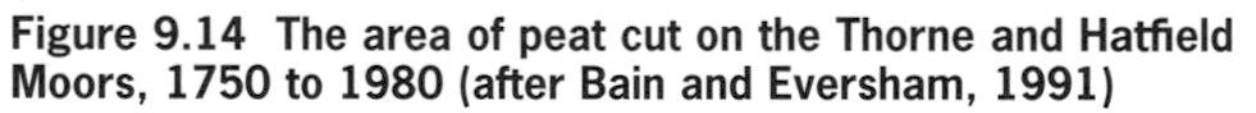

Figure 9.14 The area of peat cut on the Thorne and Hatfield Moors, 1750 to 1980 (after Bain and Eversham, 1991)

Figure 9.15 Uncut peat, Ireland

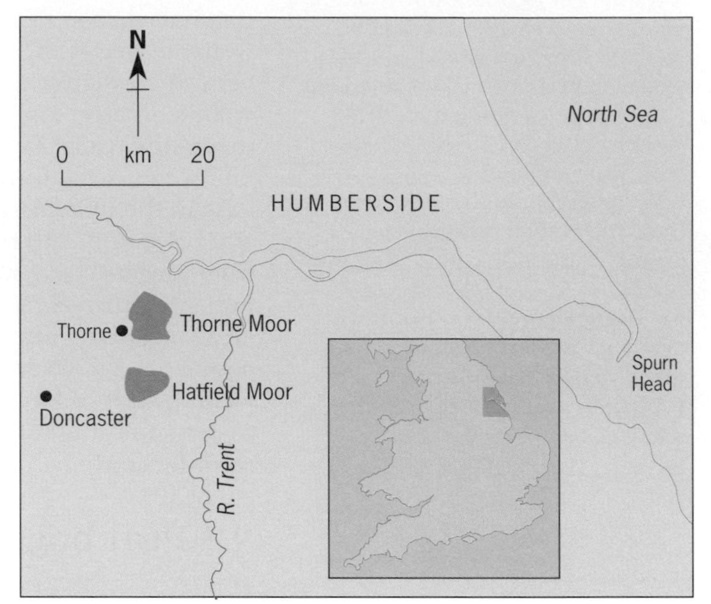

Figure 9.12 The Thorne and Hatfield Moors, Humberside

Table 9.2 Details of peat extraction on Thorne and Hatfield Moors

Technique	Area of patches (acres)	Length of cuttings (chains)	Distance of bare peat from intact vegetation (metres)	Mean times between re-cutting
Hand-graving	50–100	<10	5	10+years
Pre-1975 machine	1–200	10–20	10–20	2 years
Post-1975 machine	up to 500	to 60 (T) to 150 (H)	100–200	1 year
Milling	500–1000	50–160	500–2000	1-2 months

?

7a Study Figure 9.13 and explain how and why the use of peat has changed since 1750.
b Study Figure 9.14 and describe how the size of the cut area has changed since 1750.
c What is surprising about this when you compare it to the tonnage taken from the bog as shown in Figure 9.12?

8a Using Table 9.2 explain why, although the tonnage has decreased, the area of cutting has increased.
b Using Figures 9.15–9.17 describe the impact that each of these techniques might have on the landscape.
c Which technique would produce the largest area of bare peat?
d Which technique would mean that the remaining bog vegetation would be furthest from the bare peat?
e How would the time between re-cutting affect the growth of the bog plants?

Figure 9.16 Cutting peat by hand, Lancashire

Figure 9.17 Peat milling by machine, Humberside

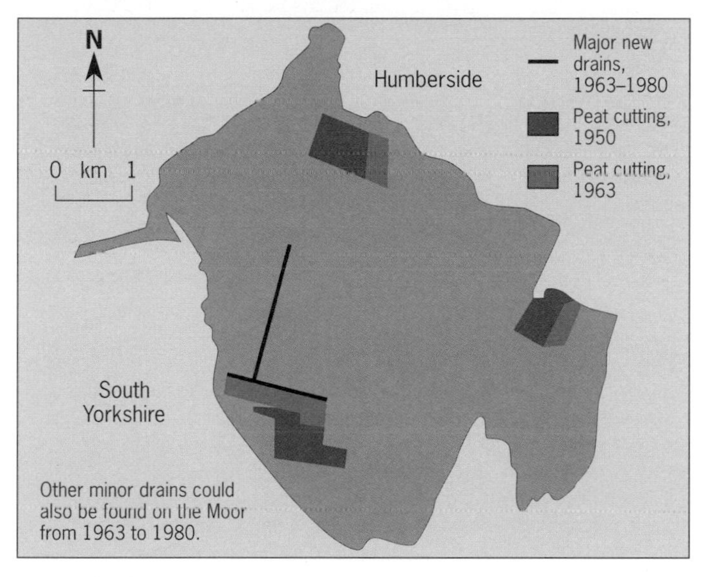

Figure 9.18 Areas of peat cutting and draining at Thorne Moors, 1963 to 1980 (after Bain and Eversham, 1991)

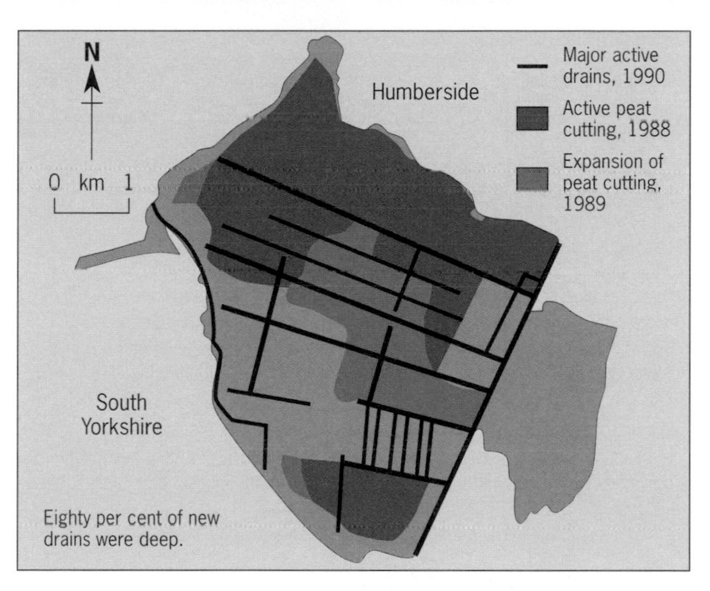

Figure 9.19 Areas of peat cutting and draining at Thorne Moors, 1989 to 1990 (after Bain and Eversham, 1991)

Planning permission for peat extraction at Thorne was granted in the 1950s. There is no means of cancelling the original planning permission before it runs out, unless compensation is paid. Hence, the local authority are reluctant to withdraw permission because it would be very expensive for them to do so.

However, pressure from several interest groups began to increase when milling became widespread (Figure 9.20). Since 1992, the peat company on Thorne Moor has agreed not to mill any vegetated parts of the site but will continue to mill the areas which are already bare.

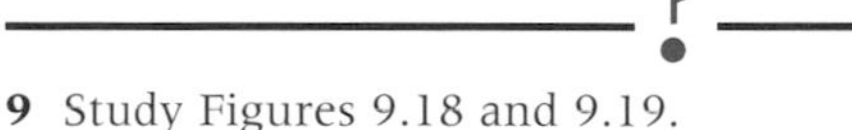

9 Study Figures 9.18 and 9.19.
a What has happened to the number of drainage channels?
b How would drainage damage the balance of the peat bog ecosystem?

10 What do you think the effect of the peat company's decision will be on the peat bog ecosystem?

11 What do you think about the use of peat for horticulture? Devise a table to show costs and benefits of peat extraction.

"Both Moors are SSSIs. They are also listed as Nature Conservation Review Sites, the most important SSSIs. Parts of the bog cut by hand in the 17th and 18th centuries still support rare plants and animals. These have been designated as nature reserves but are affected by water loss from the system through drainage channels. We are pleased that the peat company have handed over areas of the moor which are not currently milled. However 65 per cent of the 8000 acres will still be milled and drained."

"In February 1990, drainage work in the south-east corner of Thorne reduced the water table by 60cm. The summer of 1990 was dry like the one before, and the moors became very dry. Water levels in the NNR fell drastically, although the peat company had been pumping water into the area. A drier area is more likely to catch fire. Succession begins as the peat bog dries out, with more birch invading."

"Peat is not scarce on a world-wide scale. Bogs are not original habitats. If peat extraction stopped, large areas could dry out, resulting in scrub invasion. Small islands left in the peat will encourage recolonisation. If 50cm of peat is left at the bottom of workings, regeneration will take place. We have pledged not to seek planning consent for new workings in existing or candidate SSSIs."

"Peat alternatives are expensive and do not have the same qualities as peat. Peat holds and releases water, retains air space, has low bulk density, reasonable nutrient retaining capacity, is relatively sterile, free from toxins and salts, has a low to neutral pH, uniform texture and encourages fine root formation, which eases transplanting."

"For the first time in history, ten conservation groups are working together to protect a habitat. Peat bogs are biological indicators, a genetic resource, a refuge for rare species, a hydrological filtering system, a living archive, a carbon sink for carbon dioxide, a wilderness and part of our international heritage. Less than 10 000 ha of natural lowland peat bogs remain in the British Isles. Sixty-five percent of peat extraction in the UK is on SSSIs. There is no evidence that rehabilitation of bogs is possible. Few people are directly employed in the peat industry."

"*Sphagnum baticum* is a rare species of bog moss known at only six sites in the British Isles. Bog rosemary is a rare plant of the heather family only found on peat bogs. As the peat bog dries out, these become increasingly threatened. The moors are the richest sites for invertebrates in the North of England. They are the richest acid peatlands in the British Isles. In May 1991, the results of an invertebrate survey showed continued extinctions amongst this group. Although birds have declined in number they are still of international importance. Nightjars still make up I per cent of the British total on Thorne and 1.3 per cent on Hatfield. The Moors are also important archaeological archives since peat preserves pollen, insect remains and human artefacts."

"Peat bogs have been cut for centuries. Peat cutting is controlled by planning permission. Designation of SSSIs protects the site, as does the voluntary code of practice recently introduced by the Peat Producers Association. Twenty-two thousand hectares remain and the best areas are not threatened. The unmilled areas of Thorne and Hatfield are now intact and in the safe hands of English Nature."

"One thousand people are employed in peat extraction in the British Isles. Thirty thousand jobs are linked to packing and transport of peat, £500 million are linked to peat sales. Most of these jobs could be transferrred to the production of peat alternatives."

"Peat is normally cheaper and more generally used than peat alternatives. Production processes are expensive due to low demand but the cost will decrease as consumer demand increases. The true price of peat should include the cost of the lost landscape and wildlife resource. Peat alternatives use waste products, which reduces waste disposal problems."

"There are many peat alternatives!
(a) Coir, a waste product from the coconut industry, can hold air and water and is good for potting and soil improvement.
(b) Composted bark and wood wastes from the timber industry can be used for potting and seed composts.
(c) Composted garden waste collected at local authorities can be used as a mulch to keep soil moist and reduce weed growth.
(d) Composted straw and sewage is used for soil improvement. Home made compost from kitchen scraps is a valuable soil food and conditioner."

Figure 9.20 Quotes from interested parties in the peat debate

?

12 Match the quotes from Figure 9.20 to the interested parties below. One of them has *two* quotes.
a The peat company
b Representative from the Thorne and Hatfield Conservation Forum
c Hydrologist (water scientist)
d Peat Producers Association
e 'Green' economist
f Local job centre
g Peatlands Campaign Group
h Horticulturist (garden cultivator) in favour of peat alternatives
i Horticulturist against peat alternatives.

9.4 Blanket bogs

Where rainfall is much higher, bogs can spill out of their basins to blanket surrounding land in peat (Fig. 9.21). Blanket bogs (Fig. 9.22) are more common than lowland peat bogs and have been forming since the end of the last ice age.

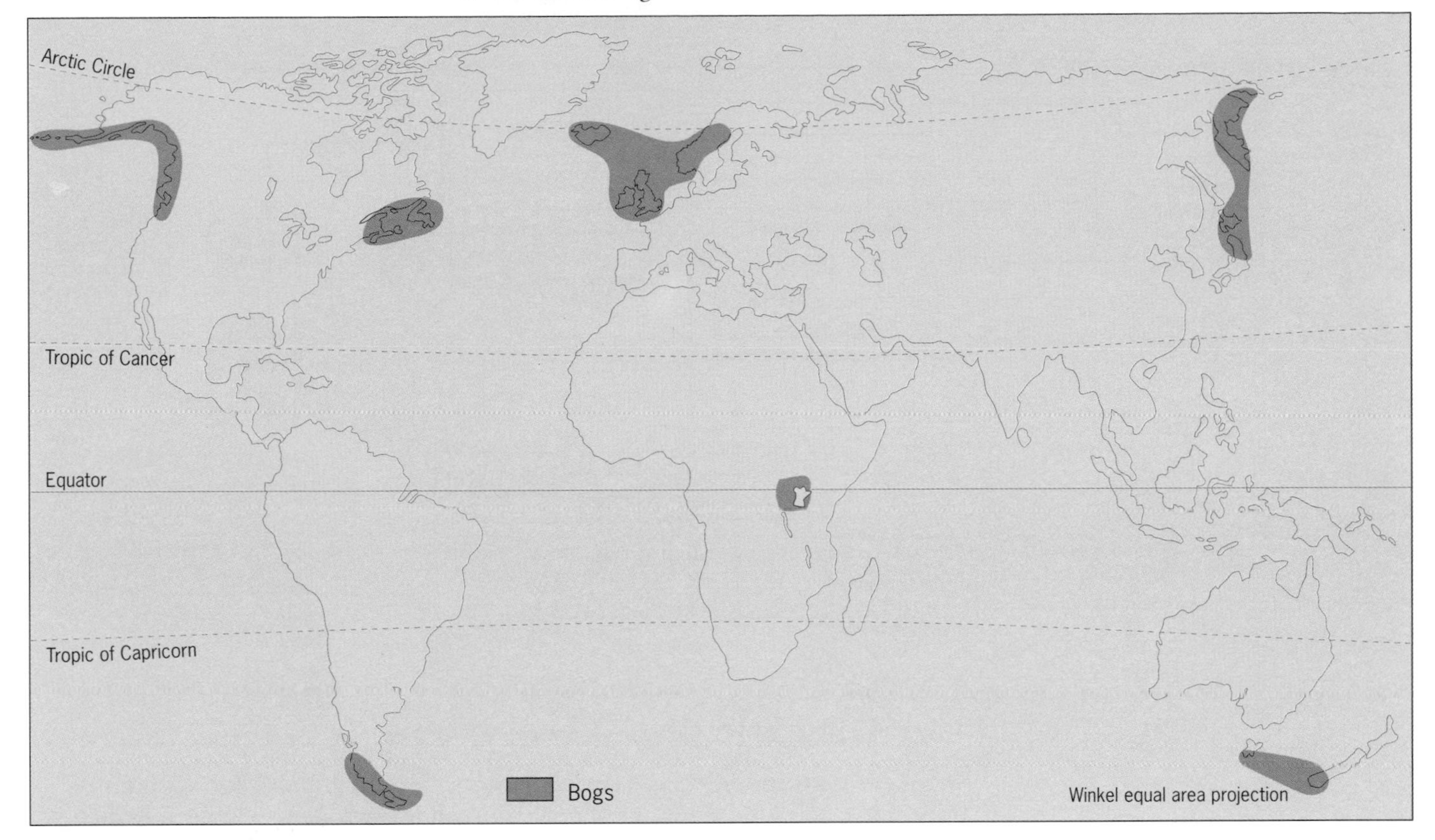

Figure 9.21 Distribution of blanket peat bogs (after English Nature, 1988)

Figure 9.22 Peat bog in the Flow Country, Scotland. Large-scale afforestation has threatened this habitat.

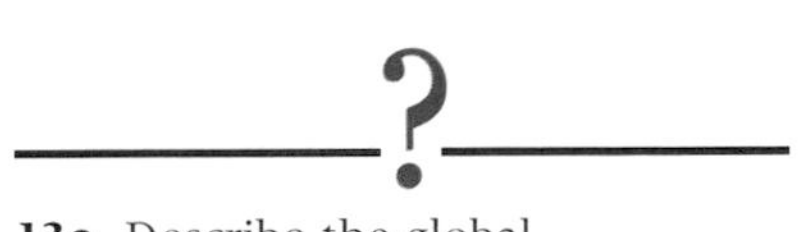

13a Describe the global distribution of blanket bog as shown in Figure 9.21.
b Using an atlas, describe the main climatic characteristics of these areas.

14 Essay: Discuss the view that the remaining peat bogs in the British Isles should become protected areas where all peat extraction is banned.

Summary

- Swamps and marshes are very productive due to inputs of nutrient-rich waters. Alkaline inputs from river systems lead to the formation of neutral communities. These are often very diverse, since many wetlands are ecotones between terrestrial and water-based ecosystems.
- Acid peat bogs form where there is excess rainfall (blanket bogs) or where water is trapped in a basin. Waterlogging and limited nutrient inputs create acidic conditions. Few decomposers exist in such an environment and so nutrients are not recycled.
- Peat bog communities are highly specialised but have limited productivity.
- People have managed wetlands to prevent succession. Where succession has occurred at a sustainable level, diverse communities exist.
- Intensive drainage of wetlands has contributed to widespread loss of wildlife but has had significant economic benefits for farmers.
- Wetlands can act as natural cleansing systems, reducing excess nutrients and pollutants from watercourses.
- Peat extraction can damage wetland areas such as peat bogs.
- Hydrological changes have one of the greatest effects on wetland management.

10 Cultivated land and wildlife in the EU

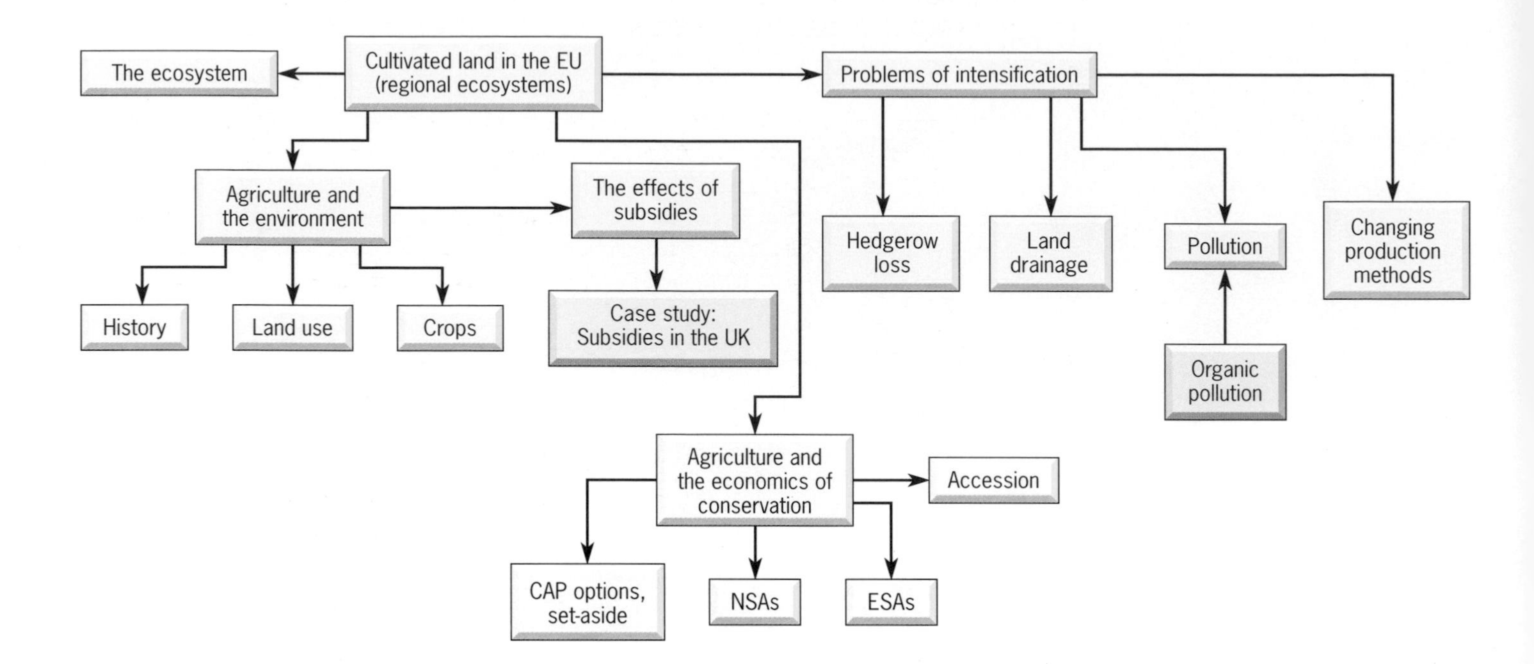

10.1 The agricultural ecosystem

Some **ecosystems**, such as extreme deserts, are not suitable for agriculture. However, the higher productivity of most other ecosystems (Table 1.1) means that people can consider them for growing food. Cultivated land has a mean net **primary productivity** of 100–3500 g/m^2/year. Land cultivated in areas of tropical rainforest or tropical seasonal forest is therefore ten times less productive than the natural ecosystems (Table 1.1). Similarly, in areas cleared of woodland and scrub, cultivated land is less productive. Only when we compare it with temperate grassland does agricultural land appear to be as productive.

Natural **ecosystems** can determine the type of agriculture which takes place in an area. However, once established, agriculture often leads to either removal of the climax vegetation or prevents **succession** towards it. Agricultural ecosystems are therefore **plagioclimax communities**.

Agricultural ecosystems are open systems. Their **inputs** are eventually transferred as **outputs** to natural ecosystems (Fig. 10.1). The extent of this has increased as technology has progressed and farming practices have become more **intensive**.

10.2 Agriculture and the environment

Before the Second World War

Before 1939, land-based activities such as farming were determined mainly by physical factors such as precipitation. Throughout Europe distinctive agricultural areas matched climatic zones (Fig. 10.2). Some areas were too wet, too high, too infertile or too dry for agricultural activity. Many of these areas were important wildlife sites. Between these sites was cultivated land which was **sustainably managed** with a relatively low level of technology. Wildlife was able to adapt to this level of use, producing few threatened **species**.

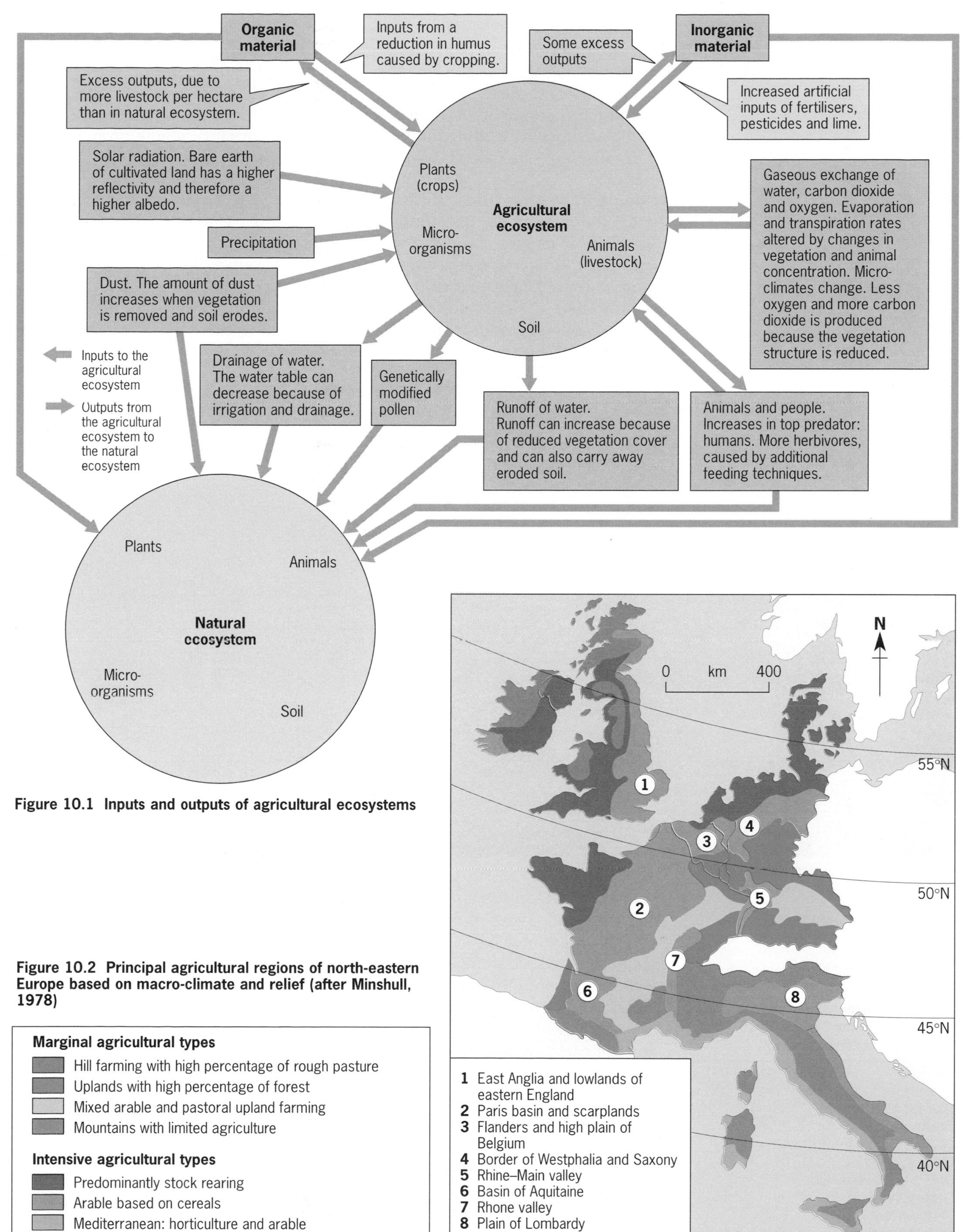

Figure 10.1 Inputs and outputs of agricultural ecosystems

Figure 10.2 Principal agricultural regions of north-eastern Europe based on macro-climate and relief (after Minshull, 1978)

After the Second World War

After 1945, government policy throughout Europe changed agriculture radically. The severe food shortages during the Second World War meant that agriculture was an important topic for the EEC when it was established in 1957. The aim of creating a common market in food was reinforced by the governments wanting to build an efficient farming industry which would produce a high proportion of the Community's food needs and thus reduce imports.

Agricultural policy and habitats

The UK did not join the EEC until 1973 (now known as the EU), but successive UK governments were strongly committed to the modernisation of agriculture. Their agricultural policies encouraged farmers to produce higher yields from more and more land under cultivation. This has been speeded up by increases in mechanisation, use of chemicals and by new breeds of crops and animals. Thousands of hectares of moorland, downland and wetland have been ploughed and 'improved' for intensive agriculture (Fig. 10.3).

Figure 10.3 Intensive agriculture in Essex

Specialisation has also occurred. Many mixed farming systems have been abandoned in favour of specialist arable or livestock farms. The variety of crops and livestock has been reduced. Government pressure on farmers to maximise production has had a dramatic effect on the UK's countryside and wildlife. Several important **habitats** have disappeared since 1945 (Table 10.1).

Nature reserves and land designated by the EU and member states as important wildlife sites have been able to protect small habitat areas. To prevent these areas from becoming isolated wildlife parks, where extinctions become more possible due to reduced immigration rates of species, the land between also needs to be available for wildlife. This is the land which sustains European economies and is therefore affected by policy change.

Table 10.1 Examples of habitat destruction in the UK, 1945–90 (*Source:* T. O'Riordan, 1987)

Habitat	Lost or damaged (%)
Lowland meadows	82
Chalk downlands	79
Lowland bogs	60
Lowland marshes	51
Limestone pavements	43
Lowland heaths	39
Upland woodlands	28
Ancient woodlands	25

Land use

Agriculture is the largest single use of land in the EU and occupies 130 million ha. This is nearly 60 per cent of the total area of the 12 member states. There is a great variety of farming systems among the nine million farm holdings in the EU (Table 10.2).

Table 10.2 Land use in European countries in 1990 (*Source:* EC, 1991)

	Belgium	Denmark	Greece	Spain	France	Ireland	Germany	Italy	Luxem-boug	Nether-lands	Portugal	UK	EU 12
Total area (000 ha)	3052	4 309	13 196	50 479	54909	7028	24 862	30 128	259	4148	9 207	24 414	22 5987
Areas under timber (000 ha)	617	493	5 755	12 511	14810	327	7401	6 434	89	330	2 968	2 297	53 998
Utilised agricultural area (000 ha)	1	2 809	5 741	27 110	30581	5697	11 868	17 215	127	2019	4 532	18 447	128 080
Arable land (000 ha)	711	2 560	2 925	15 560	17 753	1029	7 281	8 917	56	897	2 906	6 589	67 371
Permanent meadows and grasslands (000 ha)	633	N/A	N/A	6 650	11380	4666	4 375	4 877	69	N/A	671	1 205	N/A
Permanent crops (000 ha)	16	11	1 168	4 900	1 218	2	184	3 323	2	37	865	57	11 886
Number of farms in 1987 (thousands)	78.8	86	703.5	1 539.9	911.8	216.9	670.7	1 974	3.8	117.3	384	242.9	6 929
Average farm size in 1987 (ha)	17.3	32.5	5.3	16	30.7	22.7	17.6	7.7	33.2	17.2	8.3	68.9	16.5

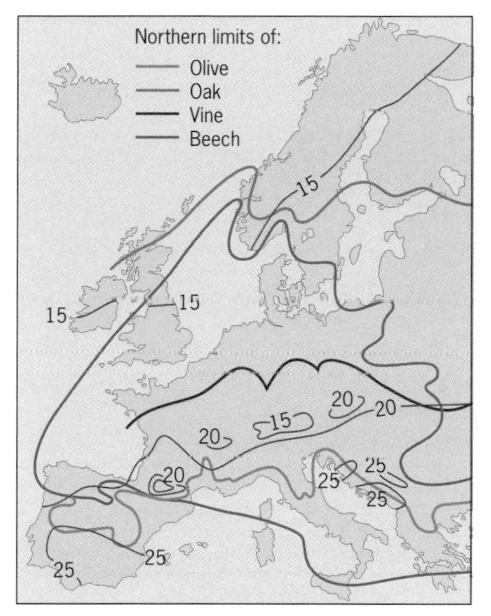

Figure 10.4 July temperatures (°C) and vegetation limits for Europe (after Speck and Carter, 1979)

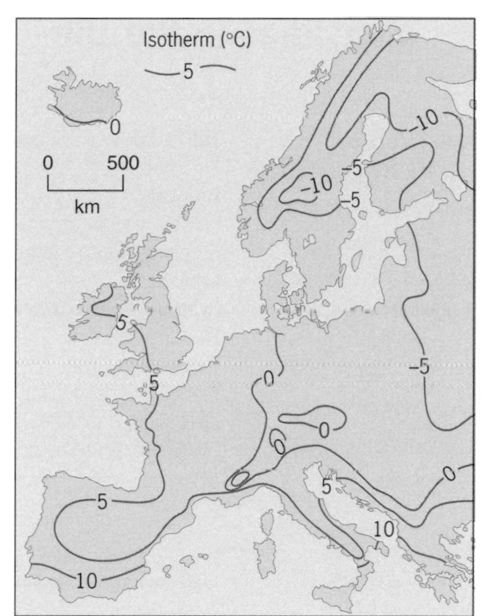

Figure 10.5 January temperatures (°C) for Europe (after Speck and Carter, 1979)

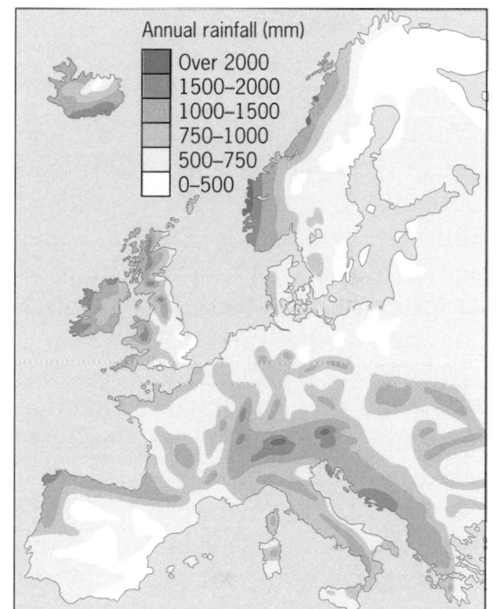

Figure 10.6 Annual precipitation (mm) for Europe (after Speck and Carter, 1979)

Crops

Some crops (such as cereals) are important throughout the EU, while others have more limited distribution patterns. Beef and dairy farms are common in northern countries while wine, olive and fruit farms are found mainly in the south (Table 10.3).

Table 10.3 Area (000 hectares) devoted to farming in 1988 for European countries (*Source:* EC Farm Structure Survey, 1990)

Crop	Belgium	Denmark	Germany	Greece	Spain	France	Ireland	Italy	Luxem-bourg	Nether-lands	Portugal	UK	EU 12
Cereals	300	1 600	4 900	1 300	8 200	9 300	300	4 900	0	200	900	4 000	35 900
Sugar beat	117	73	403	36	231	488	31	256	0	131	N/A	205	19 701
Rape seed	4	200	385	0	8	865	4	23	1	7	0	348	1 845
Sunflower seed	0	0	14	42	894	912	0	170	0	0	49	0	2 081
Olive oil	0	0	0	655	2 093	17	0	1 196	0	0	317	0	4 278
Fresh fruit (not citrus)	12	8	52	283	1 163	251	2	960	0	24	259	49	3 063
Citrus fruit	0	0	0	53	259	2	0	179	0	0	33	0	526
Vegetables	30	20	44	141	460	250	3	406	0	64	92	134	1 644
Wine	0	0	93	85	1 396	983	0	978	1	0	372	0	3 908

?

1 Using Table 10.2, list those countries that have mainly arable agriculture, and mainly permanent meadows and grasslands.

2 To what extent does the distribution of arable and permanent meadows and grassland coincide with physical factors shown in Figures 10.4–10.6?

3 Compare the average farm size in each of the EU countries. How do you account for the variations?

The effects of subsidies

Many of the intensification measures adopted by farmers were encouraged by subsidies and grants. The price support system in the EU has held product prices above world price level. Farmers were guaranteed a market for their produce and have increased output in response to favourable prices. The stability provided by the Common Agricultural Policy (CAP) also gave farmers the confidence to invest in modern intensive systems. There has been little incentive to reduce production, resulting in massive surpluses of dairy produce, beef, cereals and wine.

The CAP has consistently absorbed about two-thirds of the total EU budget. Individual member states also operate their own support programmes. In relation to its contribution to GDP, public expenditure on agriculture in the UK is several times more than that for any other industry.

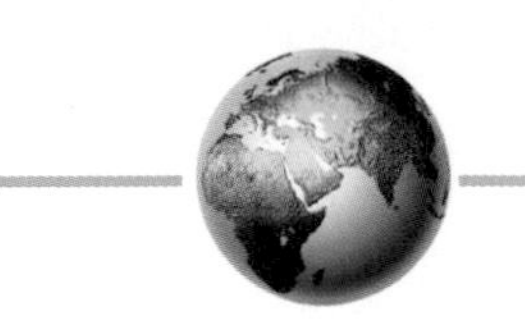

Subsidies in the UK

Upland farms

Within the EU some of the **Less Favoured Areas** are in the uplands (Fig. 10.23). One-third of Britain's agricultural area is in the uplands; a large proportion of this is rough grazing. Rainfall is high and the growing season short, which makes the production of arable crops impossible. Grass is the primary crop and is used for animals that can withstand the harsh conditions.

Sheep are the major grazing animal in the uplands. There is a lower number of sheep per unit area (stocking density) than on lowland farms because the pasture is less productive. The average stocking densities are 8.6 ewes per hectare in uplands, compared with 14 for lowland pasture.

Sheep graze specific plants, leaving others uneaten. Heavy grazing can upset the balance. They prefer sheep's fescue over the fine and creeping bent, and the unpalatable mat grass. Overgrazing has occurred if mat grass dominates. Black-faced sheep graze heather but white-faced sheep do not.

Table 10.4 Economics of sheep farming for an upland farm, 1990

Animals	
Lambs reared per ewe:	1.3
Stocking density (ewes per ha):	8.65
Income £	
Average price per lamb sold:	34.50
Wool per ewe:	2.40
Costs £	
Annual cost of buying ewes and rams to replace old animals, per ewe in the total flock:	11.50
Supplementary costs per ewe (extra feed, vet, transport):	17.90

?

4 Use the figures in Table 10.4 to evaluate the economics of sheep farming on Ty Coch Farm, a mid-Wales hill farm of 52 hectares, run totally as a sheep farm. Calculate the following:
a Total number of ewes on the farm.
b Income for a year from the flock.
c Costs for a year of maintaining the flock.
d From **b** and **c** above, calculate a figure known as the 'gross profit'. This is the profit received from the sheep farming operation without taking into account the fixed costs such as repairing barns and buying machinery such as a tractor.

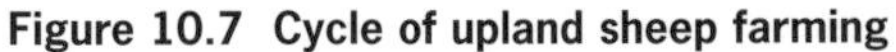

Lambs are born in late spring and remain with their mothers on uplands until late summer, when they are weaned.
Females remain on the hills for future breeding and income from their wool.
Lambs (mostly rams) are fattened before slaughter on lowland farms for four to twelve weeks.

Figure 10.7 Cycle of upland sheep farming

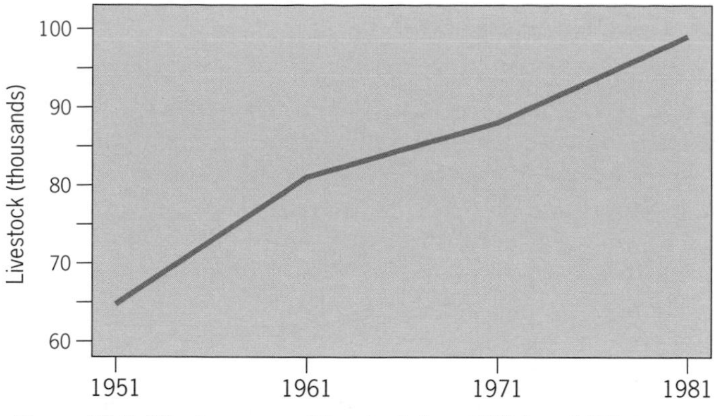

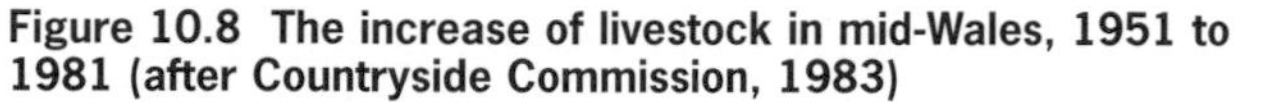
Figure 10.8 The increase of livestock in mid-Wales, 1951 to 1981 (after Countryside Commission, 1983)

Figure 10.09 Liming in the uplands, Shropshire

Figure 10.10 Improved upland fields, Scotland

Figure 10.11 A lowland sheep farm, Sussex

Table 10.5 Economics of sheep farming on a lowland farm, 1990

Stocking density (ewes per ha):	14
Lambs reared per ewe:	1.65
Average price per lamb:	£36

?

5 Sheep farming in the uplands qualifies for grants. In 1990 the ewe premium grant gave £ll per ewe, which was available to all sheep farmers. Ty Coch farm also qualified for a further grant due to the poorer nature of its land, the Hill Livestock Compensatory Allowance (HLCA), of £7.60 per ewe.
a Calculate how much Ty Coch farm received in grants in 1990.
b Calculate a gross profit for the Ty Coch farming operation which includes the HLCA grant aid.
c Comment on the figures you have calculated for income with and without grants. Bear in mind that the figures are a gross profit, without the cost of maintenance of equipment etc., and that the farm supports a family of four.

6a Comment on the trend shown in Figure 10.8. Use the economics of Ty Coch farm that you have explored in questions 4 and 5.
b Refer to Section 7.4 and Table 4.10. Explain what changes to the vegetation might occur within the heaths, moors and woodlands of the uplands when HLCA payments are made.

Improved land

In their efforts to raise stocking levels, and encouraged by MAFF and CAP policies, farmers have improved large areas of upland. If the land being improved is heather moorland, then the process will involve ploughing up heather, treating the soil with lime to neutralise its natural acidity (Fig. 10.9) and sowing the land with grass, which is then heavily fertilised.

Lowland farming

Lowland farms have better climatic and soil conditions and so tend to have more diverse farming operations than upland hill farms (Fig. 10.11). It is equally useful to look at the economics of lowland sheep farming and compare it with the economics of the upland farm.

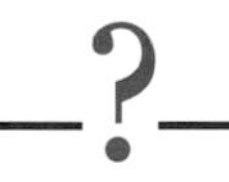

7a Describe the visual changes of the land improvement on the landscape in Figure 10.10.
b What do you think the visual effect of this land improvement would be if it were extended to cover the whole hillside?
c How do you think land 'improvement' of this type affects both numbers and diversity of wildlife? (see Section 6.3).

8 Why do you think there has been pressure to make HLCA payments more compatible with wildlife?

9a Use Table 10.5 to explain how the lowland farm is able to achieve a higher stocking rate, a higher rearing success and a better average price than Ty Coch Farm.
b In the light of these figures, what are the economic and social advantages of paying the HLCAs?

10.3 Problems of intensification

EU grants and subsidies have encouraged farmers to increase the inputs into their farming systems in return for greater **outputs**. This intensification has not only changed the nature of the upland landscape through overgrazing, but hedges have also disappeared, water courses have been polluted by agrochemicals and waste products and land has been drained.

People can add artificial inputs such as fertiliser into agricultural systems. These prevent **negative feedback** mechanisms from controlling the **dynamic equilibrium** of the ecosystem.

Hedgerow loss

Figure 10.12 Soil erosion from hedgerow removal, Cambridgeshire

Hedgerow loss became an issue during the 1970s and early 1980s as farmers became aware of the soil erosion associated with their removal (Fig. 10.12).

Hedges at the base of **food webs** are significant features of the agricultural landscape. They consist of lines of shrubs, sometimes with trees. Their functions are to be a barrier for livestock, to mark a boundary, or provide shelter.

Between 1984 and 1990, 85 000 km of hedges, over one-fifth of all hedges in the UK, were lost because of building work, agriculture and neglect. Only 3500 km of new hedges were planted. Over this period, derelict (unmanaged and unused) hedgerows increased by 75 per cent. Farmers were encouraged by grants to rip out hedges to increase the size of their fields. They also removed what may be considered to be an unnecessary maintenance cost.

In 2000, another threat to hedges emerged, prompted by EU policy changes. Existing laws allow an uncropped area around the edges of fields, thereby enabling hedgerows to survive. However, the European Court of Auditors claims British farmers are receiving subsidies for land that is not used for growing crops. It wants new laws that force farmers to make maximum use of field space to grow crops. On some farms reduced subsidies could affect income. A 400-acre farm would lose £2000 in annual subsidies. Already some farmers are pulling up hedges. A small change like this could have a devastating effect on the British landscape.

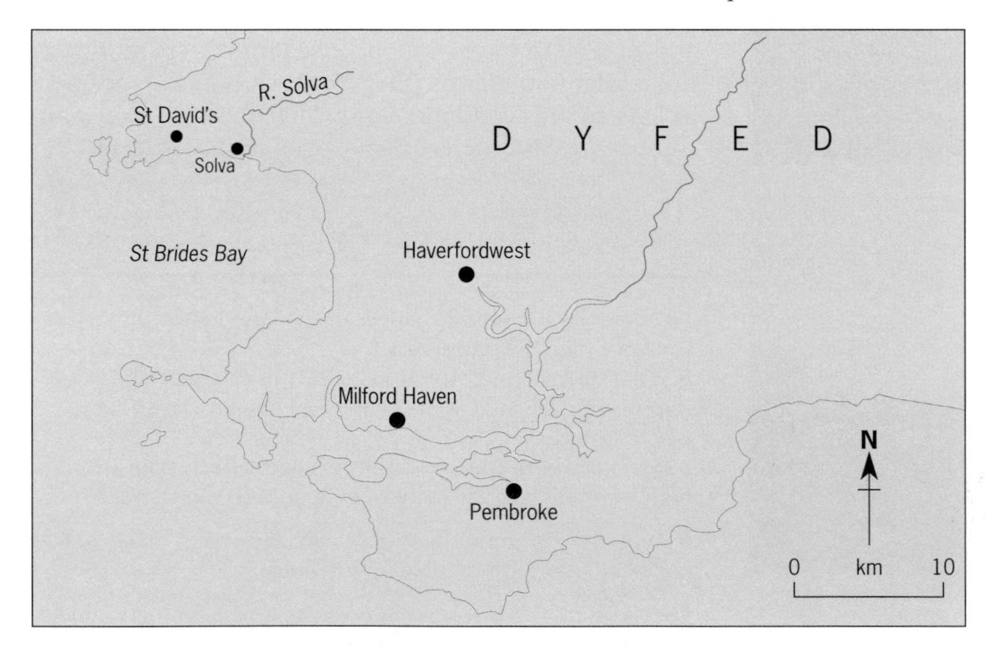

Figure 10.13 The River Solva, Wales

Land drainage

Increasing the agricultural area by draining the land was a popular option during the 1960s, 1970s and early 1980s. Grants were available from the UK government and the EU for farmers to alter the water table and change water courses to drain the land. Entire wetland and water-based ecosystems have been changed as a result (see Chapter 9).

The River Solva

Work was carried out in the 1970s and 1980s on the River Solva (Fig. 10.13) to deepen the channel. This would lower the water table and allow wet fields next to the river to be drained, thus improving their potential as grazing land. Another reason for the work was to reduce occasional flooding that occurred in the village of Solva downstream.

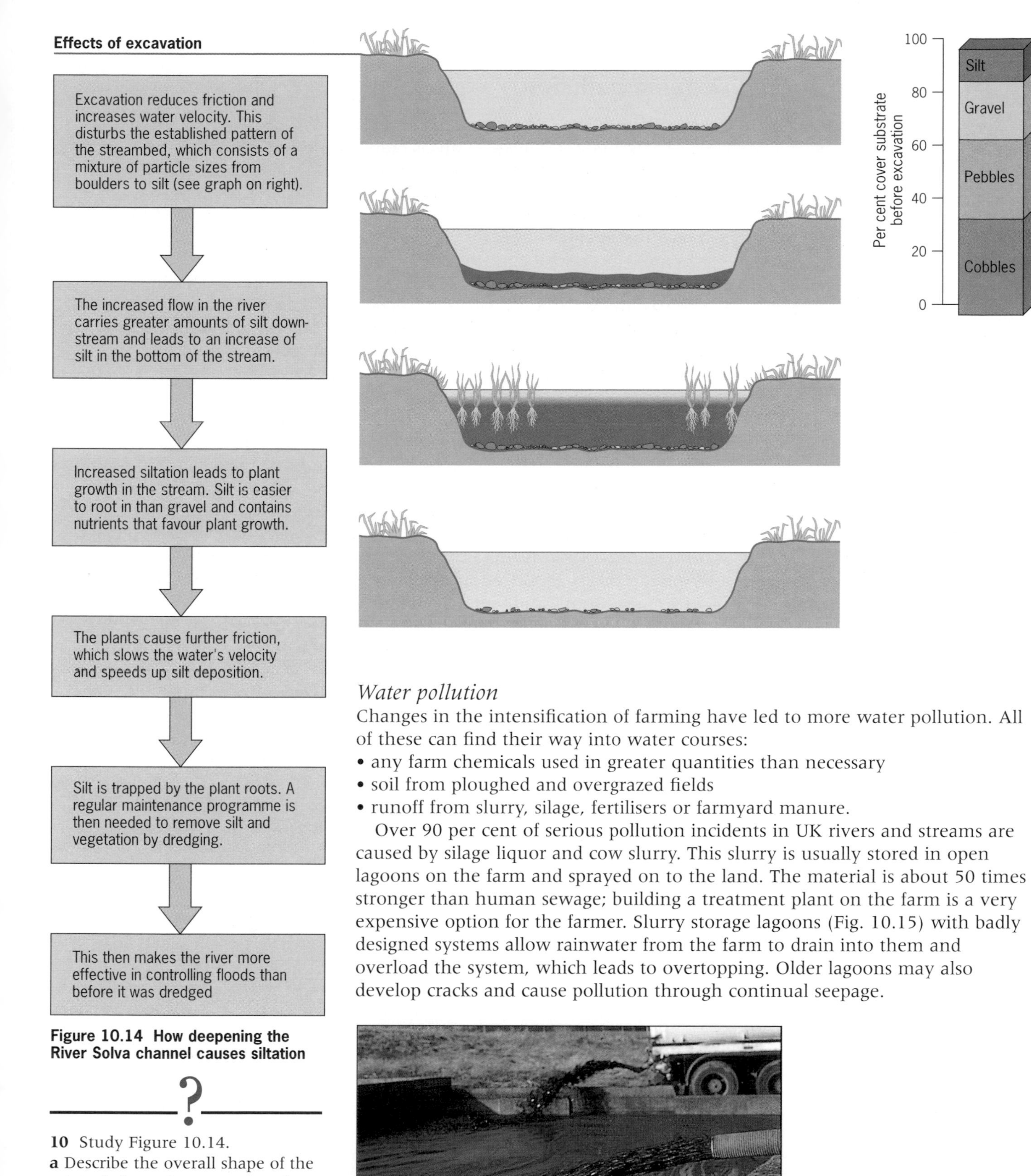

Effects of excavation
Excavation reduces friction and increases water velocity. This disturbs the established pattern of the streambed, which consists of a mixture of particle sizes from boulders to silt (see graph on right).
The increased flow in the river carries greater amounts of silt downstream and leads to an increase of silt in the bottom of the stream.
Increased siltation leads to plant growth in the stream. Silt is easier to root in than gravel and contains nutrients that favour plant growth.
The plants cause further friction, which slows the water's velocity and speeds up silt deposition.
Silt is trapped by the plant roots. A regular maintenance programme is then needed to remove silt and vegetation by dredging.
This then makes the river more effective in controlling floods than before it was dredged
Per cent cover substrate before excavation
100
80
60
40
20
0
Silt
Gravel
Pebbles
Cobbles

Figure 10.14 How deepening the River Solva channel causes siltation

?

10 Study Figure 10.14.
a Describe the overall shape of the original channel compared to the shape of the new channel after excavation.
b List the advantages and disadvantages of deepening the river channel.

Water pollution

Changes in the intensification of farming have led to more water pollution. All of these can find their way into water courses:

- any farm chemicals used in greater quantities than necessary
- soil from ploughed and overgrazed fields
- runoff from slurry, silage, fertilisers or farmyard manure.

Over 90 per cent of serious pollution incidents in UK rivers and streams are caused by silage liquor and cow slurry. This slurry is usually stored in open lagoons on the farm and sprayed on to the land. The material is about 50 times stronger than human sewage; building a treatment plant on the farm is a very expensive option for the farmer. Slurry storage lagoons (Fig. 10.15) with badly designed systems allow rainwater from the farm to drain into them and overload the system, which leads to overtopping. Older lagoons may also develop cracks and cause pollution through continual seepage.

Figure 10.15 A slurry storage lagoon, Wales

Organic pollution

The characteristics of a freshwater ecosystem are usually determined by the chemistry of the water. Phosphorus is one of the elements needed by higher plants and algae. It is also one of the least widely available in natural systems. Clear water upland lakes usually have less than 5 µg (micrograms) of phosphorus per litre. Fertile undisturbed lowland lakes may have 10–30 µg. Nitrogen is also very scarce.

A gradual increase in agricultural intensification has caused these elements to be released into water in higher concentrations than normal. This has happened where pasture or natural vegetation has been ploughed and when organic waste products enter water courses. The increased fertilisation or **eutrophication** leads to increased production of **phytoplankton** and algal growth on the water surface. This has a damaging effect on the ecosystem (Fig. 10.16).

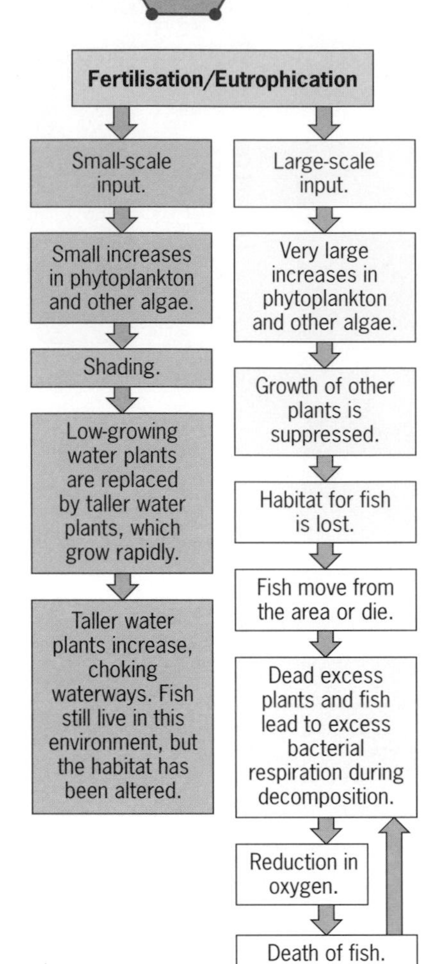

Figure 10.16 Eutrophication

Biological oxygen demand

One problem is to find a useful measure of eutrophication. Oxygen shortages can occur in water containing large amounts of organic matter. Decomposition by microorganisms uses up so much oxygen that it can affect other organisms. A common measure of organic pollution is **biological oxygen demand** (BOD). It is calculated by finding out how much oxygen is taken up by a sample of water when it is kept warm for five days at 20°C in the dark.

Using this measure, a clean mountain stream will have a value in the region of 0.05 mg per litre. Crude sewage effluent has a value of between 200 and 800 mg per litre. Silage liquor (fluid that leaks from farm silage) has a value of around 60 000 mg per litre, and slurry a range of between 800 and 60 000 mg per litre.

Animals and plants living in streams vary in their ability to tolerate polluted water. When a large quantity of effluent enters a watercourse, the dramatic increase in micro-organisms which feed on organic compounds causes a rapid fall in the oxygen levels, resulting in the death of water plants, small **invertebrates** and fish in the stretch of the stream immediately below the incident. This pollution is reduced as it moves downstream (Fig. 10.17). The dead bodies of the creatures also have a BOD, which is a **positive feedback** mechanism.

11 Using Figure 10.17, explain the data on the graph.

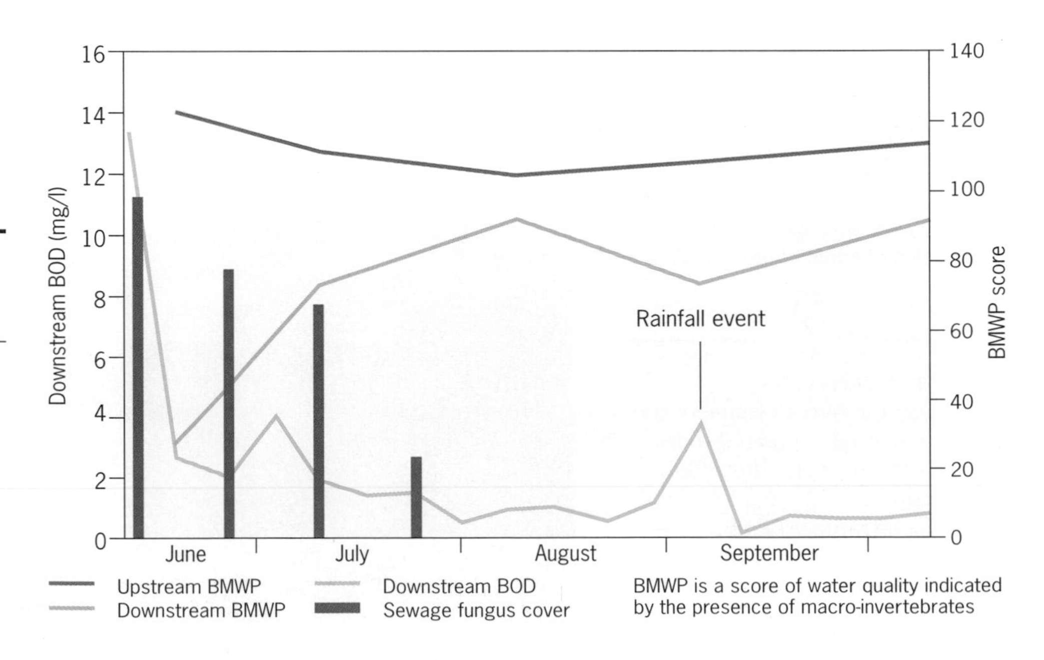

Figure 10.17 Water quality at Ponfaen Brook and its effects on sewage fungus growth and benthic macro-invertebrates (after Seager et al., 1992)

Nitrates

Nitrogen is essential for plant growth and can affect quality and yield of grain. Managing nitrates within an agricultural system is a question of balancing outputs and inputs.

There are two annual peaks of nitrate release which coincide with decomposer activity. From May to June increased soil temperatures cause more microbe activity, but the crop takes up all that the soil releases. More nitrogen is needed at this time of the year to produce high yields of crops. During September to October deep cultivation aerates the soil. This stimulates the microbes to break down more organic matter. At this time of the year the new crop does not have extensive roots to extract the nitrogen, so there is a risk of these surplus nitrates being leached. Nitrates enter water supplies and have caused health problems among young babies (called blue baby syndrome). Water companies have to meet EU standards of 50 mg per litre in drinking water. Farmers can adopt practices to keep excess nitrates from entering water supplies (Table 10.6).

?

12a Draw a flow diagram of nitrate release.
b Using Table 10.6, mark the stages where nitrate outputs could be reduced.

Table 10.6 Possible actions to reduce nitrate outputs from farmland

- Avoid applying nitrogen during the autumn.
- Avoid leaving the soil bare in winter.
- Sow crops early in autumn so that they take up more nitrogen.
- Plough grassland in late winter to stop increases in decomposers early in the autumn.
- Avoid excessive inputs of animal manure.
- Measure inputs of nitrogen fertiliser carefully.

Changing production methods

Farmers have reduced the number of hay meadows (Fig. 10.18) as they have changed their methods of harvesting grass. Silage, which, unlike hay, can be cropped more than once a year, is the preferred method in many parts of Europe (Fig. 10.19).

A combination of hedgerow removal, land drainage, changing production methods and pollution have contributed to the loss of some of the more traditional farming systems which were compatible with wildlife (Fig. 10.20).

Figure 10.18 An Alpine hay meadow

Figure 10.19 A silage meadow, Wales

Uplands
Grass and heather are the main crops. The low productivity of the uplands means that it can only be grazed sustainably at low intensities. Farmers also crop some of the grass as hay. Hay meadows are extremely valuable for plants, insects and breeding birds. Heather is often managed specifically to encourage birds such as grouse.
Lowland heathlands
Heathlands, largely in north-west Europe, have been partly created by grazing, which has prevented woody shrubs and grasses from succeeding to woodland. They support very specialised plants and animals.
Lowland wet grassland
Some of the largest areas of freshwater wetland remaining in Europe comprises a varied group of marshes and wet grassland habitats. Many sites are of great conservation value. Grazing by farm livestock is the principal use of these wetlands, and the conservation interest of many sites is sensitive to the pattern and intensity of grazing and the type of stock.
Woodlands
Much woodland was cleared to make way for farms, but there are still copses and areas of productive woodland in the European landscape.
Olive plantations
Traditionally managed, they support a rich wildlife. Olives and figs are important winter food for migrating birds such as robins, blackcaps and warblers.
Dry grassland
Grassland of various kinds make up one of the most extensive forms of land use in Europe. Permanent meadows and pastures account for about 40% of all agricultural land in many countries in north-west Europe, risng to over 60% in the UK and over 80% in Ireland. In southern Europe, grassland is much less important, but still accounts for 17% of the land in Portugal, 24% in Spain and 28% in Italy. These areas are dominated by dry grass heaths and low scrub.
Mediterranean wood pasture systems
Some of the few remaining wooded pastures of Mediteranean countries are still used for low-intensity farming. Hunting on a sustainable basis, cork harvesting, timber, honey gathering and olive growing all provide additional income for farmers. This rotation protects fragile soils and creates a rich habitat for birds and other wildlife.
The maquis
Some of the most widespread semi-natural habitats in the Mediterranean basin belong to a group including the maquis, the garigue and the matorral. All three are more open habitats than climax forest, with a shrubby vegetation. Such habitat can support a wide range of species.
The deeper shaded area shows regions in the EC where intense farming is practised. The lighter areas show less intense farming in northern and southern areas of the EC.

Figure 10.20 Low-intensity farming systems (after Taylor and Dixon, 1990)

10.4 Agriculture and the economics of conservation

One possible way to make agriculture and wildlife more compatible is for farmers to receive money for any income that might be lost by farming a piece of land for conservation instead of for profit.

Common Agricultural Policy options

Since the early 1990s, the EU has accepted the need to reform the Common Agricultural Policy (CAP) to reduce intensification. In July 1991, Ray MacSharry, the Agriculture Commissioner of the EU, launched proposals to reform the CAP. One of these reforms involved a commitment to agri-environment regulations. Farmers losing subsidies would be compensated by payments to safeguard the environment. Some of the options available for all EU farmers include **set-aside grants**.

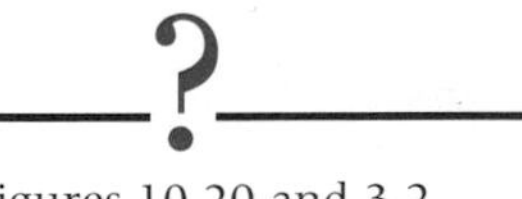

13 Use Figures 10.20 and 3.2 to make a list of arguments in favour of:

a intensification of agriculture
b wildlife conservation.

Figure 10.21 A set-aside field, Oxfordshire

Set-aside a weedy issue

British farmers are now raising 243 square miles of perennial weeds under the set-aside scheme.

A Produce Studies Group survey of five hundred English cereal growers commissioned by Monsanto found that when part fields, field edges and headlands are taken into account, 1.5 million acres are in set-aside.

Of the total, an estimated 296,000 acres are affected with couch grass, 170,000 acres with couch and broadleaved weeds, 235,000 acres with broadleaved weeds only and up to 155,673 acres are estimated to be covered by perennial weeds.

Cover crops were planted on 311,000 acres (26 per cent of national set-aside land) of which 15 per cent is down to grass.

Figure 10.22 Newspaper article detailing the extent of set-aside fields in Britain (*Source: Farmer's Guardian*, 14 May 1993)

Set-aside scheme

The set-aside scheme was introduced and funded by the EU in 1988. Annual payments (per hectare) were made to producers who reduced their arable crop area by at least 20 per cent for five years. The scheme only applies to arable land previously used to grow crops.

The new scheme, introduced as part of the CAP reform, provides payments for certain arable crops. To claim money, farmers must set land aside (Fig. 10.21). This land can be used for some non-food crops, but other money-making schemes are not allowed. It must be cared for to keep it in good cropping condition. Member states are also required to apply appropriate environmental measures to the land set aside. There are two types of set-aside: rotational and non-rotational.

Land under rotational set-aside is only eligible for payments every six years, so farmers have to move their set-aside fields around the farm. Non-rotational set-aside is able to carry on year after year. Farmers may set aside poorer land within this option.

?

14 Which type of set-aside, rotational or non-rotational, do you think would be better for wildlife? Give reasons for this.

15 Read Figures 10.22 and 10.24. Make a list of the disadvantages of the set-aside scheme for farmers.

16 Using Table 10.7, Figures 10.3 and 10.23, answer the following questions:
a Why are payments lower for set-aside in Less Favoured Areas (LFA)?
b Which areas seem to receive the highest payments?
c Explain why set-aside uptake varies throughout the EU.

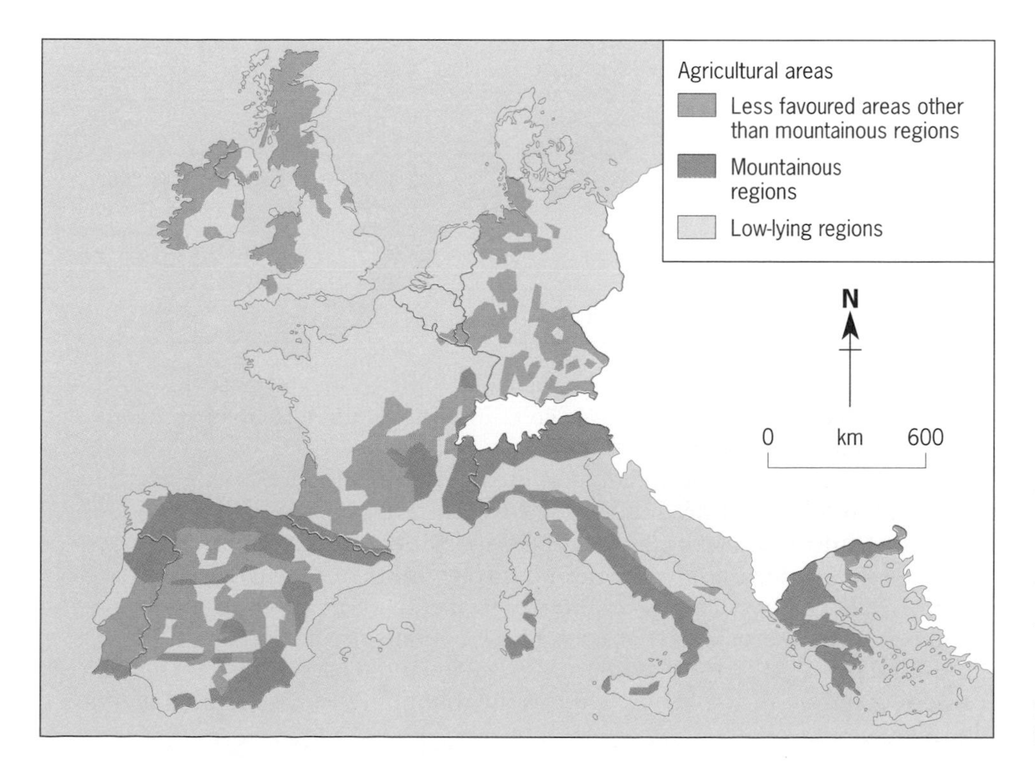

Figure 10.23 Less Favoured Areas in the EU (after Com, 1989)

Notre action vous gêne

Nous en sommes conscients

Mais notre démarche n'est pas corporatiste

Le devenir de l'agriculture et de la ruralité vous concerne tous

◆◆◆

LA REFORME DE LA POLITIQUE AGRICOLE COMMUNE EST BASEE SUR DES PRINCIPES INACCEPTABLES

◆ L'Europe risque à terme la dépendance alimentaire

Nous sommes là pour cultiver de façon rationnelle et intelligente l'ensemble du territoire, pas pour mettre des terres en jachère.

◆ Vous allez payer deux fois votre alimentation: d'abord par vos achats puis par vos impôts (directs et indirects)

Vous le savez: une baisse des prix à la production ne signifie pas une baisse des prix à la consommation.

L'EUROPE NE PEUT PAS SE CONSTRUIRE SUR LES RUINES DU PREMIER SECTEUR ECONOMIQUE ET HISTORIQUE DU CONTINENT.

Our action is causing you problems. We are aware of this. However, we are not just representing our own interests. The future of agriculture and the countryside concerns all of you.

The CAP reform is based on unacceptable criteria. If we take it to its logical conclusion, Europe risks becoming dependent on food imports.

We are here to cultivate all of the land in a rational and intelligent way, not to let fields lie fallow.

Your food bills will double: firstly through what you buy, and secondly through the (direct and indirect) taxes you pay.

You know that lowering the production costs does not mean lowering food prices.

Europe cannot be built on the ruins of its primary economic (and historic) sector.

Figure 10.24 A handbill given to the public during road blocks in southern France, June 1992

Table 10.7 Set-aside of arable land (*Source:* EC, 1991)

Member state	Premium 1990-1 (ECU*/h)	Land set aside (ha) 1988/9	1989/90	1990/1	Total
Belgium	207 } LFA 269 } 362 sandy 518 sandy/silty	339	151	250	740
Denmark	112-431 depending on yield	–	–	5 520	5 520
Germany:					
Former West	300-600 based on land quality	165 125	57 259	71 000	293 384
Former East	190-290 based on land quality	–	–	599 243	599 243
Greece	LFA: 150 non-irrigated 250 irrigated Other: 180 non-irrigated 300 irrigated	–	250	N/A	250
Spain	121 LFA 143 other 197 } 257 } irrigated 344 }	34 29	13 858	36 000	84 087
France	1 195-312 11 234-363 } 111 286-416 } groups IV 325-455 } of regions V 338-481 }	14 220	39 702	112 653	166 575
Ireland	242	1 141	438	187	1 766
Italy	380 LFA 400 other hill farms 440 plains 600 Po plain	91 617	266 366	250 752	608 705
Luxembourg	217	6	31	48	85
Netherlands	700	2 582	6 155	5 869	14 606
UK	285 LFA 314 other	51 567	50 321	30 734	132 622
Tota		360 826	434 501	1 112 256	19 075 83l

* ECU = European Currency Unit

Twenty-year set-aside

The EU Agri-environment Regulation provides payments for taking agricultural land of all types out of production for 20 years 'for purposes connected with the environment, in particular for the establishment of biotope reserves or natural parks or for the protection of hydrological systems'. The UK has a scheme called *Countryside Stewardship* where landowners agree to manage specific habitats for at least 10 years. These habitats include chalk and limestone grassland; lowland heath; waterside land; coastal land; upland; meadows and pasture; historic features; field boundaries; arable field margins; and public access.

Nitrate Sensitive Areas

To help reduce nitrate levels, the UK government set up Nitrate Sensitive Areas (NSA) (Fig. 10.25). Within NSAs, farmers were asked to follow some voluntary guidelines to reduce the output of nitrates from their agricultural systems. In return they received payments.

Environmentally Sensitive Areas

The concept of **Environmentally Sensitive Areas** (ESA) was introduced in 1985, as part of a general review of agricultural structures policy. Farmers in ESAs receive annual payments, for a period of five years, if they agree to change farming practices to protect and enhance existing landscapes and wildlife habitats such as wet grassland, moorland and heathland and chalk grassland.

?

17 Using Figure 10.25, describe and explain the distribution of Nitrate Sensitive Areas in the UK.

18 From Figure 10.25 and an atlas describe the type of habitats that are being protected under ESA schemes in the UK.

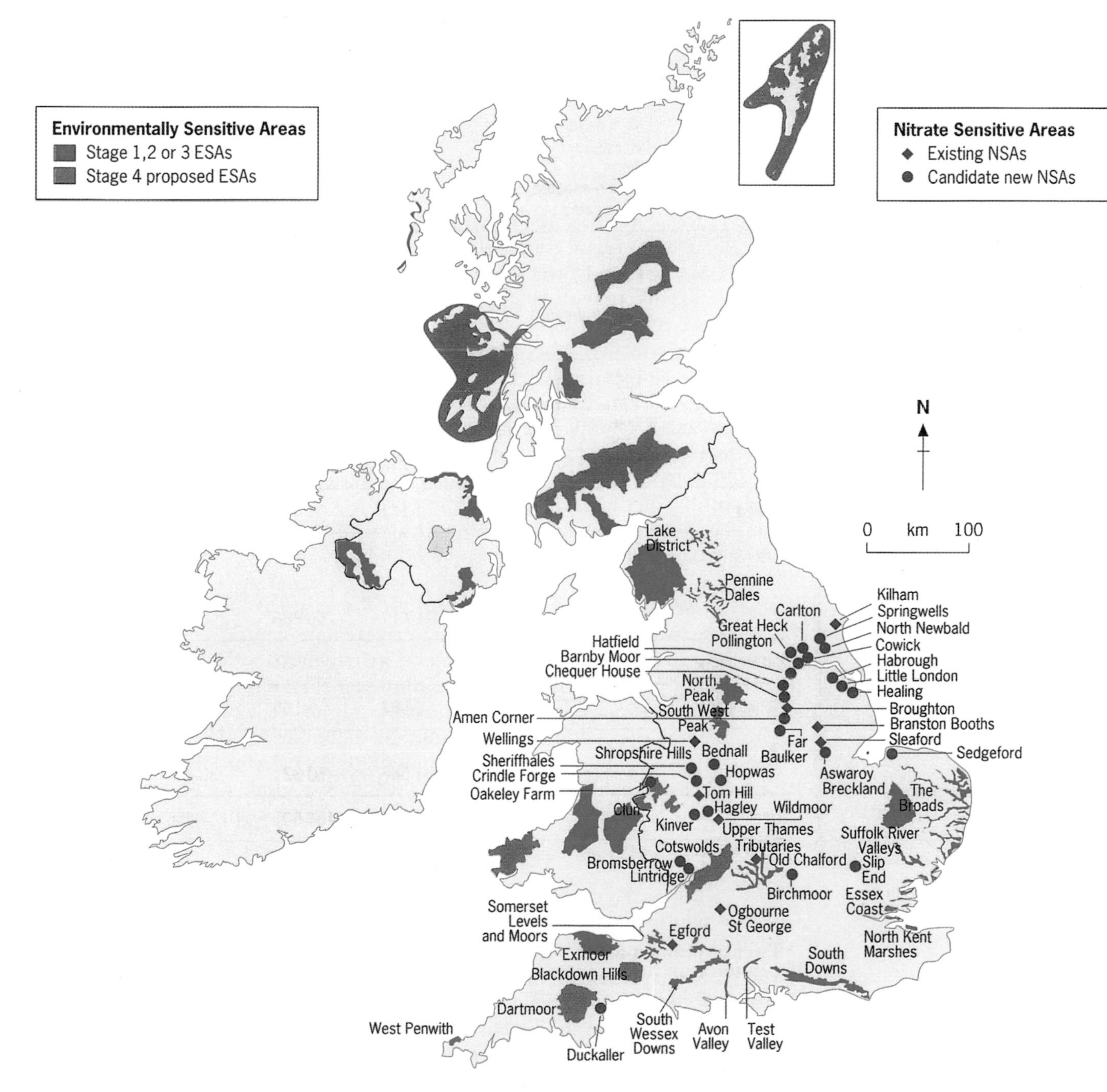

Figure 10.25 Environmentally Sensitive and Nitrogen Sensitive Areas in the UK (after MAFF, SO and DANI, 1993)

?

19 Essay: Evaluate the potential advantages and disadvantages of agri-environment regulations for:
a wildlife,
b farmers.

20 Essay: What impact will changes to the EU and new technology have on the countryside?

The scheme enables member states to introduce national aid schemes, funded in part by the EU, to support ecological and landscape management within designated areas of recognised importance.

Conservation organisations believe that many more ESAs are needed in those parts of each member state where traditional low-intensity farming systems still sustain important wildlife habitats (Fig. 10.20). ESAs could be particularly valuable in Spain, Portugal, Greece, Italy and parts of France. They could also sustain labour-intensive farming activities. This keeps rural communities in areas where there is a migration of people to towns and cities.

The UK and Germany were the first countries to begin designating suitable areas as ESAs. Habitat destruction has slowed down in some of the UK's ESAs. In Mediterranean countries, such as Spain and Portugal, there has been a call for ESAs by conservationists because many of their traditional land uses are undergoing changes due to grants from the CAP. Consequently many important sites for wildlife have been destroyed due to the new intensified farming methods. An increasing number of ESAs are now being designated throughout the EU (see Chapters 7 and 8).

10.5 Accession

Thirteen more countries are applying to join the EU: Bulgaria, Cyprus, the Czech Republic, Estonia, Hungary, Latvia, Lithuania, Malta, Poland, Romania, Slovakia, Slovenia and Turkey. Accession of these countries is complicated; they have to comply with 20 000 items of legislation to join. Many of the countries are relatively poor, with high unemployment and declining agricultural production. They are also rich in wildlife, with animals such as brown bears, wolves, lynx and varied birdlife. If these countries join, they will be able to finance major agricultural and rural development projects under the SAPARD funds (Special Accession Programme for Agriculture and Rural Development) which began in January 2000 with an annual budget of Euro 520 million. This is an opportunity to improve productivity without compromising biodiversity. Already the EU PHARE Partnership programme has funded management advice and technical support to ensure that important habitats are not lost in some of these countries.

World Trade Organisation Agreement on Agriculture

The EU is trying to influence the WTO to create a new principle of agricultural trading in the world, based on sustainable development rather than greed. It promotes the idea of multifunctionality, which recognises that agriculture is not just about producing crops for trade but that environmental, social and cultural values are all important.

The UK has tried to integrate environment and people into its Rural White Paper which was published in 1999. This will be supported by EU funding under the Rural Development Regulation and benefits people and wildlife in rural areas.

Agenda 2000

CAP was reformed in 1999. All countries must run agri-environment schemes which pay farmers and landowners grants to protect and improve the environment. Each member of the EU can run up to 8 other schemes which help hill farming; restructure agriculture; develop more profitable farming, forestry and food industries; and support rural development.

The UK continues its support for farmers through funding environmental good practice. The Rural Development Plan for England, published in 2000, states that it will reduce payments for producing crops but will increase payments for sustainable farming and forestry which leads to a successful

Benefits of GM foods	Concerns
Reduce the need for pesticides and herbicides	Limited application to poorer countries because of the expense
Expand crop growing areas	May disrupt the natural ecology of areas; could increase the competitive advantage of certain species
	Health concerns about the impact on human health and the food chain
Unethical not to produce when people are starving	Pollen travels 20 times the distance currently used around fields of GM crops
Terminator technology, which causes the seed to die after production, may prevent contamination	Terminator technology would means that poorer farmers would not be able to save seeds from provision years and would rely on companies to supply them

The England Rural Development Plan Measures
ESAs
Countryside Stewardship Scheme
Organic Farming Scheme
Woodland Grant Scheme
Farm Woodland Premium Scheme
Hill Farm Allowance Scheme (payments per head of sheep stopped, farmers now paid per ha)
Processing and Marketing grants
Rural Enterprise Scheme
Energy Crops Scheme
Vocational training for people in farming and forestry

countryside. However, ESAs have been replaced in Scotland, Northern Ireland and Wales. In each of these countries there is now one agri-environment scheme. In Scotland it is called the All Scotland Agri-Environment Scheme, in Wales, Tr Gofal, and in Northern Ireland, the Countryside Management Scheme. Already applications for funding are double the amount of money that has been allocated because the CAP spends only 3 per cent of its budget on agri-environmental protection.

Organic farming

MAFF encourages the development of organic farming in the UK by subsidising farmers who want to convert to organic farming.

GM foods

GM foods are produced from plants and animals that have their genes changed by scientists in the laboratory, rather than farmers in the field. New techniques allow a single gene for a single characteristic to be isolated and transferred from one organism to another, even between species. DNA can be transferred to increase resistance to diseases or pests, or to change ripening or nutritional value, altitude tolerance or temperature. Genes from fish that make them resistant to cold can be put into a strawberry to allow it to grow in cold conditions.

The Flowersaver tomato was one of the first products to be produced. The chemical that causes the tomato to rot was suppressed, increasing its shelf life. Patent for this tomato was introduced into the UK in 1996. In this year the EU allowed Monsanto's Roundup Ready Soya beans in food for people and for animal feeds. The soya bean can survive being sprayed with Roundup herbicide.

Summary

- The structure of the agricultural ecosystem is a simplification of a natural ecosystem.
- Agricultural ecosystems are generally less productive per unit area than the climatic climax communities they replace.
- Increasing levels of technology have enabled people to cultivate marginal areas which have previously been important wildlife habitats.
- Agriculture has always affected the landscape but the degree of change has been greatest over the last 100 years. Agricultural change has resulted in loss of wildlife habitats and changes to those that remain.
- One of the major policies to affect European agriculture is the Common Agricultural Policy.
- Post-1945 policies to increase production involved intensification, specialisation and extension of the area under production.
- Subsidies and grants within Europe have provided the incentives for farmers to change the landscape.
- Increased agricultural efficiency in Europe is starting to provide policy-makers with a means to change towards more environmentally-friendly decisions,
- Laws have been made to curb some problems such as pollution.
- Schemes such as set-aside, Environmentally Sensitive Areas and Nitrogen Sensitive Areas now aim to replace the shortfall of reduced subsidies for agricultural production.
- The uptake of environmental schemes varies throughout Europe and is linked to cultural, political, economic and environmental differences among member states.

11 Estuaries, industry and energy

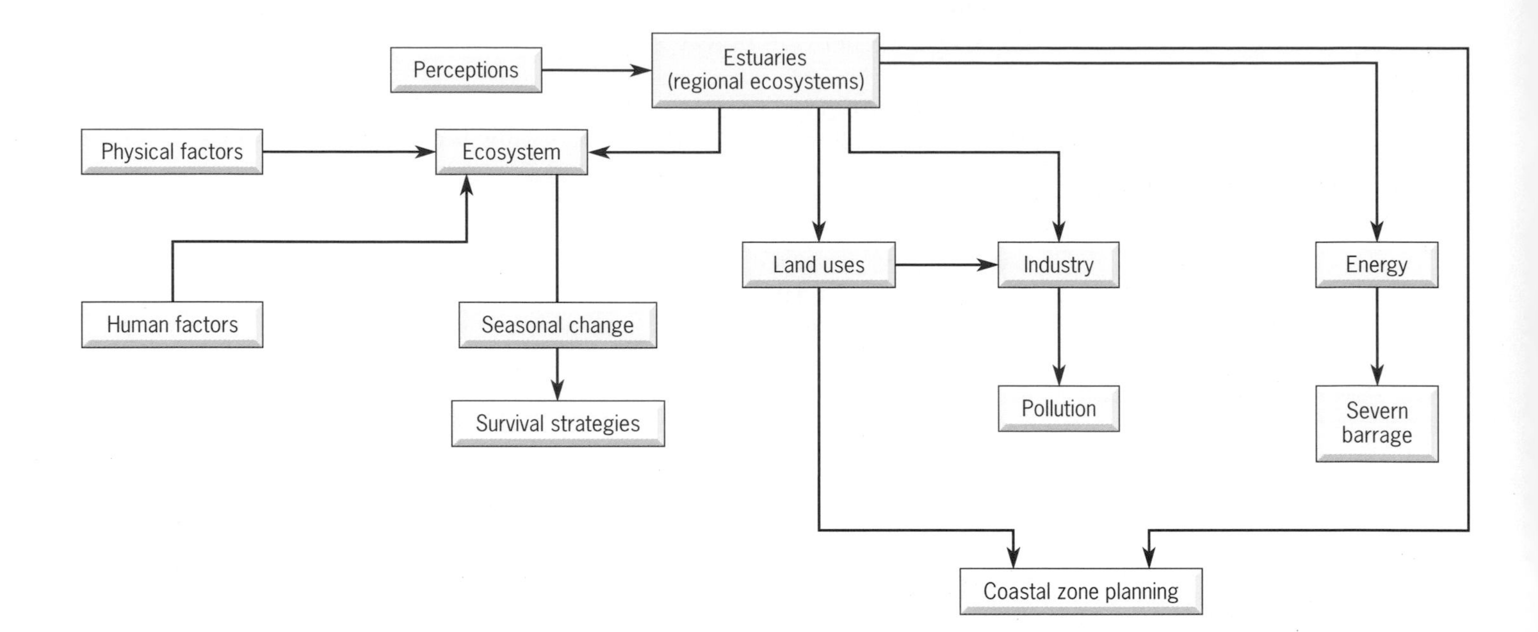

Figure 11.1 Estuary, Medway, Kent

?

la Use Figures 11.1 and 11.2 to help you write out the statements from Table 11.1 that you think apply to estuaries.
b Compare your list with a sample of people from your school or college.
c Summarise your results in a brief statement. What do your results tell you?

11.1 What is an estuary?

According to the Nature Conservancy Council's Estuaries Review of 1991, an estuary is 'A partially enclosed area of water and soft tidal shore and its surroundings, open to saline water from the sea and receiving fresh water from rivers, land runoff or seepage.'

There is competition for space along estuarine shorelines. An estuary also receives water from its catchment area. This means that an estuary (Fig. 11.1) is affected by a wide range of land uses. It is the point where recreation, agriculture, industry and energy policies all have some impact. The estuary is perhaps the **ecosystem** which has presented the greatest challenge to planners.

Table 11.1 An estuary is ...

1	a site for recreation.
2	a source of fresh farmland.
3	a site for new settlement.
4	a place to find new species.
5	a source of fish.
6	a large sewer.
7	a place to hunt wildlife for food.
8	a place to hunt wildlife for sport.
9	a place for new industry.
10	a beautiful place.
11	frightening.
12	natural.
13	a source of power.
14	a site for ports.
15	an international airport.
16	a productive ecosystem.

Figure 11.2 Aerial view of estuary, Old Hall Marshes, Essex

The aesthetic qualities of an estuary are a direct reflection of surrounding land uses. How people see an estuary now is also likely to affect their vision for the estuaries of the future. The answers to a survey summarised in Table 11.1 are a useful indication of how much people know about estuaries and their potential development.

11.2 The estuarine ecosystem

The estuarine ecosystem (Fig. 11.3) is a complex system receiving **inputs** from many sources. Thus it is one of the most highly productive ecosystems on earth.

Figure 11.3 The estuary system

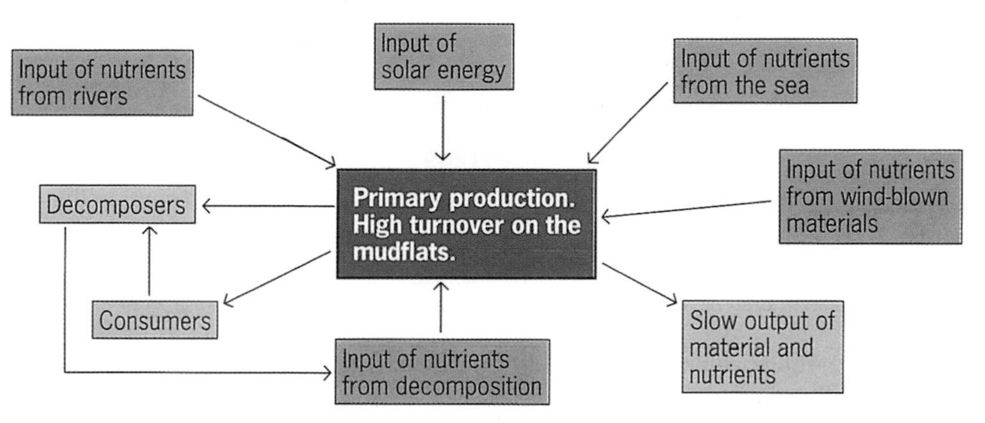

The rhythmic shifts in tide levels affect the characteristics of the estuarine ecosystem. The estuary is in a constant state of change on a daily, seasonal and yearly cycle. Tides cause daily changes which contribute to plant **zonation**. Zones are areas in which plant communities tend to be dominated by one species. This usually reflects the plants' relative abilities to tolerate salt water, to being submerged or their need for stable ground. Different plants respond to flooding in different ways, tending to be specialists for different environments (Fig. 11.4).

Complications arise where **succession** (see Chapter 6) and zonation take place at the same time (Figs 9.4 and 11.4). Succession involves variation in **species** over *time*; whereas zonation involves variation with *space*. As in any ecosystem, **competition** between the plants themselves for nutrients, water and sunlight can cause succession. It can begin due to changes either started by the plants themselves, such as by shading smaller plants, or be caused by an external factor such as a change in the environment; tidal action, for example (Fig. 11.4).

Salt marsh communities increase their gradient due to the slow build-up of sediment by species like *Spartina*. As the gradient increases, the soil drains and becomes drier. There are nutrient changes within the salt marsh. Annual species like glasswort can take nutrients, such as nitrogen, directly from sediments. Nitrogen and other nutrients would not be available to them higher up in the marsh where perennial species occur.

2 What factors shown on Figure 11.3 would contribute to relatively high productivity levels?

3a Using Table 1.1 find out how much area is taken up by estuaries throughout the world.
b How productive are they?
c How does the productivity compare with other ecosystems?
d What is the biomass figure for estuaries?
e How does the biomass compare with other ecosystems?
f Why do you think estuaries are less productive than swamps and marshes, but more productive than open oceans per square kilometre?
g Why do you think estuaries have a biomass greater than open oceans but lower than swamps and marshes?

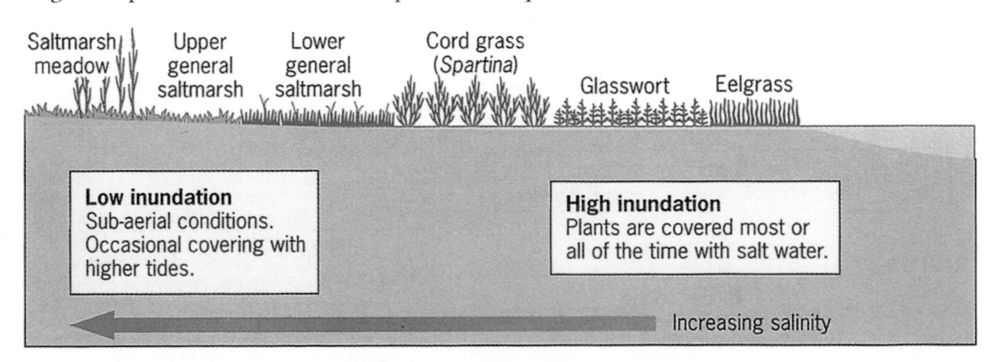

Figure 11.4 Salt marsh zonation

Figure 11.5 Changes to the Dee estuary 1648-1976 (after Dee Estuary Conservation Group, 1976)

Causes of change to estuaries

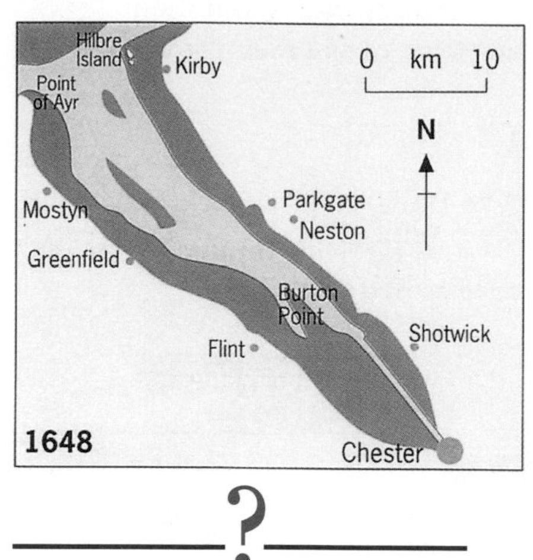

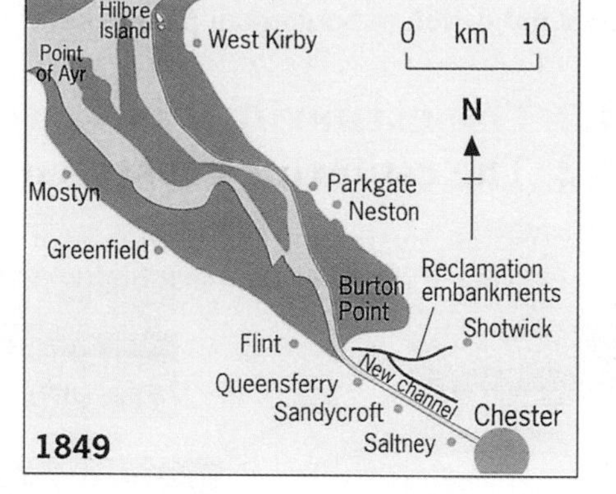

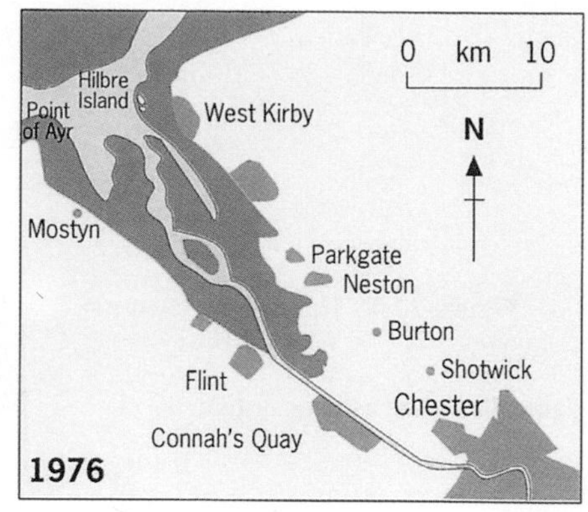

?

4a Study Figure 11.5. What do you think happened to the area of mudflats and saltmarsh between 1648 and 1976?
b What have people chosen to do to manage the natural processes?

The estuary is changing form continually, thereby affecting its **abiotic** characteristics, such as sediment inputs and current patterns. Sediments and current changes can also affect the input of the sun's energy. Turbulence and increased stirring-up of sediments causes less energy to penetrate to bottom-dwelling creatures. Currents and shifting sediments also change the nature of the estuary bottom. These, in turn, affect the distribution of **producers**, which in turn affect the population of **consumers** (Fig. 11.6).

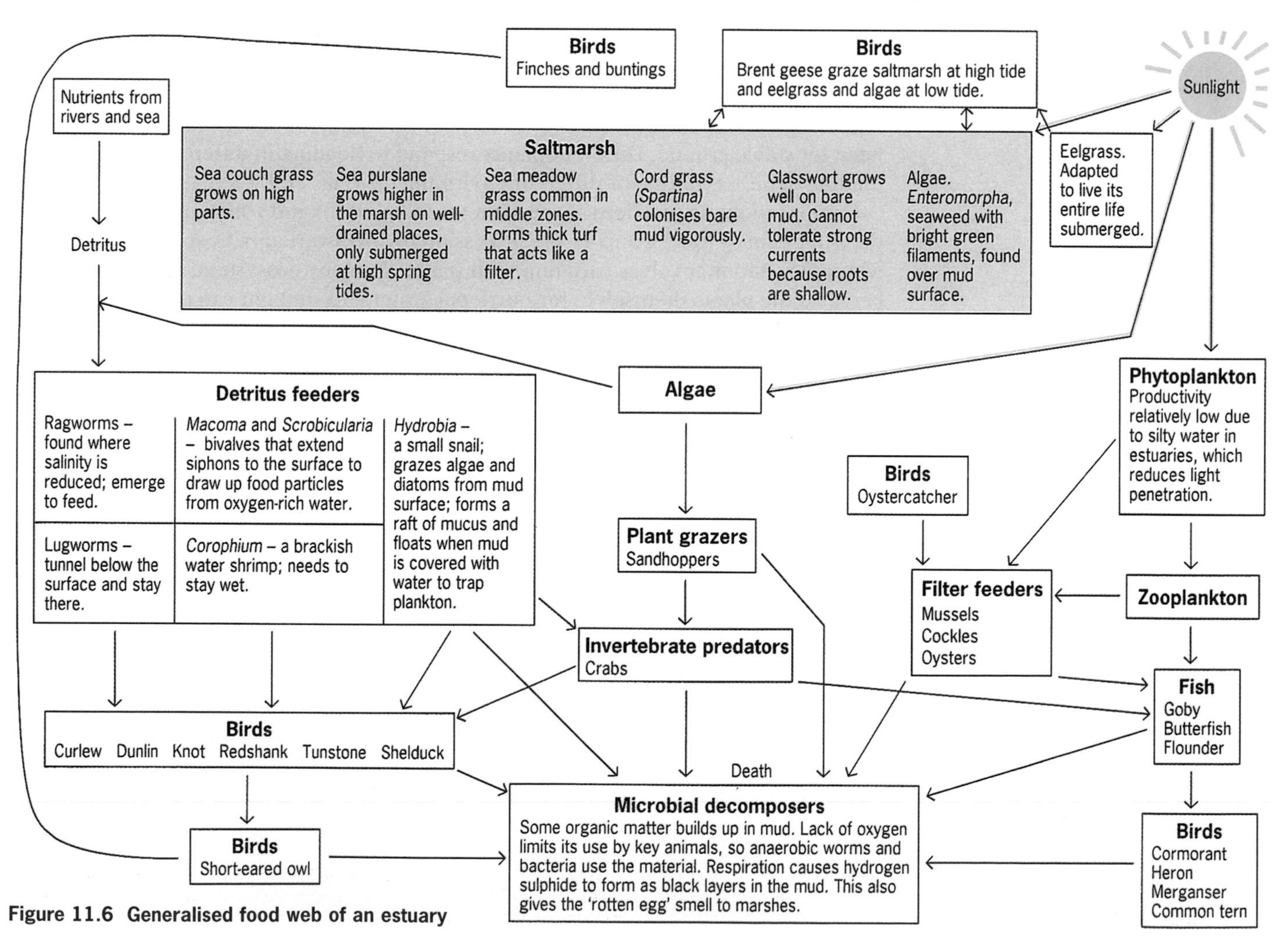

Figure 11.6 Generalised food web of an estuary

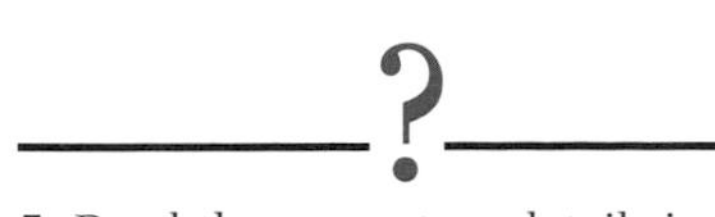

5 Read the ecosystem details in Figure 11.6.
a Identify two producers.
b Explain how the following changes might affect two **food chains**:
• increase in the number of your first producer
• decrease in the number of your second producer.

6 Using Figure 10.5, look carefully at the pattern of isotherms. Explain why you think British and other East Atlantic Flyway estuaries are so popular with migrating birds in winter.

11.3 Seasonal change

The mix of estuarine **fauna** also changes seasonally due to migratory behaviour. For instance, many fish are migratory, but some stay within the confines of the estuary for their entire life cycle, such as sea bass, common goby, thick-lipped and thin-lipped grey mullets and golden mullet. But many fish move between sea and freshwater environments. For example, sea and river lampreys and salmon move from the sea to rivers to spawn, whereas an adult eel will move from fresh water to breed in the sea.

Fish **adapt** to fresh or salt water by controlling the movement of their body fluids. Fish migrating from one environment to the other are able to osmoregulate in order to survive. Salmon increase the amount of urine they produce when they enter freshwater areas from the sea. Eels have a covering of mucus which helps to slow down the amount of water moving through their skin.

During the winter, estuaries become home to vast numbers of migratory birds. British estuaries form part of the East Atlantic Flyway (Fig. 11.7). Many of the birds use these estuaries as feeding and resting stops on their way to Africa.

Whereas predator numbers change due to migration, numbers of prey species change mostly due to reproduction and death rates. Most species grow and reproduce in the summer, producing a peak during the autumn. This is followed by a reduction in numbers during the spring and winter (Fig. 11.8). Some species migrate over short distance. For example *Macoma baltica* burrows deeper in the mud in winter, making it less available to short-billed birds.

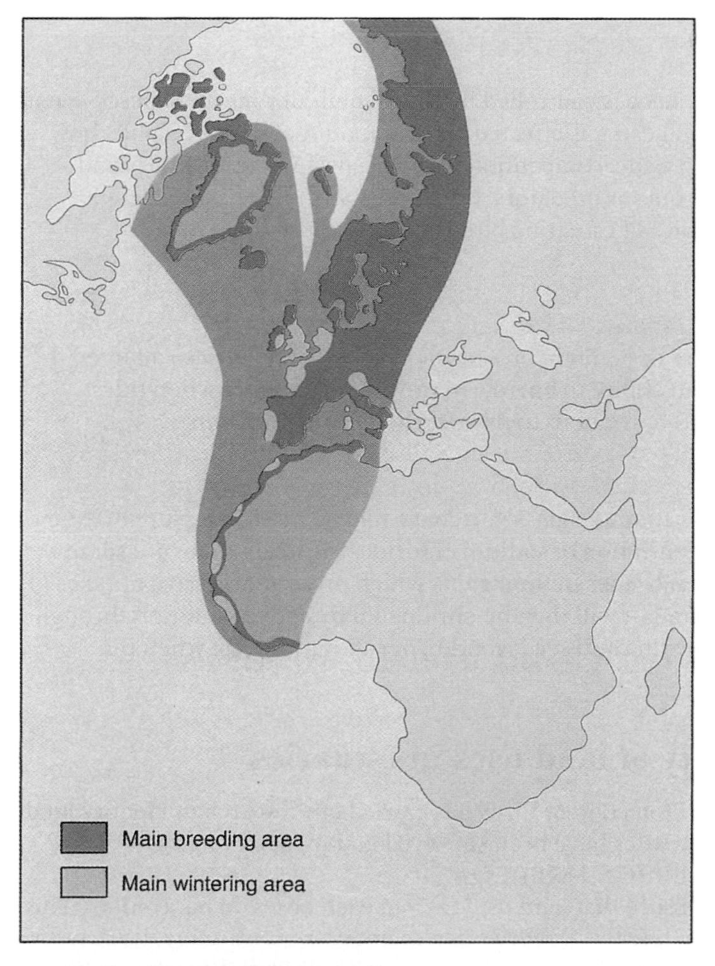

Figure 11.7 The East Atlantic Flyway (after Smit and Piersma, 1989)

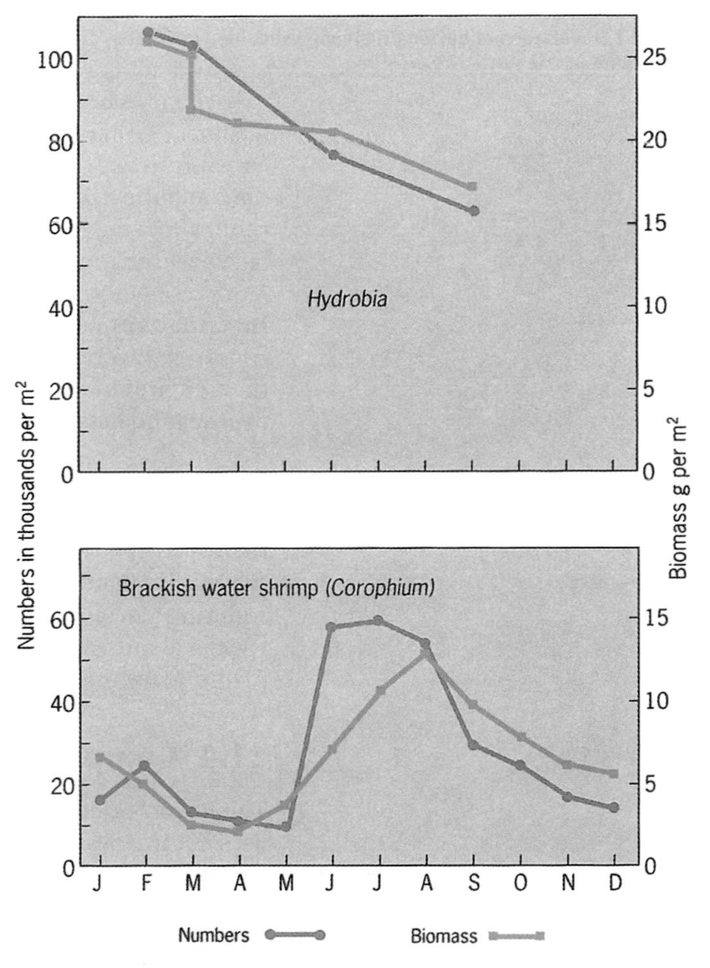

Figure 11.8 Seasonal changes in the numbers and biomass of *Hydrobia* and a brackish water shrimp, *Corophium*, in the Ythan estuary (after Milne and Dunnet, 1972)

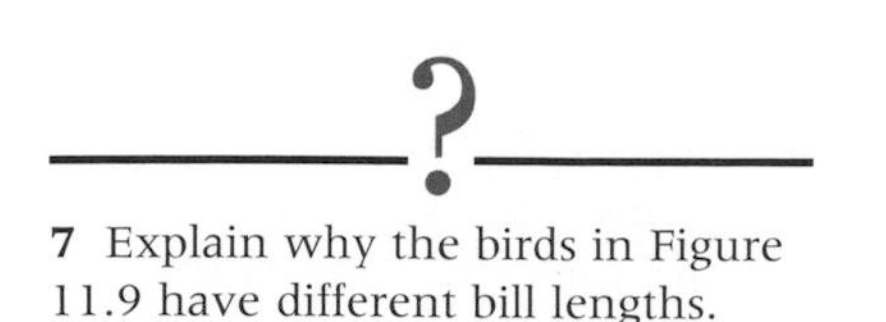

7 Explain why the birds in Figure 11.9 have different bill lengths.

Survival strategies

Birds

Birds migrate to find food or to breed in other areas. Birds are adapted to feed to ensure maximum survival. Each bird exploits a distinctive **niche** within mud and salt marsh areas. You may have noticed that birds on estuaries have different lengths of bill. The advantages of these differences are shown in Figure 11.9.

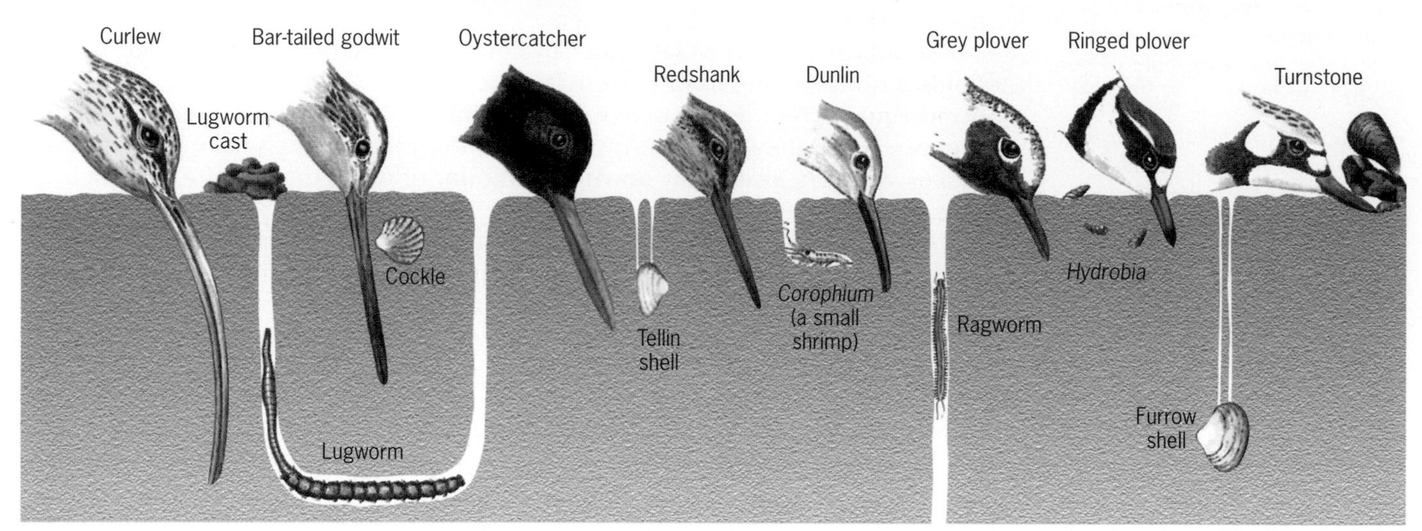

Figure 11.9 Variation of bill length among estuarine birds (after RSPB, 1988)

Each bird's feeding efficiency is controlled by the amount of time and energy spent feeding. The amount of food eaten is affected by the amount of prey available, how easy it is to find, its energy value, competition within a species, the weather conditions and the need to look out for predators. Changes in any one of these factors could affect the survival rates of estuarine birds and may cause them to move between estuarine sites.

Invertebrates

Estuarine invertebrates are dependent on sediment type. They are also adapted to their environment, with an ability to burrow to prevent drying-out when tides expose mud flats. Molluscs have shells to protect them from predators.

Plants

Most salt marsh plants, for example glasswort, have high osmotic-pressure in their tissues due to a high concentration of sodium chloride. However, there are some salt marsh plants with membranes in their roots which prevent excessive uptakes of sodium chloride. Any extra salt will then be eliminated by active secretion through special glands, as with spartina and sea lavender, or salt may be lost when the plants' leaves die.

11.4 Compatibility of land uses in estuaries

The Royal Society for the Protection of Birds has carried out research to identify land uses in estuaries. Many estuaries have been surveyed and mapped to pick out areas where land uses may conflict (Fig. 11.10).

?

8 Using Figure 11.10, classify the land uses into:
a exploitation
b recreation.

Sensible planning can ensure that land uses next to each other do not conflict. This can sometimes be difficult, because estuaries can span different planning districts, or even separate countries, as in the case of the Dee, which is partly in Wales and partly in England. Three categories of land use frequently compete with each other around estuaries: those involved with exploitation, recreation and protection.

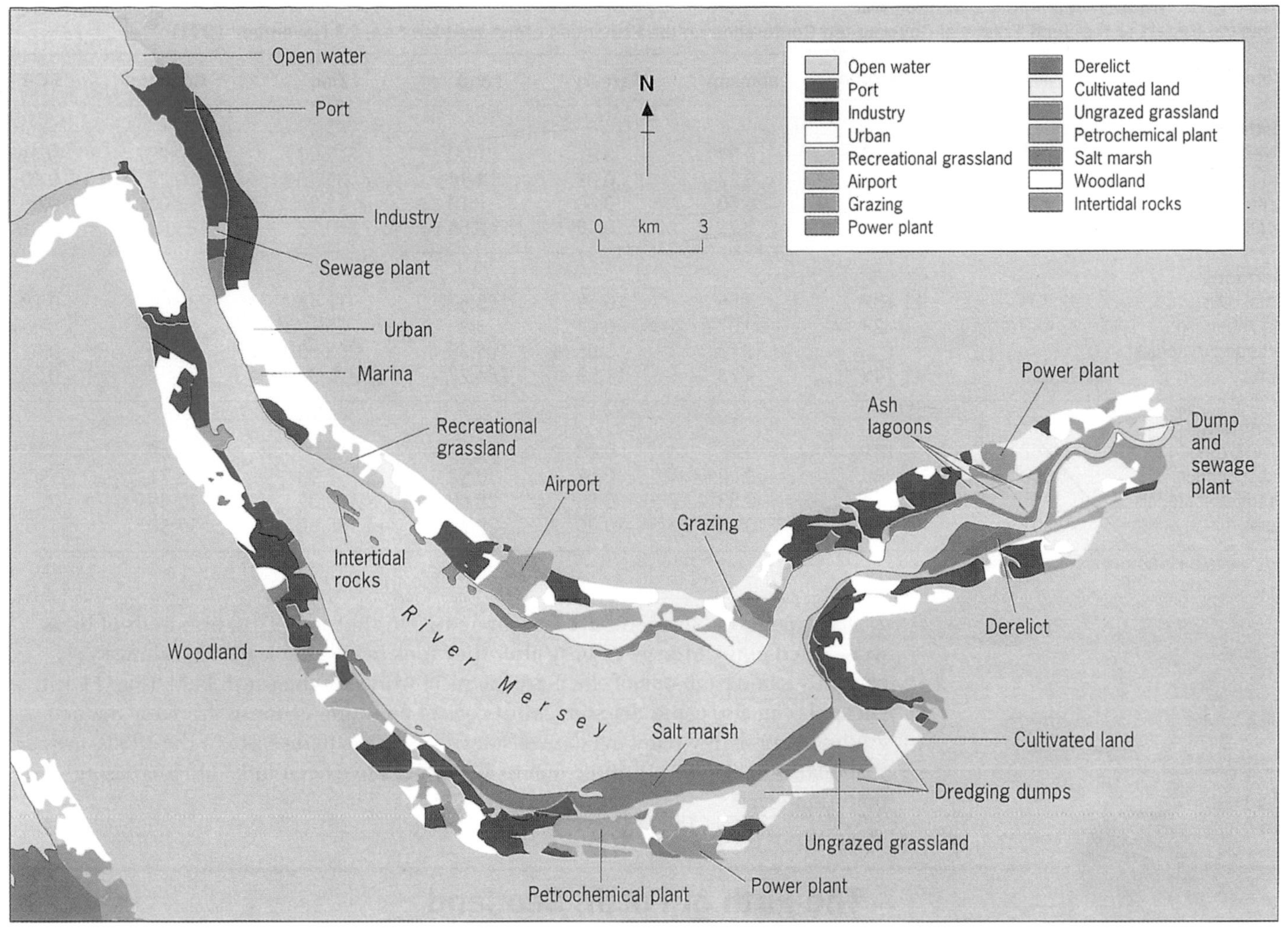

Figure 11.10 Land uses of the Mersey estuary (after RSPB, 1993)

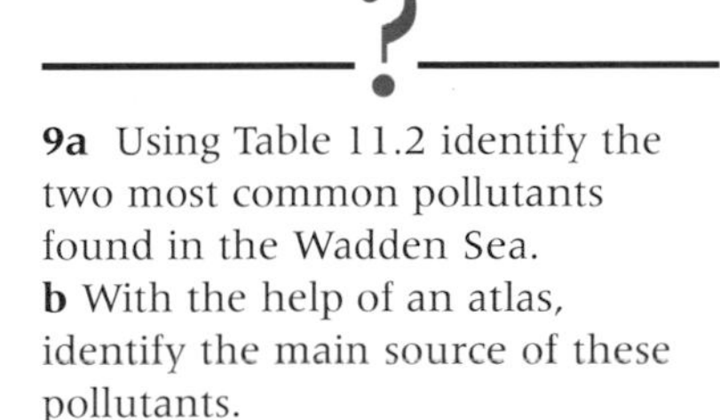

9a Using Table 11.2 identify the two most common pollutants found in the Wadden Sea.
b With the help of an atlas, identify the main source of these pollutants.
c What is the main source of the other chemicals?

10 Sewage or agricultural runoff can cause **eutrophication** (see Chapter 10), resulting in algal growth and oxygen depletion in coastal waters. Predict how this would affect the rest of the food chain. Give reasons for your prediction.

11 Using your knowledge of food chains and webs, explain why bioaccumulation of pollutants tends to reach a more toxic level in organisms at a higher **trophic level**.

11.5 Estuaries and industry

Estuaries are important sites for manufacturing industry because land is relatively cheap, flat and close to the sea for ease of transport, a plentiful water supply and a means of waste disposal.

Pollution

Pollution is a serious problem for some estuaries along the East Atlantic Flyway. For many years they have been channels for waste (Table 12.5).

Many of the mud-dwelling creatures feed on suspended matter within the water. Mussels can take in bacteria from sewage. Although the mussels are unaffected, they can transfer infections such as typhoid to humans.

Many other substances also bioaccumulate within shellfish and are then accumulated further up the food chain. These include heavy metals such as lead, mercury and cadmium and chemicals such as PCBs (Table 11.2) and organochlorides.

Organochlorides, waste products from the chlorine and pesticide industries and from paper mills, have been found to cause changes in sexual characteristics and reproductive failures in marine wildlife. Top predators are particularly affected, since the substances are virtually non-degradable in the sea and concentrate up the food chain. In Japan, at Minamata Bay, mercury entered fish through the food chain, was eaten by humans and caused illness and death.

Table 11.2 Wadden Sea inputs 1989 (tonnes)
(*Source:* Report to the Sixth Trilateral Government Conference on the Protection of the Wadden Sea, 13 November 1991)

Country	Nitrates	Phosphates	Cadmium	Mercury	Lead	Zinc	Copper	PCB
Netherlands								
Rivers/runoff	15 900	600	0.94	0.31	31.00	211.00	50.80	0.01
Ind./commun.	?	?	0.27	0.09	4.74	35.43	10.27	0.00
Dredged materials			0.00	0.00	7.10	13.80	2.00	0.00
Total	15 900	600	1.21	0.40	42.84	260.23	63.07	0.01
Germany								
Total river	192 373	11 732	8.56	10.19	175.67	1 912.48	233.72	0.18
Ind./commun.	480	23	0.072	0.045	3.3	25.1	3.15	?
Dredged materials			1.10	1.05	105.30	283.80	40.60	0.02
Total	192 853	11 745	9.73	11.29	284.27	2 221.38		277.47
0.20								
Denmark								
Rivers/runoff	6 662	181	0.08	0.03	0.80	8.00	2.40	?
Ind./commun.	2 600	285	0.02	0.01	0.92	7.90	2.71	?
Dredged materials			0.33	0.26	28.00	109.99	12.39	?
Total	9 262	466	0.43	0.30	29.72	125.89	17.50	?

Ind./commun. = industries and communities.

Figure 11.11 Advertisement promoting the Cardiff Bay development

Major pollution incidents are very dramatic. In the late 1970s, hundreds of birds were killed in the Mersey estuary after they took in organic lead. This almost certainly came from one of the petrochemical works on the south bank (Fig. 11.10). Oil spills can also cause disasters. Most coastal pollution is not so dramatic but acts gradually, probably reducing the resistance of species to disease. Grey seal deaths in northern Europe due to distemper have been linked to increasing pollution levels.

Estuary barrages

Increasingly, estuaries are also becoming important locations for service industries, because waterfront developments linked to marinas and housing developments appear attractive. In 1993, for example, a decision was made to allow a barrage to be built in Cardiff Bay and the Cardiff docks to be developed for homes and offices (Fig. 11.11). Schemes such as this have a severe impact on estuarine wildlife.

11.6 Energy from estuaries

Not only are estuarine ecosystems highly productive, but they are also a potential source of energy for people to convert to electricity. Conservationists and energy producers continue to

Table 11.3 Protective legislation for estuaries

Law	Bodies responsible	Procedure	Criticisms
Wildlife and Countryside Act 1981, SSSIs (ASSIs in NI), Sites and areas of special interest.	English Nature, Countryside Council for Wales,Scottish Natural Heritage, DOE in Northern Ireland.	Landowners must consult when they propose a potentially damaging act. Management agreements must be drawn up.	Many areas have not been designated and are threatened. Many SSSIs are damaged each year. Estuaries are the most frequently damaged. The law does not stop existing common rights (e.g. cockle digging) planning controls or private Acts of Parliament. Private Bills and schemes which receive planning permission have priority.

Fig 11.12 La Rance barrage, Brittany, France

Table 11.4 International and European legislation for estuaries

Law	Bodies	Procedure	Criticism
Ramsar Convention for the Conservation of Wetlands 1976	Governments throughout the world have adopted it.	In the UK, an area must be an ASSI or a SSSI before it becomes a Ramsar site. A site must contain over 20 000 waterfowl or support 1% of the individuals of one species.	Many areas have not been designated yet.
Special Protection Areas, EU directive on the Conservation of Wild Birds	European Union governments have to select site.	A site in the UK must be an ASSI or SSSI before it is an SPA. Sites contain rare or vulnerable species and/or regularly occurring migratory species. These sites must be protected from damaging developments.	
Special Areas of Conservation (SACs)		European Union governments have to select site. These are sites for plants and animals apart from birds.	

debate the pros and cons of tidal power. The first tidal power scheme has been operational at La Rance in Brittany (Fig. 11.12) since the 1970s, but unfortunately little baseline data was recorded to assess its impact on the environment.

Away from the shore, water generally rises and falls by half a metre. Nearer to the shore a greater tidal range occurs, since the shape of the land and variation in water depth can alter the water's flow.

A barrage can trap energy from the tides by holding back a reservoir of water. Water then passes outwards on ebbing tides and inwards on flooding tides, through turbines in the barrage to generate electrical energy.

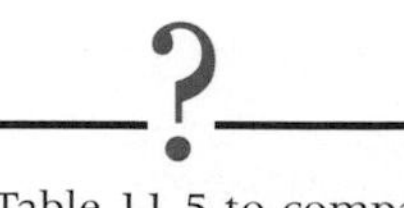

12a Use Table 11.5 to compare the tidal ranges for estuaries which have been considered for tidal power developments.
b Which has the greatest range?

13 Use Spearman's rank test (see Appendix 3) to find out if there is any significant correlation between barrage length and cost per kilowatt hour.

14 From Table 11.5 it is possible to investigate the relationships between other variables.
a Repeat the Spearman test for:
• tidal range and cost per kilowatt hour
• tidal range and annual energy output
• annual energy output and cost of energy per kilowatt hour.
b In your opinion which barrages are most likely to be built?
c What factors apart from cost may be involved in the decision-making?

Table 11.5 Potential tidal power developments on British estuaries (*Source:* ETSU for DTI, 1990)

Estuary	Barrage length (m)	Prospective capacity (MW)	Annual energy output (GWh)	Tidal range (metres)	Approximate cost of energy (p/kWh)
Mersey	1750	620	1320	8	3.6
Severn	16000	8000	17000	10	3.7
Conwy	225	30	57	8	4.0
Loughor	220	9	17	9	4.3
Morecambe Bay	16600	3040	5400	8	4.6
Solway Firth	30000	5580	10050	8	4.9
Dee	9500	800	1250	8	6.4
Humber	8300	1200	2010	6	7.0
Wash	19600	2760	4690	6	7.2
Thames	9000	1 120	1370	6	8.3
Langstone	550	24	53	4	5.3
Harbour Padstow	550	28	55	7	4.2

MW = megawatts, GWh gigawatt hours, kWh kilowatt hours.

11.7 Coastal zone planning

In 1991 the UK government set up a House of Commons Select Committee to investigate the idea of coastal zone planning (CZP). This followed the good practice of several countries who have attempted to safeguard the interests of different estuary users. CZP has been used in Australia and the USA for many years. It involves the following procedures:

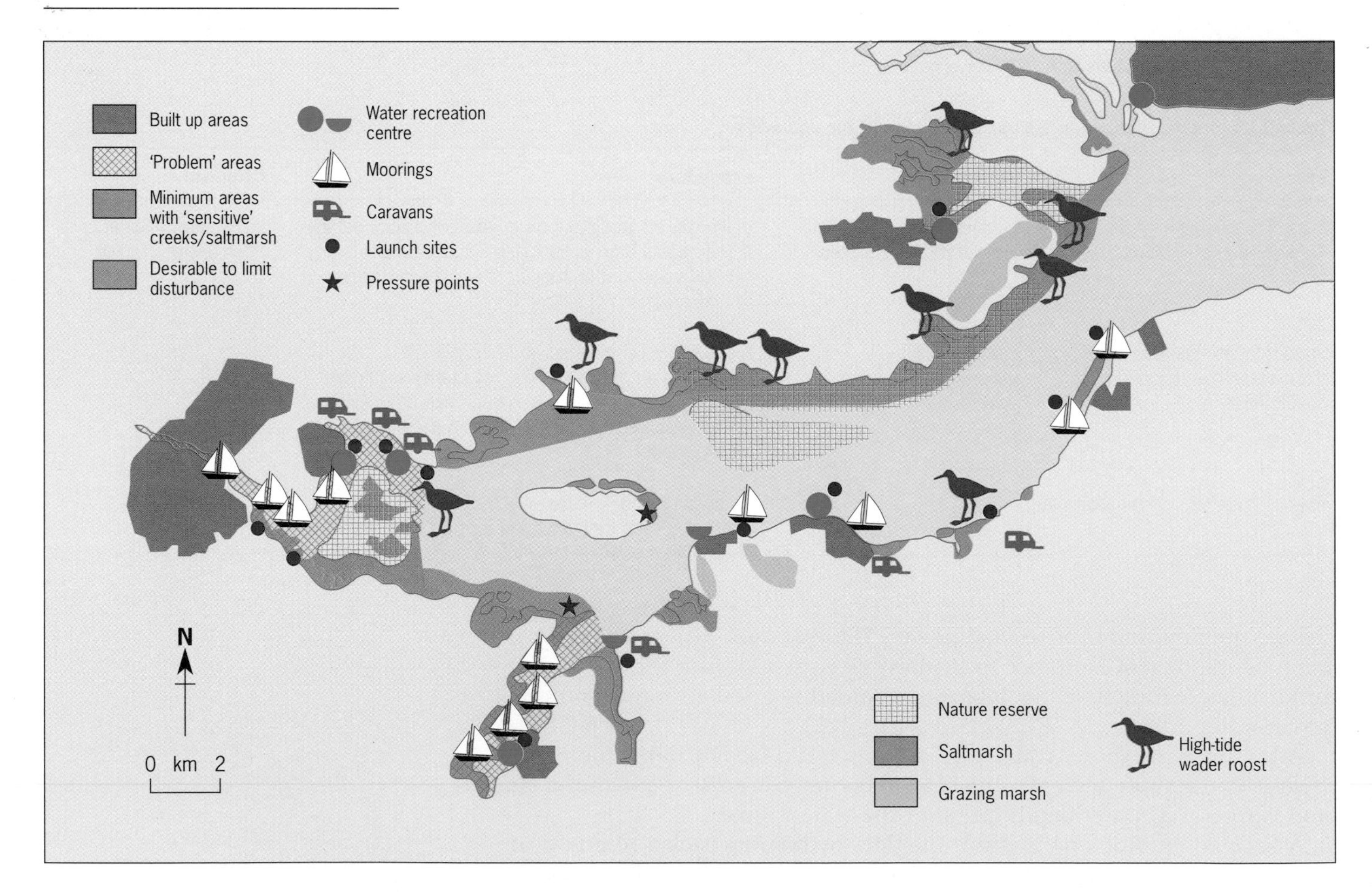

Figure 11.13 Example of a plan for the coastal zoning of an estuary (RSPB, 1993)

1 Collect data about estuarine use.
2 Produce maps to show distribution of the various uses.
3 Identify possible conflicts of use.
4 Designate zones for each use and draw up rules to limit the conflicts.

Coastal Zone Planning is a way of encouraging estuaries to be managed in a **sustainable** way. Conservationists suggest that a precautionary principle should be applied to the use of resources. This means that nothing should be undertaken if there is any doubt at all about its far-reaching effects on wildlife and the ecosystem.

?

15 A coastal zone plan has been devised for the Blackwater estuary (Fig. 11.13). Make a list of rules which may limit the conflicts:
- in the problem areas
- in the areas where it is desirable to limit disturbance.

16 Using Tables 11.3 and 11.4, explain why giving an estuary some protection is not a guarantee that it will be safe.

17 Essay: Discuss the pressures on estuarine ecosystems and suggest ways of reducing these pressures.

Table 11.6 Severn barrage: costs and benefits. These were calculated when the Severn Barrage was planned in 1990.

Costs	Monetary calculation possible?
There may be more flooding as riverbanks are weakened because of longer contact with water.	Yes
Formation of the lake and disturbance from sport may affect the energy of birds.	No
Work is needed to reduce physical damage to migratory fish in turbines. Fish will take years to get used to routes through the barrage.	Yes
Land vegetation will advance towards the sea, reducing the production of seeds and providing less food for birds.	No
SSSI may be affected by problems caused by changes in groundwater levels.	No
Salmon stocks may suffer as the adjoining river is affected by water changes.	Yes
Septic tanks in low-lying areas may need to be improved. Gravitational sewage systems could cover streets with sewage.	Yes
Inter-tidal feeding areas for birds will be lost until new inter-tidal areas can be developed.	No
Salinity of tributaries will be reduced and may require artificial inputs.	Yes
Sediment build-up in the estuary will make dredging necessary.	Yes
Toxic algae may grow in the altered conditions.	No
More stable water may prevent flushing of the estuary, leading to greater contamination of water.	No
Port of Bristol may have to pay compensation to companies whose investment was based on easier navigation conditions.	Yes
Wildlife diversity could be reduced.	No
Annual working expenses could rise.	Yes
Capital costs could rise.	Yes

Benefits	Monetary calculation possible?
The estuarine ecosystem could become more productive, leading to greater biomass and increased fishing catches.	Yes
New road link will improve access.	Yes
Jobs in construction and in tourism will be created, leading to a multiplier effect.	Yes
Barrage will enhance the landscape of the region.	No
Land and property prices could increase.	Yes
Additional generating capacity should save 8 million tonnes of coal a year.	Yes
Difficult tidal currents will be reduced, creating improved opportunities for water sports and sea angling.	Yes
Altered ecosystems may attract new wildlife.	No
There may be increases in shellfish and fish farming.	Yes
Barrage will be operated to reduce high storm surge tides and the risk of flooding.	Yes

Summary

- Nutrient inputs from both the sea and rivers ensure that the estuary is one of the most productive ecosystems.
- The estuarine ecosystem is in a constant rhythmic state of change. Tides and weather cause daily and seasonal changes.
- Estuary plants and animals form distinct zones from the mudflats to the saltmarsh, responding to variations in water cover, salinity and stability of the ground.
- It is difficult to distinguish between zonation and succession in an estuarine ecosystem because of the complex interaction between time and space.
- The estuary is an important feeding and resting area for migrating and wintering birds.
- In addition to conservation, the estuary is also an important place for manufacturing industry, port and storage facilities, agriculture, transport, recreation and energy production.
- Pollution from agriculture and manufacturing industry causes eutrophication and bioaccumulation of toxins in animals.
- Coastal zone planning is one way in which estuaries could be managed sustainably.

12 Marine resource management

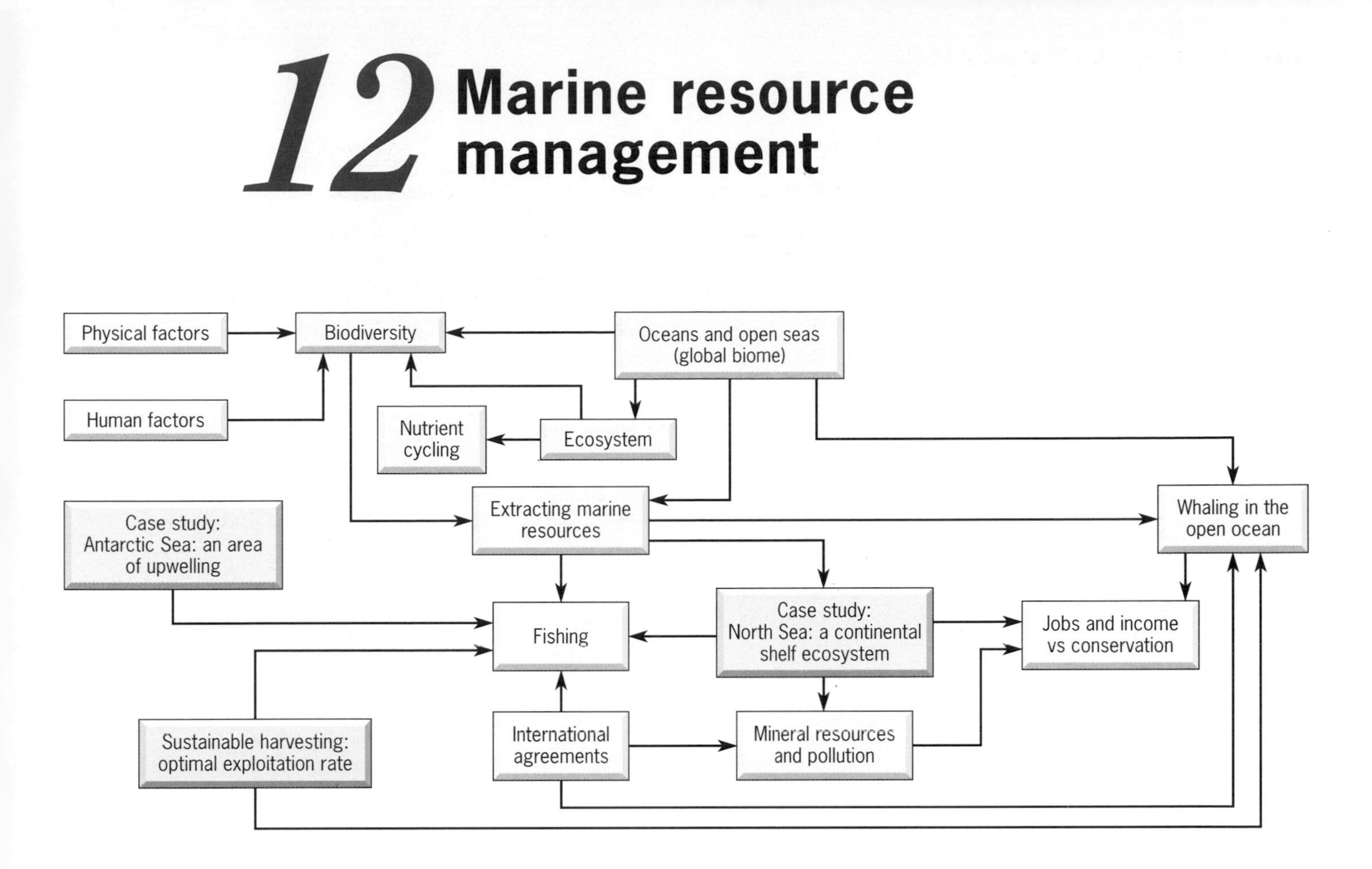

12.1 Biodiversity in the ocean and human factors

The conservation of **biodiversity**, defined as safeguarding the complex biological communities and **species** of the world, was a major issue at the 1992 Earth Summit in Rio de Janeiro. Conserving biodiversity in the seas has serious and complex problems, since information and monitoring is less easy than on land. The marine environment is more stable than the land environment. Stability increases with depth and is characterised by high species diversity. The more stable a system is, then the more fragile it is (Fig. 12.1). The effects of disturbance are more dramatic on stable environments. Species diversity tends to decrease as disturbance increases, but at the same time opportunistic species (which reproduce rapidly) invade and **compete**. For example, in the Mediterranean Sea and the Atlantic Ocean, herring gulls and blackheaded gulls have thrived on food and nest sites provided by humans and are competing with populations of terns, Auidouins's gull and the slender-billed gull.

Figure 12.1 Antarctica: a stable but fragile ecosystem

Table 12.1 Common sources of pressure on marine ecosystems

Pressure	Substance or activity involved	Major sources	Potential effects
Waste input	Nitrate and phosphate nutrients	Sewage from people	Eutrophication, agriculture, industry
	Pathogens	Sewage, agriculture	Disease and infection, shellfish contamination
	Oil	Industry, sewage, shipping, vehicles, urban runoff	Oiling of birds and animals, seafood tainting, beach contamination
	Pesticides and herbicides	Industry, sewage, agriculture, forestry	Metabolic dysfunction
	Radioactive wastes	Nuclear fuel reprocessing	Metabolic dysfunction
	Heavy metals	Industry, sewage, ocean dumping, vehicles	Metabolic dysfunction
	Plastics and debris	Litter, shipping wastes, lost fishing gear	Entanglement of wildlife, digestive interference
	Solid waste (organic and inorganic)	Sewage, ocean dumping, industry	Reduced oxygen habitat, smothering
Environment restructure	Coastal development	Dredging, industrial, residential and tourism development	Aesthetic and habitat loss, coastal erosion
Resource exploitation	Fish and shellfish harvesting, petroleum development	Harvesting activities, drilling accidents, oil seepage	Stock depletion, ecosystem changes, oil and chemical contamination, mechanical disturbance of the sea bed during construction of bore holes, extraction installations and accommodation platforms. Reception areas on coast can result in habitat destruction. Incineration of migrant birds by gas flares
	Mineral development	Extraction of sand and gravel	Destruction of fish spawning areas. Decreased water quality, coastal erosion, changes in sea bed
Atmospheric change	Carbon dioxide, and other greenhouse gases	Energy production, transportation, agriculture, industry	Sea-level rise, coastal flooding
Nitrogen oxides	Motor vehicles	Eutrophication	

Over-exploitation of food resources from the sea can cause reductions in numbers of both the food resource and creatures higher up the **food chain.** Often these numbers can be restored when harvesting is reduced. However, sometimes competition between species can prevent this from happening. In the 1930s and 1940s on the US Pacific coast, sardines were harvested. By 1955 stocks of sardine were so depleted that little money could be made, so fishing stopped. It was thought that stocks would increase, but by 1970 anchovy had filled the **niche** previously occupied by the sardine.

Table 12.1 identifies some of the most common sources of pressure within marine **ecosystems**. Most of these problems are concentrated in continental shelf areas, since these are closest to human settlements (Fig. 12.2). These

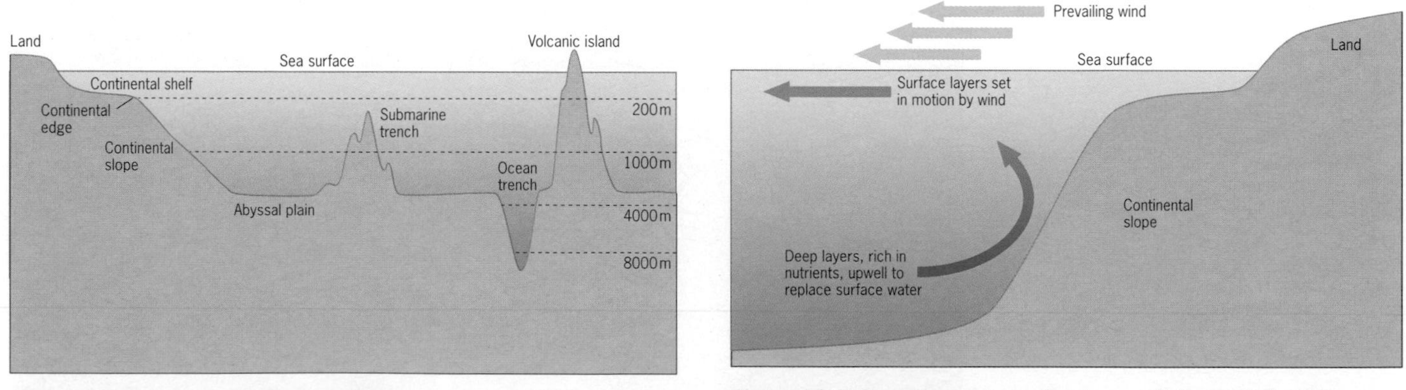

Figure 12.2 Cross-section of oceanic features (after Tait, 1972)

Figure 12.3 Upwelling occurs due to an offshore wind at the surface (after Tait, 1972)

areas also happen to be the most likely locations for oil and gas and sand and gravel resources. In addition, the relatively high productivity rates of continental shelves encourage the use of marine resources by both humans and the marine food chain.

12.2 The nature of the ecosystem

Sea water covers approximately 71 per cent of the earth's surface. In the deepest parts it is 10 000 metres but averages 3700 metres (Fig. 12.2).

Nutrient cycling

The concentration of carbon dioxide is fifty times greater in sea water than in the atmosphere, because the gas is soluble in water. The limiting factors for **photosynthesis** are therefore nutrient availability (phosphates, nitrates and trace elements) and light. As water absorbs light, primary production can only occur in the upper layers of the sea.

The overall productivity of the ocean is determined by the availability of nutrients from many sources, such as runoff from the land, decaying sea life and the atmosphere. Unlike in land systems, there are no roots to take nutrients upwards. The oceans and seas depend on currents to stir up and circulate the nutrients. Some bottom living creatures also need the water to move nutrients in and waste products out. There are two main ways in which this movement can take place: upwelling and turbulence.

?

1 From Figure 12.4 identify the most productive areas from the categories of oceans and seas, upwelling zones and continental shelves.

Upwelling

Off-shore winds set surface water in motion and cause water from deeper levels to reach the surface (Fig. 12.3).

Turbulence

This can be caused by convection where surface water cools, its density increases, it sinks and warmer water from the bottom replaces it, thereby circulating nutrients. Turbulence can also be caused by water moving over an uneven bottom such as on a continental shelf or by tidal movements. Sometimes production can be limited if turbulence occurs, since mixing can move **phytoplankton** (primary **producers**) to a level where they can no longer obtain light.

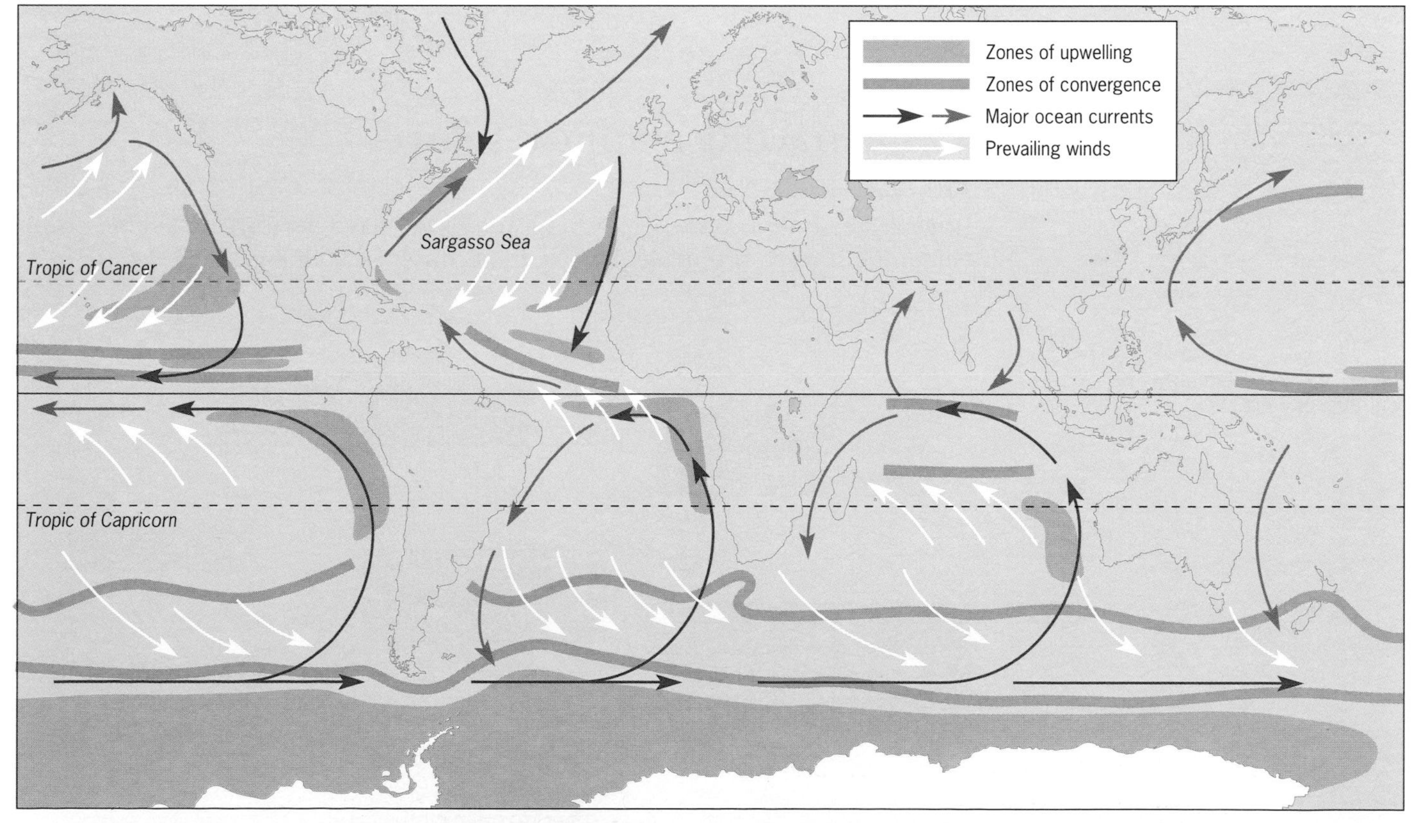

Figure 12.4 Prevailing winds and major surface currents of the oceans, showing areas of upwelling and convergence (after Tait, 1972)

The Sargasso Sea (Fig. 12.4) is a semi-tropical marine desert area because it has a slight horizontal mixing of its waters. However, its total productivity is more than some seas in temperate areas where there is a great deal of nutrient mixing in winter.

Although the Mediterranean Sea (Fig.12.4) is at a lower latitude than some temperate seas, the Mediterranean is also an area of relatively low fertility because a bottom current flows out at the Strait of Gibraltar and the inflow from the Atlantic is of surface water, which is poor in nutrients.

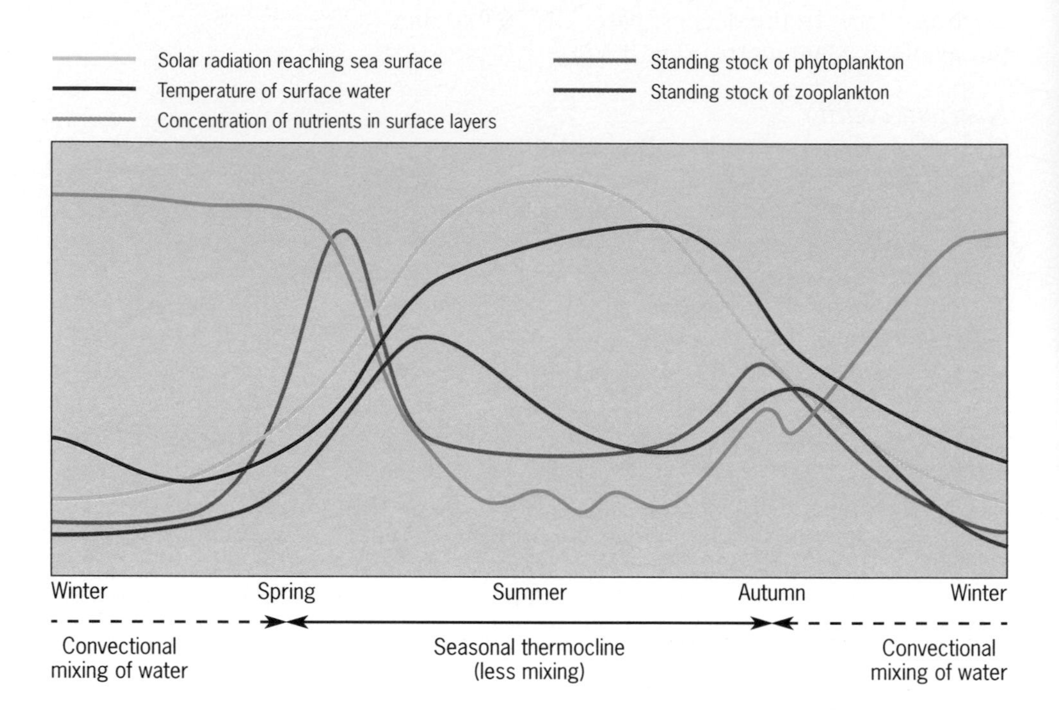

Figure 12.5 Generalised seasonal changes in temperature, nutrients, phytoplankton and zooplankton in the surface of temperate oceans (after Tait, 1972)

12.3 Extracting marine resources

Fishing

Huge increases in commercial fishing since 1945 have been partly led by demand but were also made possible by technological changes (Fig. 12.6).

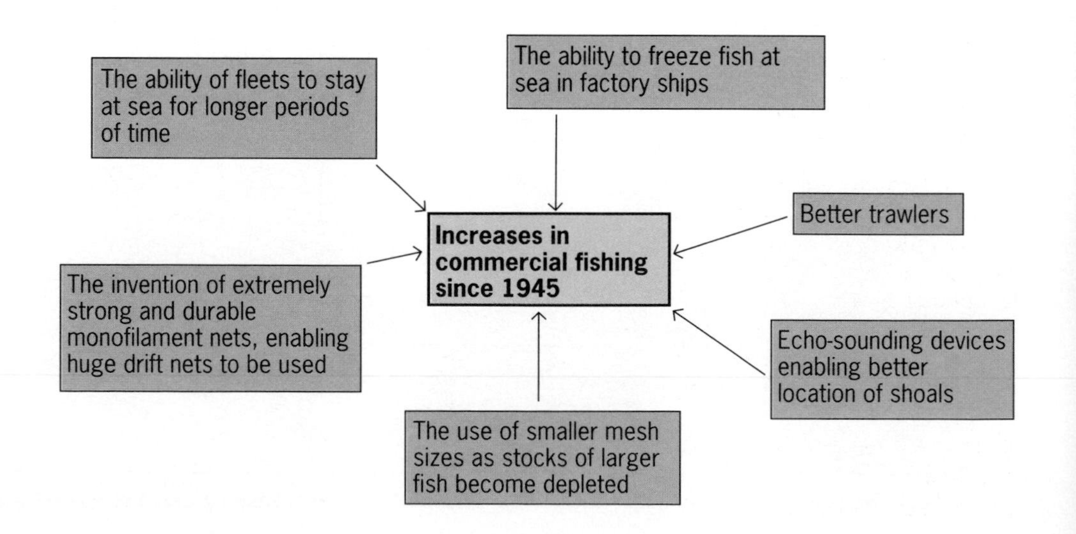

Figure 12.6 Technology and commercial fishing

Figure 12.7 A generalised food web (after Tait, 1972)

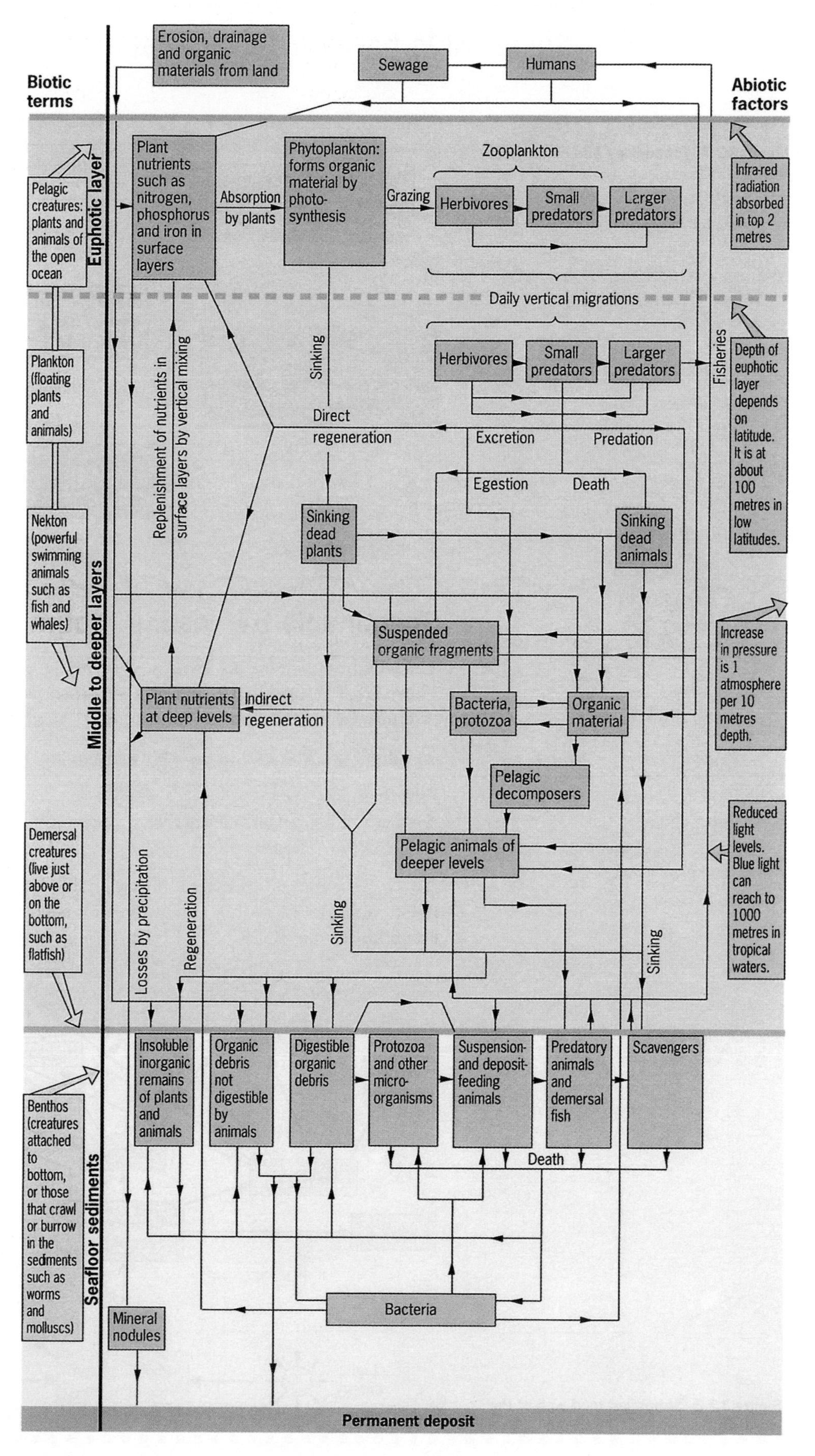

?

2a Using Figure 12.4, explain why Arctic seas are less productive than Antarctic seas.
b Using Figure 12.7, explain why a continental shelf (Fig. 12.2) is more productive than the open ocean.

3a Using Figure 12.5, describe what happens to nutrients in spring, summer, autumn and winter.
b Explain why this is so.
c When is the productivity of phytoplankton at its highest? Explain why this is so.
d When is the peak production of **zooplankton** (a primary **consumer**)? Explain why this is so.

4a From Figure 12.7, describe which living organisms are found at the surface of the ocean/sea ecosystem. Explain their distribution.
b On what might **herbivores** in the deeper levels feed? What other types of creatures are found in this zone?
c On what do creatures feed on the sea floor?

Sustainable harvesting: optimal exploitation rate

The weight of fish stocks increases as the fish grow and young fish join the stock. There is no natural increase, however, since fishing and natural death control population numbers. **Sustainable** harvesting is calculated as follows:

Weight of stock at end of year = weight of stock at beginning of year + $A + G - C + M$

Where:
A = annual recruits of young fish C = fishing loss
G = growth of existing stock M = natural death

$A + G - M$ = the natural yield. If sustainable harvesting is maintained, then $C + M$ must equal $A + G$, so that an equilibrium is kept. This is known as the optimal exploitation rate (OER).

Table 12.2 Level of exploitation of fish species in England and Wales (1= lightly exploited, 2 = under-exploited, 3 = around optimal exploitation, 4 = heavily exploited)

Species	Overall level of exploitation	Percentage of each catch species due to gill nets (1985–9 mean)
Salmon	3	70
Sea trout	2	70
Bass	3	40
Grey mullet	3	70
Sole	3	10
Plaice	3	2
Rounder	2	19
Cod	4	9
Whiting	3	1
Pout	3	1
Skates and rays	4	10
Spurdog	3	9
Herring	3	19
Sprat	3	1
Mackerel	3	2
Pollack	2	77
Saithe	3	9
Ling	2	53
Hake	4	48
Conger	3	14
Dab	2	2
Monkfish	4	13
Turbot	4	24
Brill	3	4

Figure 12.8 Simplified food web of the North Sea

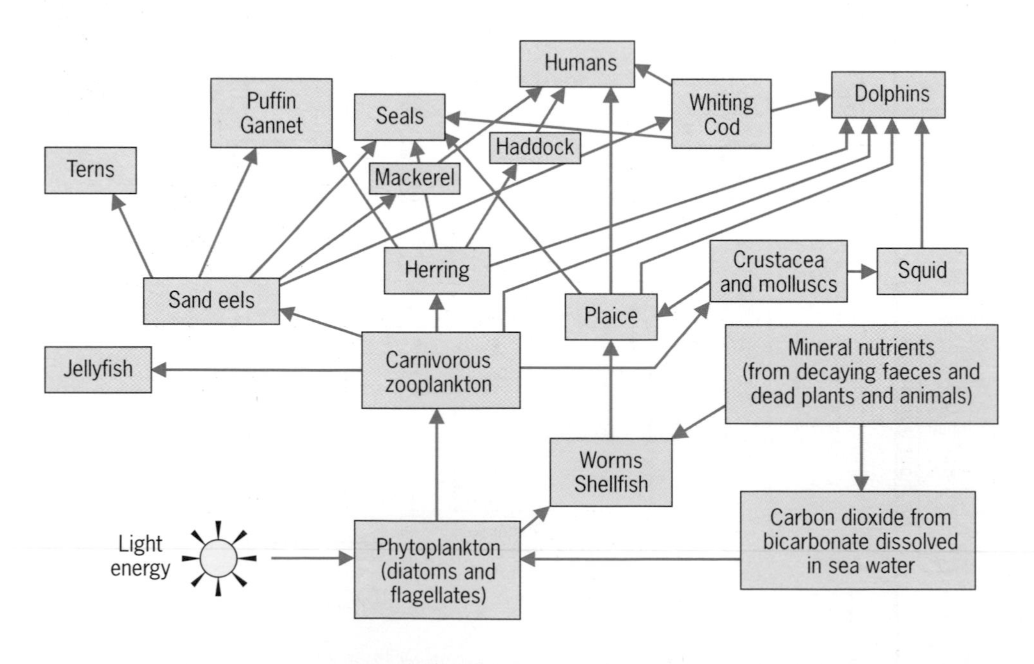

?

5 Study Table 12.2.
a Suggest reasons which you think may explain the different levels of exploitation.
b Is there a connection between gill netting and level of exploitation?

6 From Table 12.2, calculate the number of species fished at:
a below the OER
b at the OER
c above the OER.

7 Use the formula for calculating OER to suggest what needs to happen to stocks of monkfish (Table 12.2).

8 How would unsustainable harvesting of mackerel, herring, cod and haddock affect the North Sea food web (Fig. 12.8)?

The North Sea: a continental shelf ecosystem

The geology and productivity of the North Sea make it a very important resource area for all the countries which border it. Its **management** therefore requires international co-operation.

Fisheries

Addressing the International Fishery Exhibition in London in 1883, T.H. Huxley said, 'I believe that it may be affirmed with confidence that, in relation to our present modes of fishing, a number of the most important fisheries such as the cod fishery, the herring fishery and the mackerel fishery are inexhaustible'.

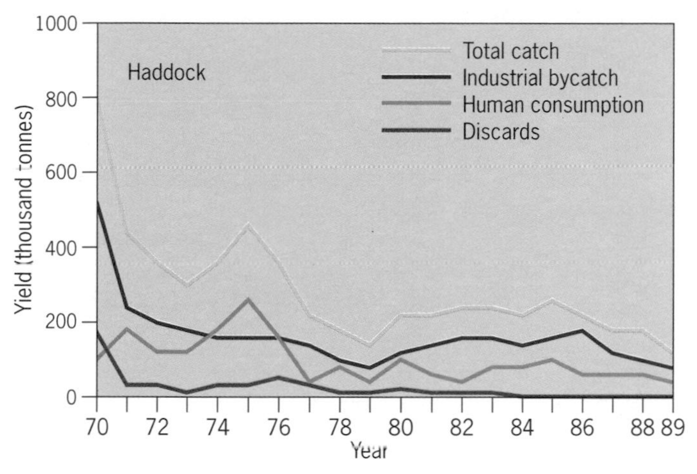

Figure 12.9 Yield of haddock from the North Sea, 1948–89 (after ICES, 1989)

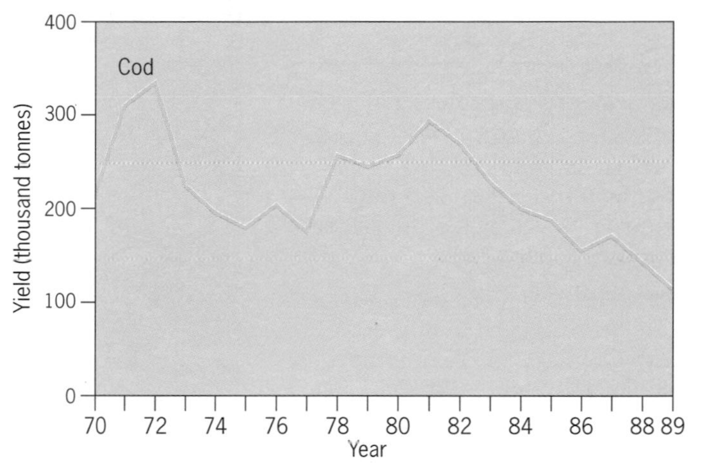

Figure 12.10 Yield of cod from the North Sea, 1970–88 (after ICES, 1989)

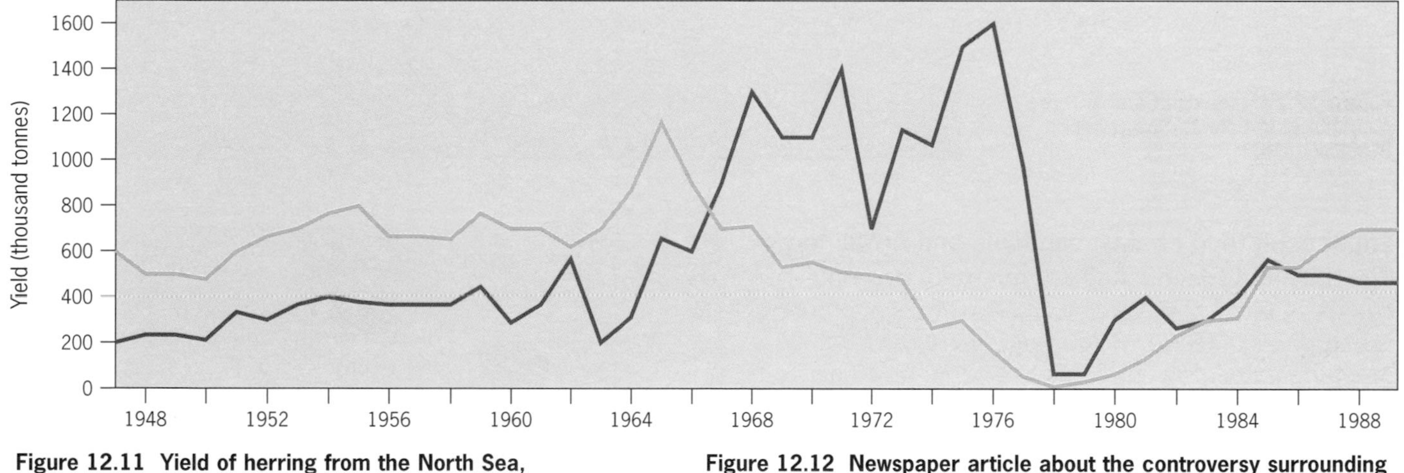

Figure 12.11 Yield of herring from the North Sea, 1970–88 (after ICES, 1989)

Figure 12.12 Newspaper article about the controversy surrounding EU fishing regulations (*Source: Guardian*, 10 December 1992)

The Government pledged to oppose proposals by the European Commission which would keep British cod and haddock boats tied up in the harbour for 10 days each month. Sir Hector Munroes said that they would be doing their 'very best' at talks to decide fish quotas for next year. 'I know they (the fishermen) feel that they have had a rough ride this year. But the issue is not as straightforward as some people might like to think.' He said that stocks of haddock were doing well, but the general picture was 'very worrying'.

For the Liberal Democrats, Jim Wallace said, 'What has angered fishermen more than anything is that we have these restrictions on British fishermen, when their European counterparts are not subjected to similar restrictions.' This was echoed by Robert Hughes (Lab. Aberdeen North), who said that there were suspicions that the policing of quotas for other countries was not as well done as in Britain.

Table 12.3 Employment changes in the UK

Year	Number of people employed in UK fisheries
1967	18 675
1977	16 337
1986	15 962

?

9a Using Figures 12.9–12.11, assess the change in yields of cod, herring and mackerel.
b How is this reflected in the number of jobs within the fishing industry (Table 12.3)?

10 Refer to Figure 12.8. What products extracted from the sea may also have an effect on the decline in yields of cod, mackerel and herring?

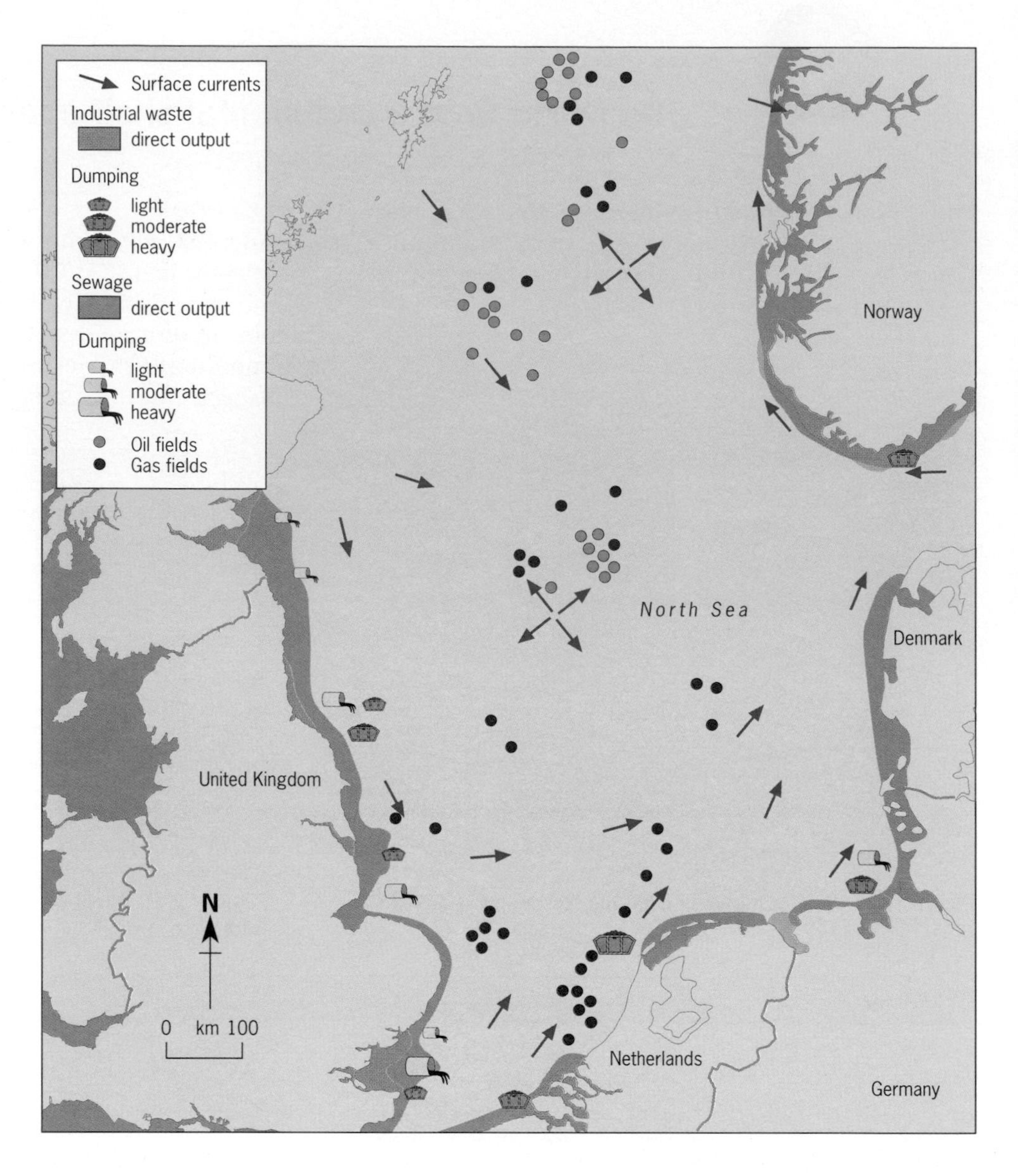

Figure 12.13 Industrial and sewage pollution in the North Sea (after UNESCO, 1985)

Impacts on food chains: sandeels and Arctic terns

The Shetland Islands hold an important breeding colony of birds: the Arctic tern. The bird's main prey is fish: the sandeel. In 1980 there were over 30 000 breeding pairs of Arctic terns in Shetland (40 per cent of the British population). Between 1983 and 1990, no young birds fledged.

The Shetland sandeel fishery started in 1974. It trawls for sandeels in inshore waters; the catch is frozen and exported for food for mink farms, for fishmeal, and oil and food for salmon farms.

A voluntary ban on sandeel fisheries has recently taken place and this has resulted in an increase in the number of terns raised in 1991. A voluntary ban obviously does not have the same force as EU law.

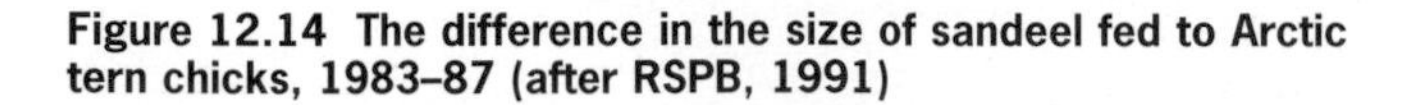

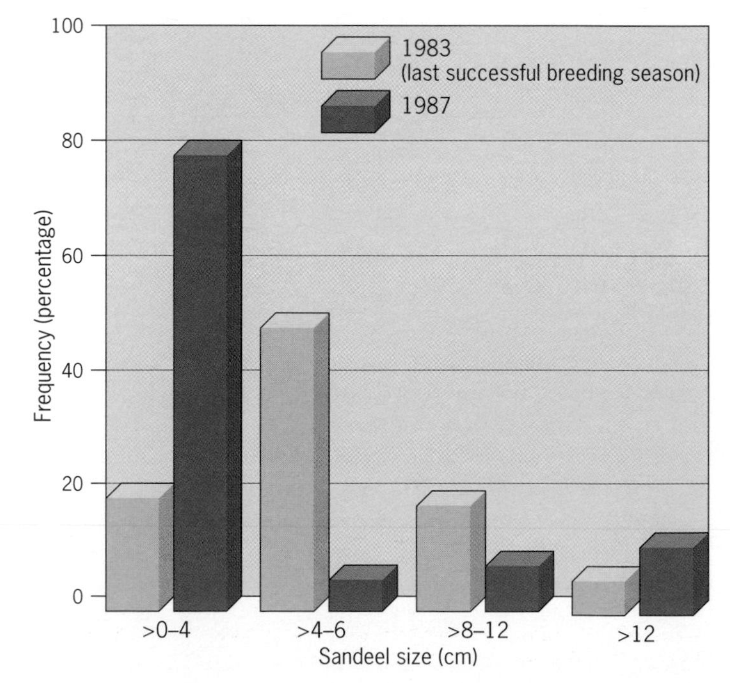

Figure 12.14 The difference in the size of sandeel fed to Arctic tern chicks, 1983–87 (after RSPB, 1991)

11a Figure 12.14 shows the amounts of food brought to the chicks in 1983 and 1987. What happened to the structure of the sandeel population between 1983 and 1989?
b Why did this happen?
c What effect would you predict this would have on the breeding capability of sandeels?
d What might be the impact on the population of Arctic terns?

12 Read Figure 12.12. What methods of management does European Common Fisheries Law use to control fishing in the North Sea?

13 Using Figure 12.13 and Figures 12.15-12.16, make a list of the possible environmental impacts of activities in the North Sea.

14a How many of the laws in Table 12.5 follow the guidelines in Table 12.6?
b For each guideline in Table 12.6, suggest an activity in the North Sea that conflicts with it.

Table 12.4 EU Common Fisheries Policy, first introduced in 1983, also signed by Norway

Main aims
- To maintain the level of fish stocks by preventing overfishing.
- To guarantee stable supplies at reasonable prices and thus to provide an adequate standard of living for the fishing community.

Mechanisms
- The total allowable catch is shared and each member state is given a quota for each species.
- Restricted and prohibited zones are established, each country having exclusive access to the waters within twelve miles of its coasts.
- Standards for fishing and gear are set, for example a minimum mesh size for nets and minimum landing size for fish, so that young stock are not caught.
- Restrictions are placed on effort, boats being required to remain tied up in harbour for certain lengths of time.
- Decommissioning schemes.

Table 12.5 Agreements and laws that apply to the North Sea

Global
- United Nations Conference on the Law of the Sea, 1982 (UNCLOS 1 11)
- International Maritime Organisation (IMO) (The North Sea is a Special Area)
- International Council for the Exploration of the Seas (ICES)
- London Dumping Convention: controls dumping and incineration of hazardous wastes at sea. Incineration to cease completely by 1994.

Regional
- Bonn Convention to combat oil pollution, 1983
- Oslo Convention signed in 1972 by 13 countries. Designed to prevent the dumping of pollutants by ships and aircraft.
- Paris Convention signed in 1974 by most of the same countries. Deals with sources of pollution on land which affect the seas.
- Convention for the Protection of the Marine Environment of the North East Atlantic, 1992. Aimed to 'safeguard human health and conserve where practicable restoration of the marine ecosystem'. It promotes the 'precautionary principle' and that of 'the polluter pays'.'

UK regulations
- Wildlife and Countryside Act 1981. Non-statutory Marine Consultation areas have been set up but no Marine Nature Reserves exist in North Sea waters.
- Sea Fisheries (Conservation) Bill aims at controlling access to fish stocks under pressure.
- Sea Fisheries (Wildlife Conservation) Act 1992 requires that fisheries have regard to the conservation of marine flora and fauna.

Table 12.6 Guidelines for conservation of the sea

1 Survey vulnerable species and set up protected areas to ensure their survival.
2 Preserve natural habitats by establishing marine parks with rules according to local conditions.
3 Prevent and control pollution along heavily developed and industrialised coasts, including diverting potentially harmful cargoes away from sensitive marine areas.
4 Restore habitats which are deteriorating and re-create sites of ecological interest.

Mineral resources and waste inputs

The North Sea is rich in mineral resources which provide income and employment for the countries bordering it (Table 12.7), but these countries also use the sea as a waste dump.

Figure 12.15 Salmon farm. Fish-farming is an important industry on the Shetland islands and particularly dependent on the high quality of Shetland's environment. All salmon harvested within an exclusion zone which surrounds the islands were destroyed in 1993. This prompted the Scottish Office to impose a fishing ban in areas to the south of Shetland.

Table 12.7 The UK's North Sea oil and gas income and employment, 1990

Income from oil and gas sales	£11 billion
Investment into other industries	£3.5 billion
Cash contributions to Scottish economy	£23.5 billion
Off-shore workers employed	36 500
On-shore support and service workers	73 000
Petrochemical and industry workers	652 000

(WL)

Figure 12.16 The Braer wreck. The Liberian-registered tanker, the MV Braer, was en route between Norway and Canada carrying 84 500 tonnes of light Norwegian crude oil when she lost power in the Fair Isle Channel. On 5 January 1993, the Braer grounded at Garth's Ness on the southern tip of Shetland spilling her cargo.

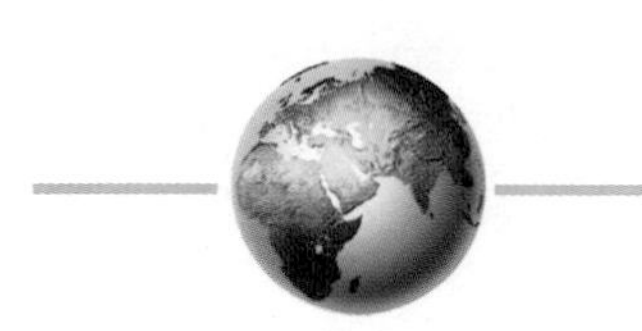

Antarctic Sea: an area of upwelling

The Southern Ocean is 36 million km^2 in extent and occurs south of the Antarctic Polar front. This is where cold water on the surface travelling north from the edge of Antarctica meets the warm water moving south from the Tropics. It sinks below the warm water. The polar front's position changes each year but it is usually 50-60 degrees south. The Polar front marks the edge of the Antarctic.

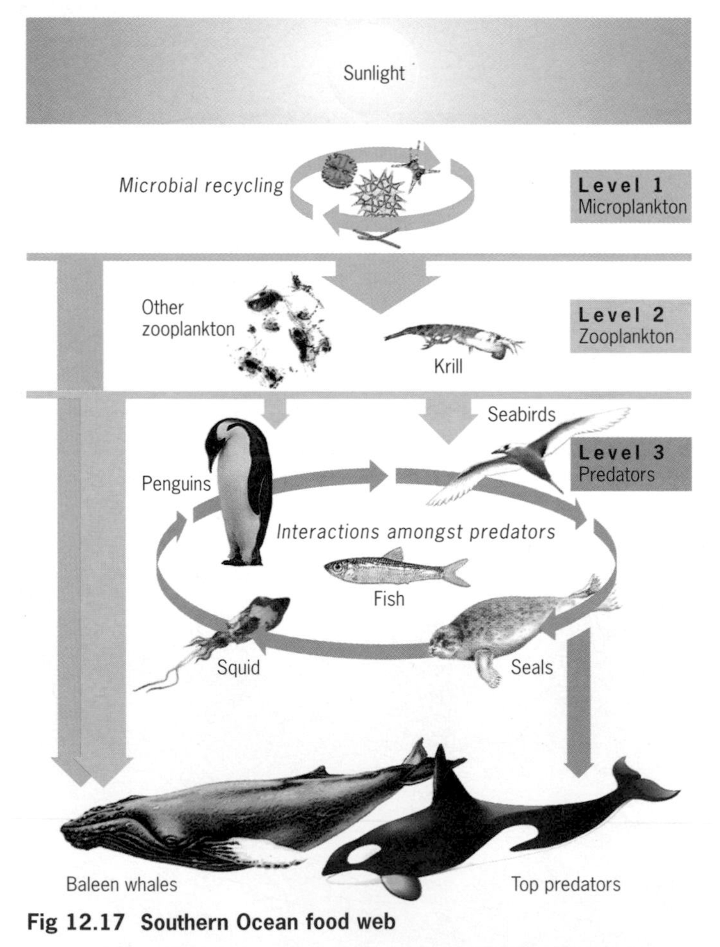

Fig 12.17 Southern Ocean food web

The seas are very rough as winds of up to 300km per hour blow from the ice caps onto the sea surface. The currents move westwards due to the Coriolis force, caused by the rotation of the earth, and are called the East Wind Drift.

Each winter 57 per cent of the ocean is covered in ice and this prompts many creatures to move further north. Wildlife in Antarctica occurs in concentrations which follow the food chain. Phytoplankton therefore determines where creatures are found. Although it is cold, rising currents in the Antarctic seas cause turbulence that supply vital nutrients. Phosphates and nitrates are high. 24-hour daylight in summer provides ideal conditions for phytoplankton but turbulence separates the patches and they are unpredictable.

Krill are next in the food chain. They are the most numerous creatures and can be so dense that they cause 'red waves'. In 2000 it was estimated that there were 62–137 million tonnes in the Southern Ocean. Harvested by eight nations, krill are cooked, peeled and made into paste with cheese; most of the catch is used for animal feed. As patches of krill depend on phytoplankton, this in turn means that predators also follow the krill patches. If krill are harvested, there will be a major impact on breeding success.

Poaching

Krill fisheries are legal under the convention on the Conservation of Antarctic Marine Living Resources. In recent years there have been many incidents of poaching. Poachers use longlines to fish toothfish; their hooks catch albatrosses and other sea birds in the process. It is estimated that 60 000 seabirds are dying from this poaching activity. Decreases of krill and fish will all have an impact on creatures further up the food chain.

12.4 Whaling in the open ocean

Whales are marine mammals, large, mostly migratory creatures of the open ocean. They have been hunted for their blubber and flesh by many nations for centuries, but a worldwide ban on whaling (apart from a number of whales used for scientific research) was agreed in 1986 by the International Whaling Commission. This ban was prompted by a decline in the numbers of most whales in the twentieth century (Fig. 12.19).

Japan admits using aid to build pro-whaling vote

Japan has admitted that it is using its overseas aid budget to persuade developing countries to join the International Whaling Commission and vote for a resumption of commercial whaling.

Japan aims to recruit four or five new countries each year and, within three years, gain enough support to overturn the international moratorium on whaling which came into force in 1986.

The commission over whaling has been deadlocked for ten years, with pro-conservation countries, such as Britain and the USA, totally opposed to lifting the ban, and Japan and Norway in favour.

Although any country can join, only 40 belong to the commission. A 75% majority is required to overturn the ban, so recruiting poor countries would enable Japan to create a pro-whaling majority and change the rules.

The Japanese plan was made public by Horoaki Kamaya, the Vice Minister for Fisheries, after trips to Guinea, Namibia and Zimbabwe. He said he had already visited Trinidad and Tobago which had understood the whaling issue.

His African trip was partly to increase the number of nations working in the International Whaling Commission and the World Trade Organisation. He said both Namibia and Guinea would join this year and there were talks with Morocco and Mauritania.

Mr Kamaya added 'we would like to utilise overseas development aid as a practical means to promote nations to join, expanding grant aid towards non-member countries which support Japan's claim.'

There have been previous allegations against Japan of using aid to buy votes, but Tokyo has always denied doing so.

The Commission yesterday described the tactic as understandable. 'In reality, getting nations who are supportive of your position to join is a sensible strategy which anyone could use,' its secretary, Roy Gamble, said.

Mr Gamble said Japan was using the same tactics at the Convention on International Trade in Endangered Species of Fauna and Flora (CITES) which is due to meet in Nairobi next April. They are trying to get a majority vote to get whales moved off the fully protected list. 'If they succeed in Nairobi it will make our next meeting in July in Adelaide a lot more interesting. But Japan's tactics enraged the WWF for Nature, Stuart Chapman,' the Conservation Officer said.

'If this new Japanese offensive goes unchecked, it could lead to the resumption of large-scale commercial whaling within three years.'

Japan is blatantly misusing development aid to buy votes at the IWC and CITES, and such steps violate the spirit, if not the law, of international treaties.

Until four years ago, Japan and Norway were almost alone in wanting to resume whaling. But a group of Caribbean countries joined and began voting a block, with Japan, Antigua, Grenada, St Vincent and the Grenadines, St Lucia, Dominica, the Solomon Islands and St Kitts and Nevis.

Allegations that these countries' £20 895 annual fees were being paid by Japan were denied, although funding for fishing fleets, fish processing plants and a conference centre were paid from overseas aid. Since the ban was imposed, Japan has continued whaling under the guise of science. A fleet sailed for the Atlantic yesterday with the aim of killing 440 minke whales.

A loophole in the 1946 UN treaty allows the sale of whale meat taken for scientific purposes, and the Antarctic catch will fetch high prices in Japan.

Hideki Moronuki, a Japanese official, said Japan was not vote buying, but that aid would go to countries which had been reluctant to join the commission for fear of damaging economic ties with the West.

He added: 'the whale population had grown so large that it is even damaging the supply of other fish.' But the IWC refuses to listen to anything Japan says.

Paul Brown
Environment Correspondent

Figure 12.18 From *The Guardian*, 11 November 1999

?

15 With the aid of the scale used in Table 12.2, estimate the level of whale exploitation in Figure 12.19:
a in 1900
b in 1980
and comment on your findings.

16 Read Figures 12.18, 12.20 and 12.21. Write a short speech from each of the following delegates to a meeting of the International Whaling Commission.
a Japanese whale hunter.
b Greenpeace activist.
c member of ASOC or WWF.
d Chairperson of the IWC.
e developing country that received aid from Japan.

17a In your view, which is the worst problem for Antarctica: fishing for krill, poaching of toothfish or whaling? Explain your decision.
b Write a reply to the letter that was sent by Greenpeace to the Japanese ship (Figure 12.20).

18 What problems could result if whalewatching is not carried out properly?

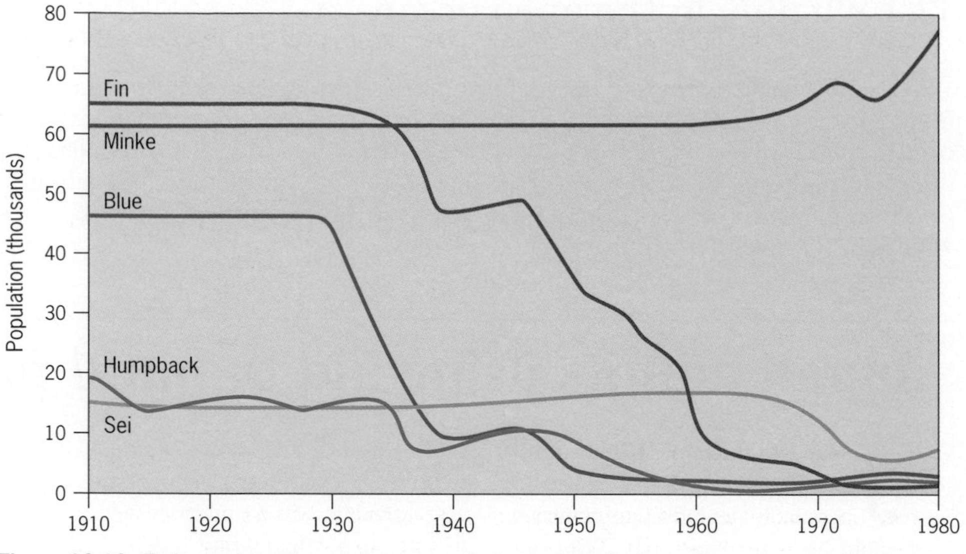

Figure 12.19 Estimated decline of the great whales before the ban on commercial whaling (after Beddington and May, 1982)

Jobs and income versus conservation

The whaling communities and the government of Japan feel that the minke whale has recovered enough for whaling to begin again (Fig. 12.18). Hunting whales for profit is a highly emotive issue: the people and governments of many countries are trying to protect them. Compared with other marine species that also suffer from over-exploitation, such as the sandeel (p. 154), whales receive vastly more attention from the media, politicians and conservation groups.

The work of Greenpeace

Greenpeace has been involved in an anti-whaling campaign for a long time and was instrumental in the outcome of the 1986 moratorium on whaling. It maintained its activities to stop illegal whaling by nations such as Norway and Japan. It has also campaigned to keep Antarctica as a World Park. It captures public attention by taking direct action to stop conservation problems. This action achieves media coverage and informs the public. Greenpeace issues press releases and lobbies governments behind the scenes to push through important legislation. In January 2000 they sent one of their ships to intercept a Japanese whaling fleet.

GREENPEACE

I demand an immediate end to all activities by your whaling fleet which threaten safety and human life. We will hold the ICR responsible for any damage, injuries or deaths that result from the dangerous behaviour displayed by your employees in the Antarctic. I urge you also to withdraw your whaling fleet from the Southern Ocean Whale Sanctuary. Again, Japan's continued whaling in the Southern Ocean Sanctuary undermines the International Whaling Commission, the United Nations Convention on the Law of the Sea and the effectiveness of international law. If all countries were to selectively ignore the international maritime treaty obligations as Japan is doing by whaling in the Southern Ocean Sanctuary, chaos would likely reign on the high seas, and render the conservation and protection of the marine environment impossible.
I look forward to a prompt response from you.

Sincerely
Thilo Bode
Executive Director
Greenpeace International
cc Sanae Shida, Executive Director Greenpeace Japan

Figure 12.20 Letter from Greenpeace to the Japanese ship

Greenpeace uses wall of water to stop illegal Antarctic whaling

Southern Ocean/Amsterdam

Greenpeace today used a wall of water to prevent the harpoonist from shooting whales on day 20 of its campaign to stop illegal Japanese whaling in Antarctica.

Using a water pump mounted on the stern of an inflatable boat, activists sprayed an eight metre wall of water directly in front of the harpoon of the catcher vessel, Toshi Maru No. 25, blocking the line of sight and preventing the killing of whales.

For 20 days Greenpeace has taken all possible peaceful action to stop Japan's illegal whaling and force the country to honour their international legal obligations. 'It is high time that the governments of the world took diplomatic action to force Japan to obey international law and stop illegal whaling in the Antarctic', John Bowler, campaigner on board the Greenpeace vessel MV Arctic Sunrise, said today.

Greenpeace has repeatedly called on national governments to demand that the Japanese government cancel its illegal Antarctic whaling programme. However, only Britain, the USA, Australia and New Zealand have made soft diplomatic moves to pressure Japan to abandon the whaling programme.

Japan's Antarctic whaling programme is in violation of Articles 65 and 120 of the United Nations Convention on the Law of the Seas (UNCLOS – adopted in 1982), which requires all states to cooperate with the International Whaling Commission (IWC) in the matter of whale protection. Each year the IWC passes a resolution condemning Japan for whaling in the Sanctuary and calls on it to stop its whaling programme.

The Southern Ocean Whale Sanctuary was declared in 1994, and the area has been off limits to commercial whaling ever since. However, Japan has a target to kill 440 Minke whales inside the Sanctuary this year. The whale meat produced during the hunt will be sold on the open market in Japan.

Greenpeace has been tracking the fleet since 20 December 1999 and has, since then, successfully disrupted illegal whaling on eight occasions. The activists currently inside the inflatable are Curtis Barnett (Australia), Aaron Barbetti (Australia), Richard Pearson (Australia).

Figure 12.21 Greenpeace press release, January 2000

Whale watching

Whale watching offers considerable financial reward to tour operators involved in this business. As ecotourism gathers momentum it could offer the Japanese and other nations a viable way of using this resource to stimulate income in a sustainable way. The IWC issues guidelines for the industry (Fig. 12.22).

Figure 12.22 IWC guidelines

Manage the development of whale watching to minimise the risk of adverse impacts

- Take care to regulate numbers of viewing platforms, their size and frequency. There may need to be closed seasons when the whales are breeding and could be affected most.
- Data needs to be obtained about the numbers, characteristics and distribution of the whales.
- New whale watching operations will need to monitor their programmes and check for problems.
- Research about the acoustic environment is important.
- All personnel must be trained carefully on the biology and behaviour of the whales. The public must be informed and supportive as a result of the activities.

Design, maintain and operate platforms to minimise the risk of adverse affects to whales

- Boats and other equipment must be designed, maintained and used to reduce impacts on the whales.
- Ship operators must be aware of the acoustic characteristics of the species and reduce the need for disturbing sound.
- Operators need to be able to keep track of whales during encounters.
- Allow the whales to control the nature and duration of interactions.
- Operators must know about whale behaviour and know when they are disturbed.
- Operators need to approach the whales at appropriate angles and within certain distances, depending on the species. Sudden changes in direction or speed must be avoided.
- Operators should welcome but not cultivate friendly whale behaviour. Whales should be able to detect the ships at all times so operators must avoid being too quiet. Young whales and nursing mothers must be approached very carefully.

Summary

- Oceans and seas can be sub-divided into areas of open ocean and continental shelf.
- The productivity of oceans and seas depends on nutrient availability and light. Seas and oceans near lower latitudes tend to have greater inputs of light, and those in areas of upwelling and continental shelves tend to have more nutrients.
- Nutrient cycling depends on the action of currents and wind.
- Oceans and seas are very stable systems, which makes them very fragile.
- Continental shelves are the areas most threatened by human actions. They are often the sites of oil and gas exploration, areas for waste disposal, as well as being particularly productive systems for food resources (fish).
- Human actions can affect the interspecific competition between species.
- As technology has increased, some species which are harvested have been unable to sustain population levels.
- Overfishing has been controlled by policies in the EU. However, there are difficulties in working out correct quotas and keeping a balance between conservation and economic needs.
- There is currently a worldwide ban on whaling to help depleted populations recover. Some nations believe that the populations of some whales are back at the level where sustainable harvesting is possible.

13 Cold deserts

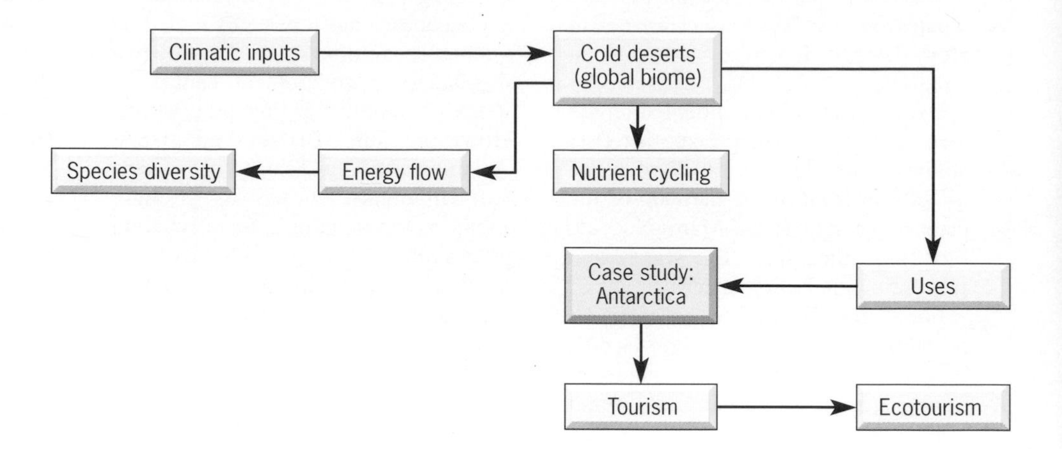

13.1 Introduction

Extreme deserts of rock, sand and ice make up 24.0 million square kilometres of the land surface. These are areas of such low net primary productivity per unit area that they are of limited agricultural use. A number of factors contribute to this low level of productivity. The common factor which links hot and cold deserts is their lack of water.

?

1a Using Figure 13.1, describe the amount of solar radiation received by polar areas.
b Explain why polar areas receive reduced solar radiation inputs.

13.2 Ice and rock deserts of polar regions

Energy inputs

If polar regions were permanently snow-free, their rocks, vegetation and ocean surfaces would absorb enough energy to provide warm temperate summers. However, polar areas are covered in ice and the albedo is high. Up to 90 per cent of the solar energy is reflected back into space (dark rocks and blue seas absorb 80–90 per cent of the radiation). Snow and ice use energy in vaporising and melting. Air temperatures cannot rise above freezing point because solar energy is absorbed by the ice for melting. The ice caps therefore cause the polar regions to lose more heat by radiation than they gain from the sun. Their energy budget is balanced by the input of warm air and sea currents from lower latitudes.

In the northern hemisphere, the North Atlantic Drift carries heat from the tropical Atlantic directly to the Polar basin; Siberian rivers carry the warmth of Central Asia to the Arctic. The coldest part of the northern hemisphere is north-east Siberia. In the southern hemisphere some of the polar islands are warmed by the sea but the interior continent of Antarctica has little warming influence. While some of the Arctic has tundra vegetation with a higher level of productivity, the Antarctic continent consists entirely of ice and rock desert. Only 4 per cent of Antarctica is not covered by ice and is the only part of the continent which can support life.

Table 13.1 Climatic data for some stations in Coastal Antarctica (Holdgate 1970)

Station	Latitude	Longitude	Mean annual temperature	Mean summer temperature	Maximum temperature	Precipitation
Hallet	72° 18'S	170° 18'E	-15.5°C	-2.4°C	8.3°C	12cm
Scott	77° 18'S	166° 45'E	-20.8°C	-5.6°C	5°C	15cm
Port Martin	66° 40'S	141° 24'E	-11.8°C	-3.2°C	–	–
D'Urville	66° 40'S	140° 01'E	-10.9°C	-3.3°C	–	–
Wilkes	66° 33'S	110° 34'E	-9.9°C	-1.5°C	8.0°C	36cm
Mirny	66° 33'S	93° 01'E	-10.2°C	-1.5°C	5.0°C	85cm
Mawson	67° 36'S	62° 53'E	-10.6°C	-1.8°C	7.3°C	–
Showa	69° 00'S	35° 35'E	-11.4°C	-2.6°C	6.5°C	–

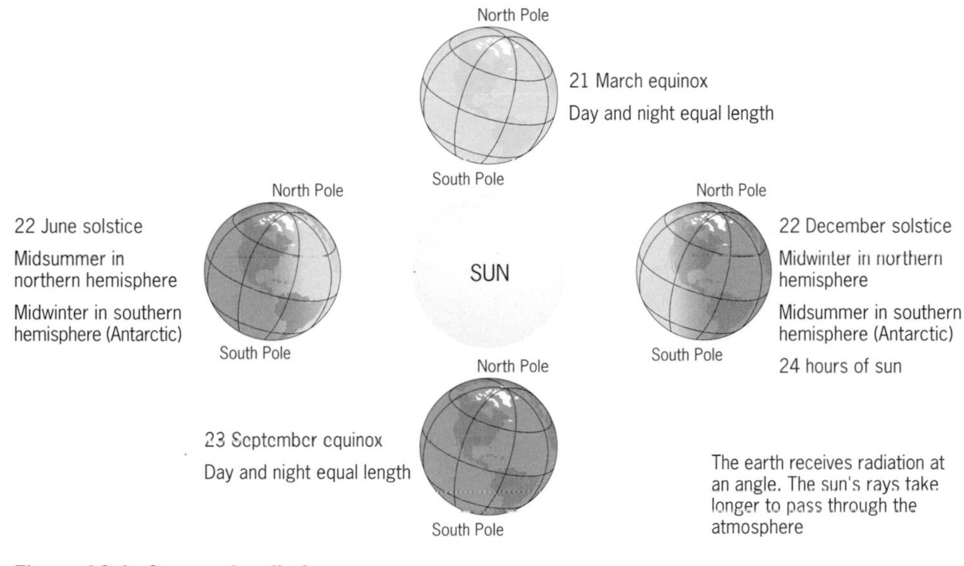

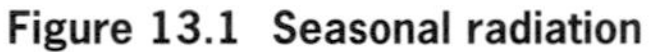

Figure 13.1 Seasonal radiation

The simplicity of Antarctic and Arctic ecology is due to both the harshness and newness of the environment. Large areas have emerged from the ice during the last two to four million years and some are still emerging. Before this there were no ice-covered seas, frozen soils at sea level nor solid frozen lakes. Within this frozen environment soils develop slowly because frost action produces rock debris quicker than plants can colonise it. The climatic climax community for polar deserts is a mixture of mosses and lichens.

Colonisation

- Ahumic soils containing small amounts of water between clay and silt particles allow some plants to establish. Some areas are too dry for soils to mature beyond ahumic level.
- Plant cells add organic material to ahumic soils.
- Chemicals released by the plant material react with the rock to release minerals which can stimulate plant growth. Blue-green algae and nitrogen-fixing bacteria begin to settle. They release nitrogen in solution for the use of other plants.
- Organic buffers develop which neutralise excess acidity or alkalinity and provide a stable chemical environment for plants and small animals.
- A growing amount of vegetation holds water, protects the soil from sun and frost, and creates a stable physical environment.
- Nematodes, tardigrades, rotifers, mites and springtails move in to feed on living and dead plant materials. They circulate minerals and increase the productivity of the developing soil.

?

2 From Figure 13.2, describe and explain the distribution of ice-free areas.

3 Using Table 13.1, describe the relationship between latitude and temperature. Explain this relationship.

4a Using Figures 13.3 and 13.4, describe how the permafrost (permanently frozen ground) varies from the coast to the ice plateau.
b Describe where melting occurs.
c Explain the distribution of permafrost and areas with melting.

5a Describe the distribution of heavy moss, moss tufts and vegetation which grows in rock cracks.
b What factors appear to influence the distribution of vegetation?

6a Using Figure 13.3, explain why darker rock creates higher temperatures.
b Why do older soils contain more salt?

7a Refer to Figure 13.4. Which areas will have the saltiest soils? State your reasons.
b What is the relationship between bacteria and salt shown in Figure 13.3?

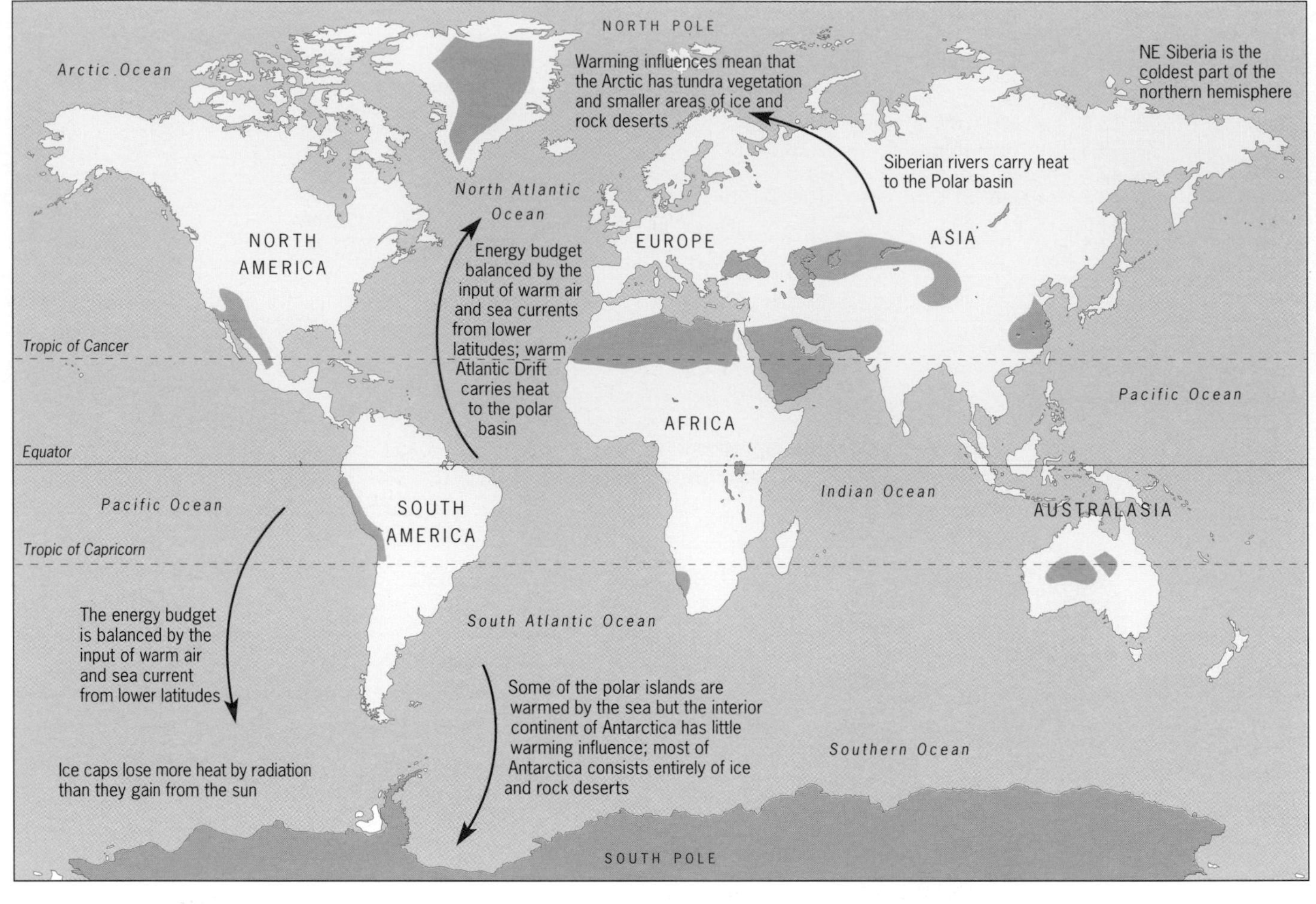

Figure 13.2 Energy budget for hot and cold deserts

?

8a What factors limit colonisation in Antarctica?

b Why is biomass and productivity of Antarctic ecosystems so low?

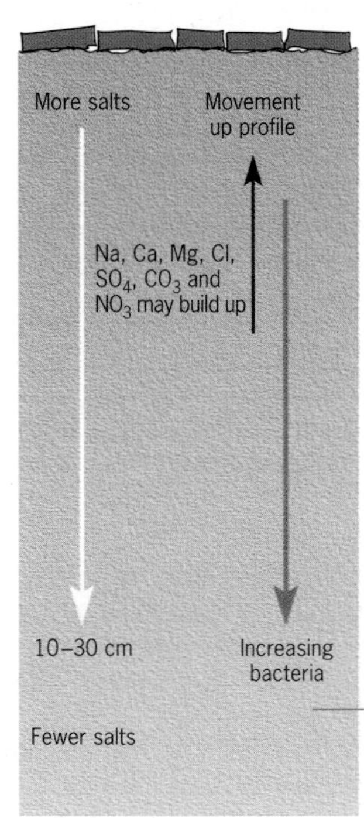

Near the coast there is more fauna and micro-fauna in upper layers; some ornithogenic soils from penguin colonies much more favourable
<1% organic content
Often no life in top layers
Base rich pH 7–9
High salinity of some soils may be a limiting factor
Older soils have high salt content
There is an inverse relationship between the age of soil and biological diversity
More salts
Movement up profile
Na, Ca, Mg, Cl, SO_4, CO_3 and NO_3 may build up
10–30 cm
Increasing bacteria
Fewer salts
Ice cemented permafrost
Dark desert pavement covering soil surface with a positive heat balance allows creation of a micro-environment milder than air temperature suggests. The colour of rock is important; dark basaltic rock is warmer than lighter coloured limestones. Temperatures can rise to 30°C on some rocks. Microclimates above freezing are advantageous for wildlife. Some types of rock determine a particular flora. Limestone creates a special flora.
Small holes in the soils hold water, allowing some plants to establish.

Figure 13.3 Antarctic soil profile

13.3 Resource use in Antarctica

Table 13.2 Activities advertised in the travel index by Antarctic tour operaters in North America (CM Hall, 1992)

Activity	Number of times mentioned, 1997–90
Airlines	3
Archaeology	1
Backpacking	6
Birdwatching	4
Camping	3
Chauffeur	1
Childminding	2
Conservation	4
Cruises	25
Cultural	1
Dogsledding	8
Ecology	2
Equestrian	1
Fishing	1
History	3
Horseriding	1
Marine biology	2
Mountaineering	13
National parks	1
Nature	18
Overland	2
Photography	1
Research	2
Safari	1
Seminar	3
Ski	6
Snowmobile	8
Spa	1
Volcano	1
Trekking	1
Yachting	6

Antarctica is not 'owned' by any nation, although Argentina, Chile, Ecuador, Great Britain, France, New Zealand and Norway claim territory. These claims are not recognised by the international community. The status of the land and its resources are governed by the Antarctic Treaty which was signed on 1 December 1959.

Tourism

Tourism is the only activity which uses the natural resources of the Antarctic continent commercially. Increasing numbers of tourists are visiting the Antarctic (Table 13.3). The IAATO is an organisation that co-ordinates travel and tourism in Antarctica. It issues guidelines to tour operators, such as that cruise ships should contain no more than 400 passengers. Cruises are the most popular form of tourism; 9857 people visited in this way between November 1998 and March 1999. There were 12 flights in 1998–99: 15 people visited the Patriot Hills, nine went to the Emperor penguin colony at Dawson Lambton Glacier, 11 went to the South Pole and 34 climbed Vinson Massif and the Elsworth mountains. All these activities were run by IAATO member, Adventure Network International.

From 1976 to 1980, 4000 people flew over the Antarctic as part of an overflight. However, an air accident on Mount Erebus in 1979 highlighted the problems of tourist activity exceeding the capacities of air traffic control communications and search and rescue activities in the Antarctic. The flights began again in 1994 and since then, 13 000 people have used them.

Most Antarctic cruises begin from Punta Arena (Chile), Puerto Williams (Chile) or Ushuaia (Argentina), because Drake passage can be crossed in 48 hours as opposed to ten days from Hobart (Australia) and Christchurch (New Zealand). Australia is trying to improve transport to the continent for its scientists by establishing an Antarctic air link based on blue-ice runways.

Table 13.3 Numbers of tourists to the Antarctic

Date	Numbers	Date	Numbers
1989 – 1990	2 581	1997 – 1998	9 473
1990 – 1991	4 824	1998 – 1999	10 026
1991 – 1992	6 495	1999 – 2000	13 906

Table 13.4 Elements of the Antarctic Treaty system (CM Hall, 1992)

Element	Characteristics	Geographical area
Antarctic Treaty, 1961	Provides for the management of Antarctic resources; establishes agreed measures, SSSI and other measures for the management of Antarctic resources	Northwards to 60°S latitude
Agreed measures for the Conservation of Antarctic Fauna and Flora, 1964	Plants and land-breeding seals, birds and invertebrates are protected; establishes the agreed measures, SSSIs and other measures	Northwards to 60°S latitude
Scientific Committee on Antarctic Research SCAR	Coordinates, initiates and promotes scientific activity in Antarctica	Northwards to the Antarctic convergence and sub-Antarctic islands
Convention for the Conservation of Seals, 1972	Protects Ross and fur seals; establishes seal reserves and sealing zones	From the the sea ice zone northwards to 60°S latitude
Convention for the Conservation of Antarctic Marine Living Resources, 1980	Applies to all marine organisms except whales, which are covered by the International Convention for the Regulation of Whaling; provides for the establishment of marine sanctuaries	From the sea ice zone to the Antarctic convergence
Convention of the Regulation of Antarctic Resource Activities	Yet to be agreed by all members; a policy of voluntary restriction on minerals; exploitation depends on progress towards a satisfactory minerals plan	
The Protocol on Environmental Protection to the Antarctic Treaty adopted in 1991; came into force in 1998	Antarctica is designated as a natural reserve devoted to peace and science. Mineral activities stopped for at least 50 years, except for scientific mineral research. All activities are to be carried out to limit adverse environmental impacts.	

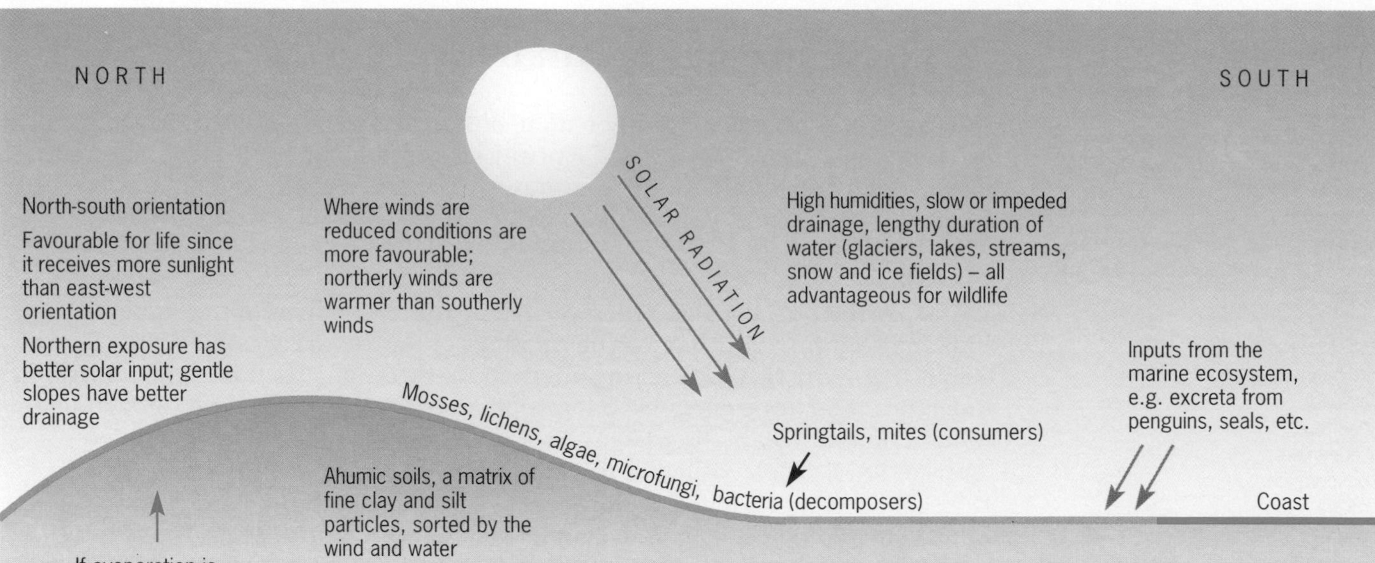
NORTH
SOUTH
SOLAR RADIATION
North-south orientation
Favourable for life since it receives more sunlight than east-west orientation
Northern exposure has better solar input; gentle slopes have better drainage
Where winds are reduced conditions are more favourable; northerly winds are warmer than southerly winds
High humidities, slow or impeded drainage, lengthy duration of water (glaciers, lakes, streams, snow and ice fields) – all advantageous for wildlife
Inputs from the marine ecosystem, e.g. excreta from penguins, seals, etc.
Mosses, lichens, algae, microfungi, bacteria (decomposers)
Springtails, mites (consumers)
Coast
Ahumic soils, a matrix of fine clay and silt particles, sorted by the wind and water
If evaporation is dominant, more salts are moved up the profile

Some lichens are symbiotic. Algae inside them trap nitrogen from the atmosphere

Lichens

Penguins and guano

Mosses

Figure 13.4 The Antarctic terrestial ecosystem

?

9a From Table 13.2 describe the activities advertised to potential tourists.
b Which of these would require
• overflights
• ship-based facilities
• onshore facilities?
c Using Table 13.2 and Figure 13.5, identify five activities that could lead to the most and five that could led to the least environmental damage. Give reasons for your choice.

10 Using Table 13.5 and Figure 13.5, write a paragraph to say why the Project Oasis Development was rejected.

11 To what extent do the principles of ecotourism listed in chapter 3 apply to the development of Antarctica?

Green Flag International is a non-profit making company formed in response to the growing demand for conservation advice from tour operators and the travelling public. It rates holiday tour operators and resorts using the following check list:

- consideration to landscape, wildlife, cultural heritage
- efficiency
- waste disposal and recycling
- interaction with local communities in terms of goods and services
- sympathetic building and architecture.

Table 13.5 Environmental impacts of tourism in the Antarctica (C M Hall, 1992)

Type of activity	Nature of impacts	Infrastructure characteristics
Overflights	Fall-out from engines; noise disturbs wildlife; low overflights cause panic stampedes or penguins to desert nests	No requirements from land-based facilities
Ship-based	pressure placed on regularly visited sites; oil spills; wildlife disturbed; potential to introduce bird and plant diseases; exotic flora introduced; cruises coincide with peak breeding times	No requirement for permanent land-based facilities
Onshore	Increased demands for free land and fresh water supply; disposal of sewage and rubbish; frequently visited sites degraded; wildlife disturbed; potential to introduce bird and plant diseases; exotic flora introduced; damage to heritage sites	All-weather airstrip provided capable of handling large commercial aircraft; accommodation facilities; possible tourist facilities with scientific bases

A major commonwealth study of air transport between Australia and Antarctica has found that a direct intercontinental air link is feasible and would significantly benefit Australia's Antarctic programme.

The report released today by Environment Minister, Gilbert Hill, lists 12 options as feasible and efficient, and recommends that four of these be subject to further practical investigations and market testing through a competitive tender process. The four shortlisted options are based on use of ice and snow landing surfaces.

'The government's principal concern is to increase the efficiency of our science programme by getting scientists and other personnel to and from Antarctica more efficiently than we currently do,' senator Hill said.

This study opens up the prospect of rapid and more flexible deployment of research teams to and within Antarctica and greatly improves capacity to support important airborne and remote area research in Antarctica.

It is proposed that existing areas of blue ice or snow are used as landing areas instead of the alternative conventional gravel runway. These ice and snow airstrips would be located well away from wildlife habitats, minimising the potential effects of air traffic on seals, birds and vegetation.

'A full environmental impact assessment in accordance with Commonwealth legislation and Antarctic Treaty obligations will be undertaken before any final decision,' senator Hill said.

This application is likely to be more successful than one proposed in 1992 because it fits in with the international communities plans for Antarctica to be a place for science rather than commerce.

Antarctic air link a step closer

In 1992 Helmut Rhode and partners proposed to build a year-round integrated tourism, environmental and scientific development, known as Project Oasis, in the Vestfold hills near Darwin station in the Australian Antarctic Territory. The visitor and accommodation facility would provide 344 visitors, 70 researchers an 174 staff. Up to 16 000 people per year could use the facilities and it was proposed that two Boeing 747 flights would operate between Davis and the Australian mainland. The facilities would be floated complete from the Australian mainland and a hovercraft-based transport system would be used for passengers, freight, search, rescue, recovery and medical evacuation.

Project Oasis was opposed by conservation groups because the project competed with the fauna and flora for the ice-free area. Within the Australian sector, the ice-free area is less than 0.3 per cent. However, the development was the direct result of the growing demand for eco-tourism from the public, the tourist industry and some conservation organisations.

The Antarctica and Southern Ocean Coalition, ASOC, is concerned about an increase in tourism. Most tour operators profess their concern for protecting Antarctica's environment and their desire to create ambassadors to the seventh continent. That tourism has doubled since the Protocol signing is worth noting. Environmental considerations take lower precedence than travel and tourism.

The ASOC is concerned that tourist numbers will increase. Interest in unusual activities, such as scuba diving and kayaking, is increasing in Antarctica. At an IAATO meeting in July 1999, several people warned that there will be further growth, particularly if large ships with over 400 passengers are allowed and if Russia supplies Antarctic-worthy ships.

The ASOC is also worried about tourists landing at Antarctic Specially Protected Areas, ASPAs. One firm visited an Emperor Penguin colony that had not been visited before. This meant that scientific data could no longer be collected as this area had been affected by humans. In 1989–90 there were 164 landings but by 1998–99 there were 858, a 400 per cent increase.

Guidance is given before before visiting sites. For example, visitors to Mawson station are advised that 'each person must ensure that they are appropriately equipped, and are required to clean their boots, clothing and survival gear to guard against soil and other materials being transferred to or from the area. Weather conditions can change extremely rapidly; a careful eye should be kept on the weather at all times and immediate action taken to ensure that large numbers of people are not housed ashore during bad weather. Please go to the toilet before leaving the ship as this helps to minimise use of sewerage facilities ashore.'

Figure 13.5 Press release, 1 September 1999

Table 13.6 Advantages of tourism in Antarctica

- Cities from which tours to Antarctica are launched can gain economically, e.g. Hobart.
- Commercial tourism could subsidise government-specific operations and justify claims to Antarctic territory.
- Damage by tourism insignificant compared to that caused by construction and refuse of Antarctic bases, harvesting of marine life, or the effect of oil and mineral exploration and extraction.
- As people gain knowledge of the area, they are more likely to protect it.

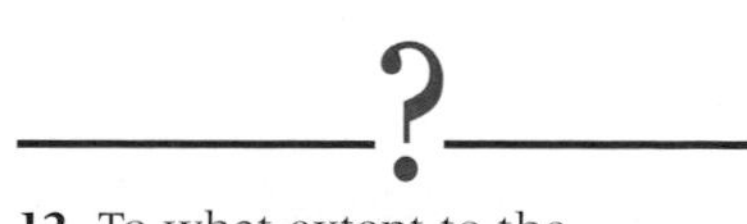

12 To what extent to the economic advantages of eco-tourism conflict with the principles of eco-tourism in Antarctica?

13 If you went to Antarctica, which activities from Table 13.4 would you undertake? Give reasons for your choice.

14 If you had the opportunity to go to Antarctica, would you go? Give reasons for your decision.

Summary

- Cold deserts only have land-based life on areas which are not covered permanently by ice.
- Colonisation is limited by the poor soils and lack of water.
- The climatic climax community is one of lichens and mosses.
- Antarctica is not owned by any countries and is protected from exploitation by the Antarctic Treaty.
- Tourism is increasing rapidly and presents a challenge to sustainable management of Antarctica.

14 Coral reefs: ecosystems under pressure

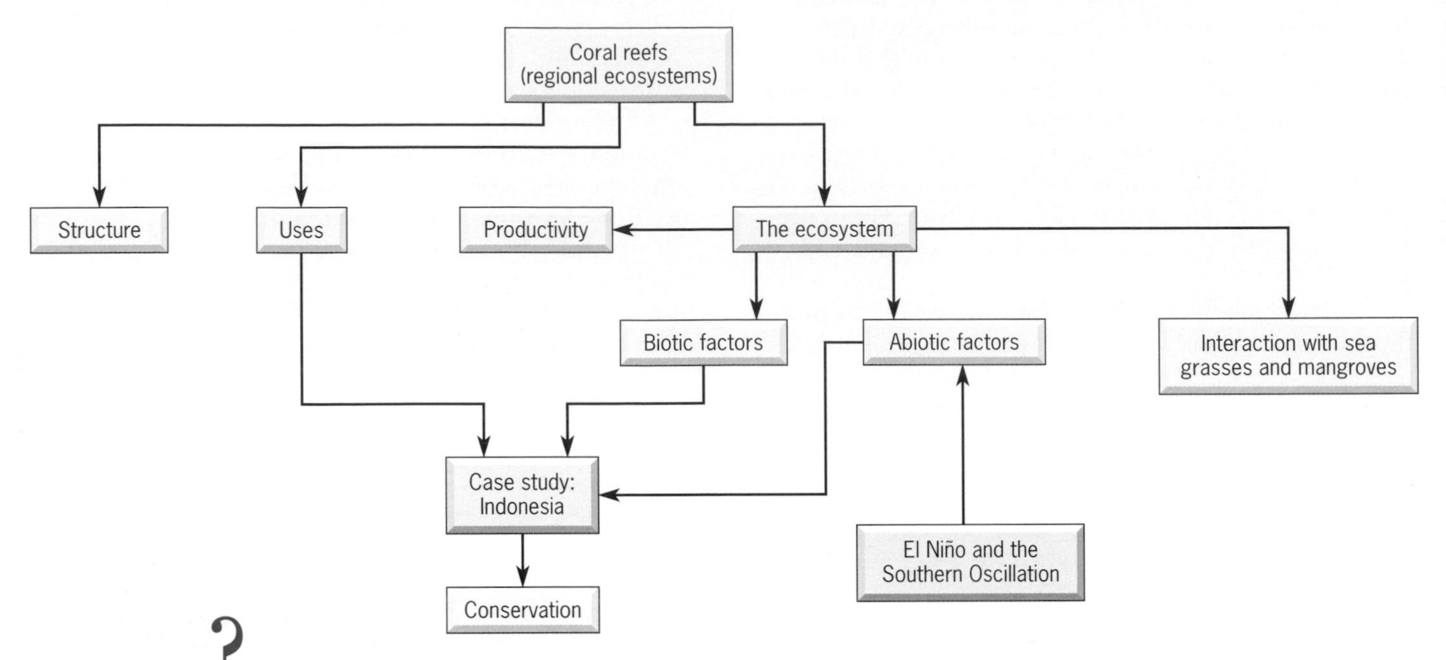

?

1 Use Figure 14.4 to identify the following features on Figure 14.2: fore reef, windward reef, reef flat, lagoon, leeward reef.

2 From Figures 14.1, 14.2, 14.5, 14.6, 14.7, list some of the attractions of coral reef areas for tourism.

14.1 Introduction

Within the area of tropical warm water which could be occupied by coral reefs (Fig. 14.3), reef communities are rare. Very often, reefs are small and separated from each other. The Great Barrier Reef off the coast of Australia is 1500km long and the largest reef of all. However, it is broken up into 2900 different reefs.

Reefs grow in mainly three ways: as fringing reefs, barrier reefs or atolls (Fig 14.2). No two reefs are the same, but they tend to follow the same general structure (Fig. 14.4).

Figure 14.1 Coral reefs are rich, thriving habitats

Figure 14.2 An atoll behind a small island, Maldives, Indian Ocean

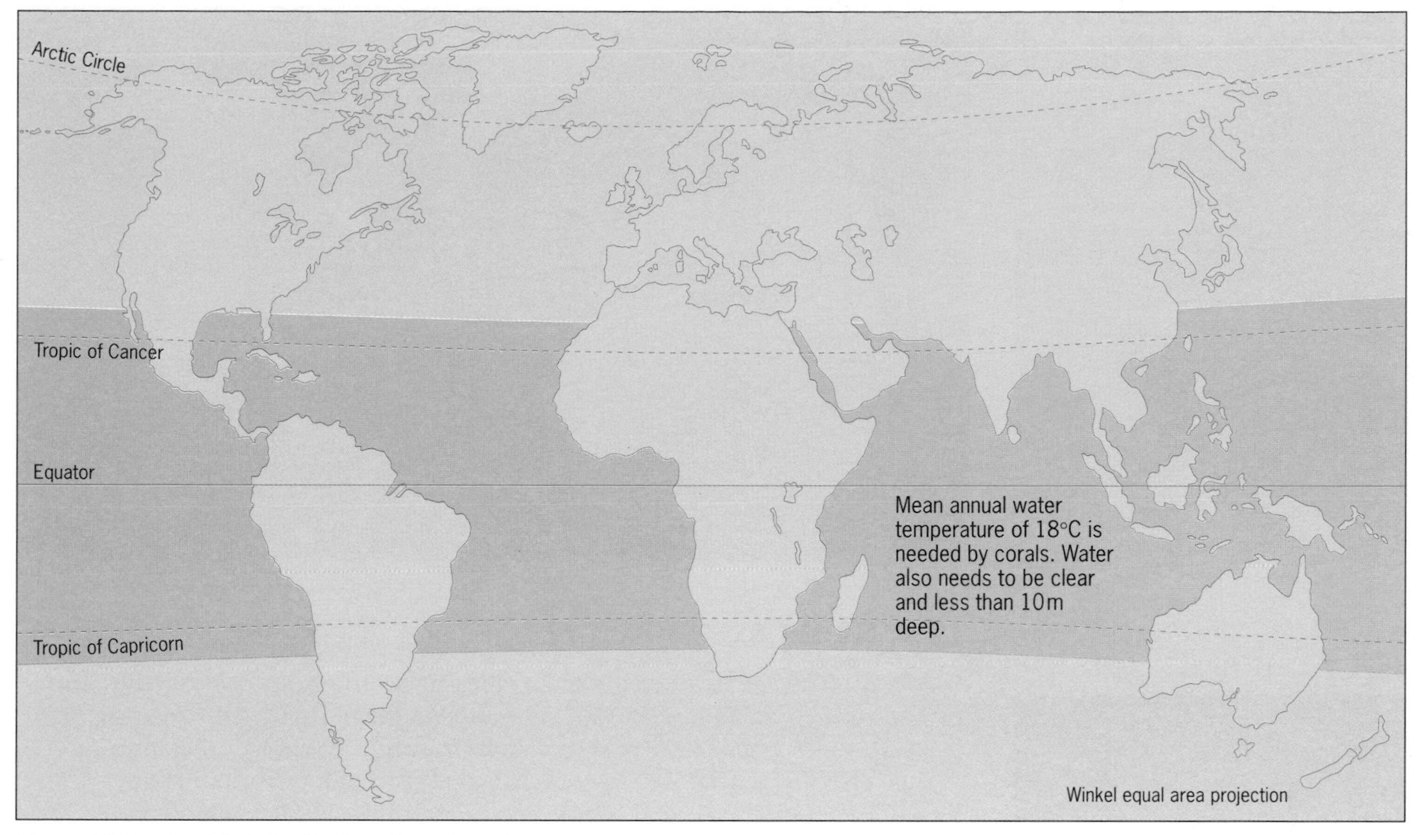

Figure 14.3 The global distribution of coral reefs

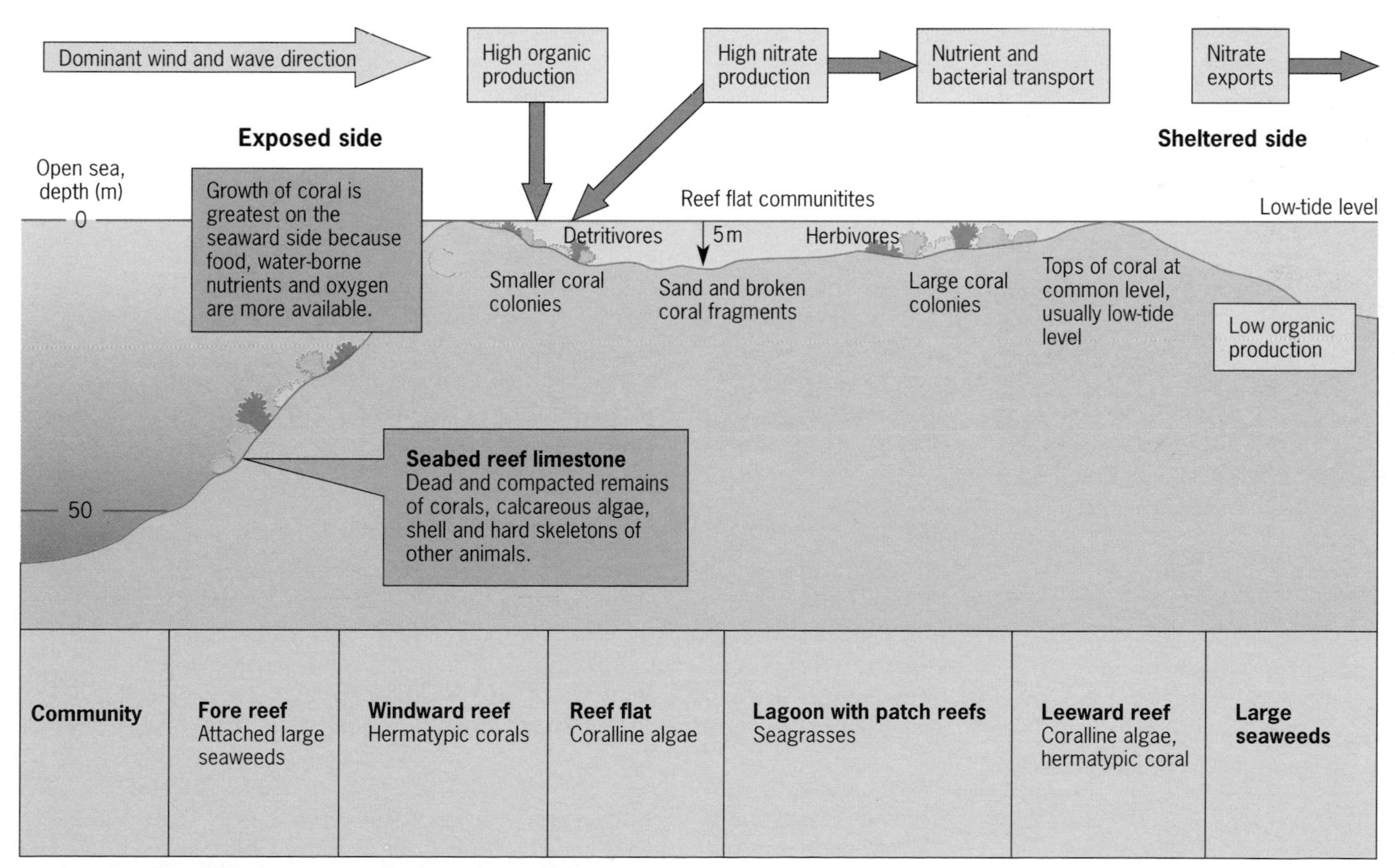

Community	Fore reef	Windward reef	Reef flat	Lagoon with patch reefs	Leeward reef	Large seaweeds
	Attached large seaweeds	Hermatypic corals	Coralline algae	Seagrasses	Coralline algae, hermatypic coral	

Figure 14.4 Generalised reef structure and communities

Figure 14.5 People reef-walking at low tide, Queensland, Australia

Figure 14.6 Green turtle, Australia

14.2 Reefs as resources

People use reefs as a source of food, collecting marine souvenirs and aquarium fish, mining coral, and as sites for sub-aqua diving and reef walking. Some of these activities can be managed in a **sustainable** way, while others cannot.

The fisheries contribution of reef and adjacent shallow-water habitats is 9 per cent of the annual commercial fish landings worldwide. It accounts for 8–20 per cent of the Philippines' catch and 18–25 per cent of the catch in Sabah, Malaysia.

Removal of living materials from a coral reef community on a large scale can cause serious damage. Coral reefs are very sensitive. Like tropical rainforests, they depend on very tight cycling of nutrients.

Over-exploitation of resources such as fish, green turtles (Fig. 14.6), oysters and giant clams (Fig. 14.7) can affect the entire community structure. The number of fish **species** in a coral reef is very large, and monitoring yields of them all is very difficult.

Increasingly, coral reef **ecosystems** are affected by tourism and recreation. Activities such as sub-aqua diving and reef walking (Fig. 14.5) can easily damage the fragile ecosystem.

Members of the reef community respond differently to different levels of **stress**. Reefs tend to be gradually changed from healthy reefs to sick reefs and then to dead reefs. Certain algae can continue to flourish when the corals are dead. This degradation is more common than destruction of complete reefs by coastal engineering or coral mining.

Figure 14.7 A giant clam, Australia

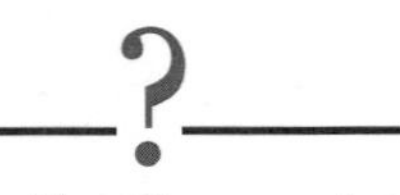

3 Draw a spider diagram of the uses of coral reef areas.

4a From Figure 14.3 and Table 3.1, estimate the GNP range of most countries that have coral reefs
b Draw a spider diagram of the pressures on coral reef environments in those areas.

14.3 The nature of the ecosystem

Different physical environments create different coral communities. Extremes of depth and wave action cause **zonation** to occur in reef communities (Fig. 14.8) both vertically and horizontally. However, interspecific **competition** controls the distribution of many species, particularly in the middle zones away from the extremes of depth and wave action (Fig. 14.9). This is evident from the various survival strategies of individual coral species.

Productivity

Coral reefs are one of the most productive natural communities (see Figure 1.8). Marine productivity is higher in areas around islands because of the factors shown in Figure 14.10 which increase nutrient inputs.

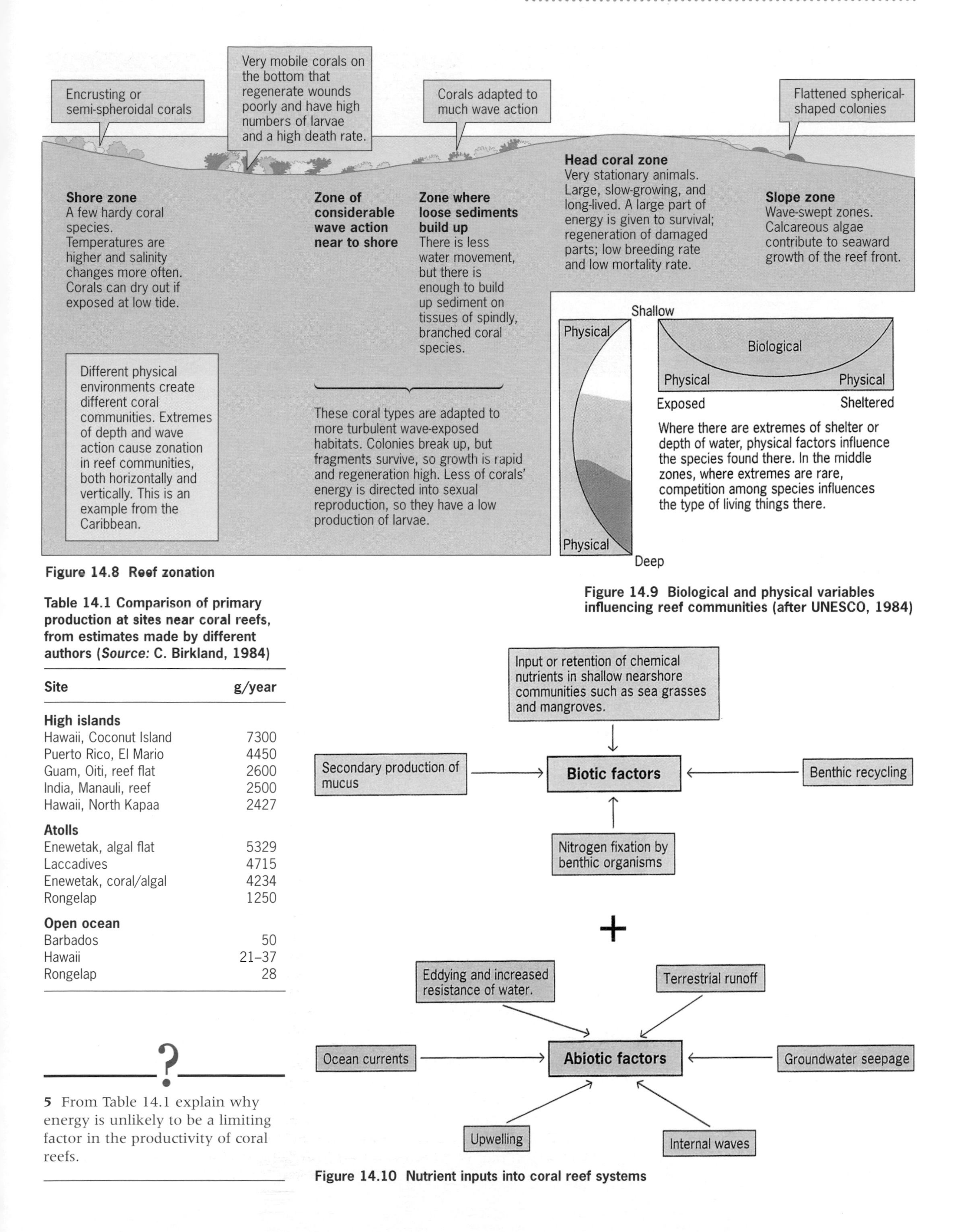

Encrusting or semi-spheroidal corals
Very mobile corals on the bottom that regenerate wounds poorly and have high numbers of larvae and a high death rate.
Corals adapted to much wave action
Flattened spherical-shaped colonies
Shore zone
A few hardy coral species. Temperatures are higher and salinity changes more often. Corals can dry out if exposed at low tide.
Zone of considerable wave action near to shore
Zone where loose sediments build up
There is less water movement, but there is enough to build up sediment on tissues of spindly, branched coral species.
Head coral zone
Very stationary animals. Large, slow-growing, and long-lived. A large part of energy is given to survival; regeneration of damaged parts; low breeding rate and low mortality rate.
Slope zone
Wave-swept zones. Calcareous algae contribute to seaward growth of the reef front.
Different physical environments create different coral communities. Extremes of depth and wave action cause zonation in reef communities, both horizontally and vertically. This is an example from the Caribbean.
These coral types are adapted to more turbulent wave-exposed habitats. Colonies break up, but fragments survive, so growth is rapid and regeneration high. Less of corals' energy is directed into sexual reproduction, so they have a low production of larvae.
Shallow
Physical
Physical
Deep
Biological
Physical
Physical
Exposed
Sheltered
Where there are extremes of shelter or depth of water, physical factors influence the species found there. In the middle zones, where extremes are rare, competition among species influences the type of living things there.
Input or retention of chemical nutrients in shallow nearshore communities such as sea grasses and mangroves.
Secondary production of mucus
Biotic factors
Benthic recycling
Nitrogen fixation by benthic organisms
+
Eddying and increased resistance of water.
Terrestrial runoff
Ocean currents
Abiotic factors
Groundwater seepage
Upwelling
Internal waves

Figure 14.8 Reef zonation

Figure 14.9 Biological and physical variables influencing reef communities (after UNESCO, 1984)

Table 14.1 Comparison of primary production at sites near coral reefs, from estimates made by different authors (*Source:* C. Birkland, 1984)

Site	g/year
High islands	
Hawaii, Coconut Island	7300
Puerto Rico, El Mario	4450
Guam, Oiti, reef flat	2600
India, Manauli, reef	2500
Hawaii, North Kapaa	2427
Atolls	
Enewetak, algal flat	5329
Laccadives	4715
Enewetak, coral/algal	4234
Rongelap	1250
Open ocean	
Barbados	50
Hawaii	21–37
Rongelap	28

?

5 From Table 14.1 explain why energy is unlikely to be a limiting factor in the productivity of coral reefs.

Figure 14.10 Nutrient inputs into coral reef systems

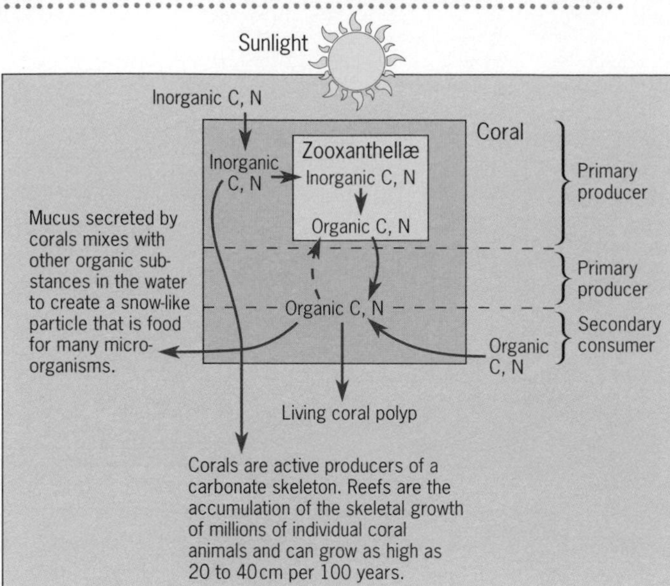

Figure 14.11 Nutrient cycling of carbon (C) and nitrogen (N) in a coral (after UNESCO, 1983)

Figure 14.12 Coral polyps feeding at night

?

6 Study Figure 14.11. Why are the trophic levels difficult to separate in coral reef communities?

7 Explain why algal blooms would be more likely to form around high islands rather than atolls.

Biotic factors

Phytoplankton in coral communities is not responsible for high productivity. The high productivity is linked to zooxanthellae (single-celled algae) in **symbiotic** relationship with coral. They produce three times the primary production of plankton. Zooxanthellae contribute sugar and other nutrients to the coral cells by **photosynthesis**. In return, the coral polyp provides the algae with a protected environment and it receives carbon dioxide (Fig. 14.11). During the day, most corals contract their tentacles and this allows the zooxanthellae to photosynthesise more efficiently. At night, the tentacles come out and feed on passing **zooplankton** (Fig. 14.12).

The most diverse reefs are the oligotrophic systems. These are systems with little nutrient enrichment. The pathways of nutrients and energy within the coral reef system are very short, particularly in corals themselves. Nutrient recycling is therefore extremely important. Nutrients are in short supply for fish on coral reefs. Fish that browse on coral structures or coral mucus need extra small worms, crustacea or fish eggs to obtain essential nutrients. At least 45 species of coral-reef fish eat fish faeces in addition to other foods.

?

8a On a copy of Figure 14.13: Locate the Java Sea, Tahiti, Okinawa and Tokelau. Use an atlas to help you.
b Write the date of bleaching in each area (Table 14.2), then draw arrows to join them in sequence.
c Compare the pattern on your map with the normal South Equatorial current in Figure 14.13.
d Comment on the relationship between ENSO and climatic effects on coral reef areas.

Abiotic factors

Internal waves, eddying and upwelling

Around islands, internal waves appear to be more important as nutrient sources than terrestrial inputs. Eddies can bring together up to 40 times more plankton than in adjacent water. A greater nutrient **input** can cause different communities to form. Near to high islands there tend to be more suspension feeders such as sponges.

Some inputs from the land can harm reef communities. A sudden input of additional nutrients can favour the growth of species which affect the equilibrium of the reef communities. Heavy rains following excessive droughts in South-East Asia and Australasia caused an increase in nutrients, followed by a population explosion of the crown-of-thorns starfish which feed on coral. Algal blooms and red tides occur around islands but rarely around atolls. They can compete with corals for space and light.

El Niño and the Southern Oscillation

In 1983, there was a drought in Australia, floods occurred in Ecuador, 17 million sea birds and 18 species virtually disappeared from Christmas Island, and hundreds of young sea birds died in Guguan, Marianas Islands. Their populations were affected by decreased productivity around coral reef communities and other ocean communities. This decrease in productivity was caused by a phenomenon known as El Niño (the Christ Child) which occurs in the equatorial Pacific Ocean at irregular intervals (two years or ten) but usually around Christmas-time. El Niño is a warming of the ocean surface, caused by upwellings of warm water, and is part of the larger and more complex phenomenon called the Southern Oscillation which causes climatic change in the equatorial eastern Pacific. The 1982–3 El Niño and Southern Oscillation (ENSO) led to the deaths of some corals (Table 14.2).

The phenomenon of coral bleaching has also become an issue in recent years, following changes caused by El Niño. Coral bleaching happens when the zooxanthellae die and the coral loses its source of nutrients from photosynthesis (Fig. 14.10). Once the zooxanthellae disappear, the entire coral reef ecosystem starts to die.

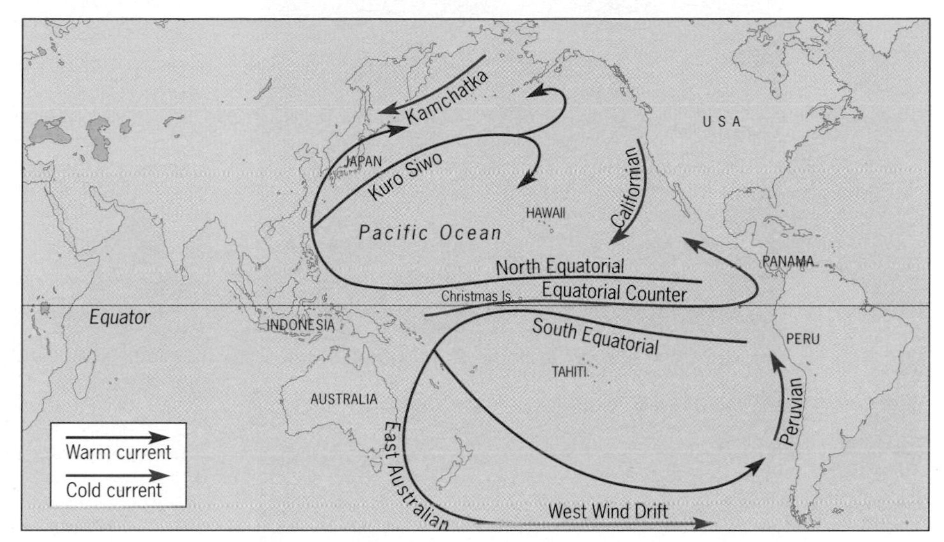

Figure 14.13 Main currents in the Pacific Ocean

Table 14.2 Biological observations on coral reefs worldwide and possible El Niño and Southern Oscillation associated phenomena during 1982-3 (*Source:* B. Brown, 1987)

Site	Biological observation	Date of bleaching	ENSO phenomenon	Date of ENSO
Eastern Pacific	Bleaching and death of reef-building corals over a wide area			
W. Panama		Feb–June 83	Surface sea-water warming	Feb–June 83
Gulf of Panama		June–Oct 83	Upwelling limited	July–Aug 83
Galapagos Islands		Feb–June 83	Surface sea-water warming	Oct 82
			Sea-level rise due to ceased south-east trade winds	Feb–June 83
Java Sea (Palau Pari, Kepulauan Seribu)	Bleaching and death of reef-building corals on reef-flats	April–May 83	Severe drought;	June 82
			intensification of drought	Dec 82–Feb 83
			and warming of sea water	April 83
French Polynesia (e.g. Tahiti)	Coral death on shallow reefs attributed to crown-of-thorns starfish as well as bleaching	1982	Sea pressure at 3mbar below normal.	April–Aug 82
			Five times normal rainfall.	Nov 82–Feb 83
			Five hurricanes recorded, two intense hurricanes.	Jan–April 83
Southern Japan	Extensive bleaching and death of corals in shallow waters			
Okinawa		Aug 83	Sea surface temperatures abnormally high (31°C)	June–Aug 83
Yaeyama Iriomote		Aug 83	Severe drought	June–Aug 83
Tokelau Islands	Extensive death of corals in shallow waters	Early 83	Fall in sea level	Early 83

Figure 14.14 Mangrove community, Trinidad

Interaction between coral reefs, sea grasses and mangroves

A living reef which grows faster than it is eroded provides a natural barrier along coasts. This helps to dissipate wave energy and creates a lower-energy environment near to the shore which soon becomes **colonised** by sea grass (Fig. 14.15), which in turn gradually **succeeds** to mangrove swamp (Fig. 14.14). Sea grass and mangrove swamp therefore depend to a great extent on coral reefs (Fig. 14.16).

Inputs of sediments produced by soil erosion, destruction of mangroves and sea grasses or dredging operations reduce the amount of light in a system as well as interfering with feeding and respiration. Corals use up valuable energy in cleansing themselves of the sediment.

Figure 14.15 Sea-grass community

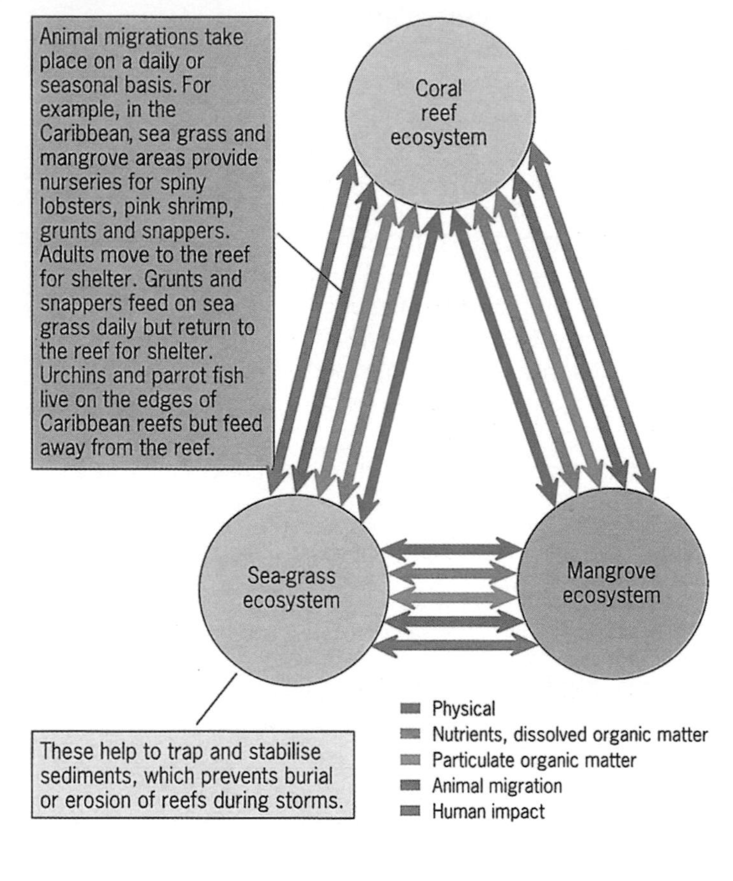

Animal migrations take place on a daily or seasonal basis. For example, in the Caribbean, sea grass and mangrove areas provide nurseries for spiny lobsters, pink shrimp, grunts and snappers. Adults move to the reef for shelter. Grunts and snappers feed on sea grass daily but return to the reef for shelter. Urchins and parrot fish live on the edges of Caribbean reefs but feed away from the reef.
Coral reef ecosystem
Sea-grass ecosystem
Mangrove ecosystem
These help to trap and stabilise sediments, which prevents burial or erosion of reefs during storms.
Physical
Nutrients, dissolved organic matter
Particulate organic matter
Animal migration
Human impact

Figure 14.16 Interactions among coral reef, sea-grass and mangrove ecosystems (after UNESCO, 1983)

Indonesia

Indonesia has a coastline of 81 000km and 13 700 islands (Fig. 14.18) and a mixture of ringing, barrier and atoll reefs. Damage to coral reefs is wide-ranging (Table 14.3).

Legislation to prevent over-exploitation has been in force since 1912. Certain methods of collecting pearl oysters and sponges were stopped. Licences were issued to pearl-oyster fishermen and separate areas were set aside for traditional oyster divers. Licences have also been issued to collectors of ornamental fish. Several reefs have been marked as protected marine areas. They number 225 sites and cover 3 per cent of Indonesian waters. Indonesia has some problems in providing trained staff to manage the coral reef reserves.

The Kepulauan Seribu (Thousand Islands) and Jakarta Bay Islands and their reefs are close to the city of Jakarta, the capital of Indonesia. Their short distance from Jakarta means that they are greatly affected by tourism, fisheries and extraction of raw materials.

The reefs are influenced by both human and natural factors. Separating causes and effects of changes to the reefs is important for the future **management** of the area. Adjoining islands have been divided into five zones to reflect their distance from the coastline (Fig. 14.17). Zones 1 to 3 are in Jakarta Bay; zones 4 and 5 are of the Thousand Islands group.

Physical factors

Winds are an important physical factor that influence the currents responsible for moving nutrients and high-salinity water into the area around Jakarta Bay. Inputs of fresh water close to the coast can cause salinity levels to be lower than that of open waters. During the west monsoon, cold salty water moves into the Java Sea from the China Sea (Fig. 14.25). This is responsible for improved coral growth in the northern islands and is shown in the preferred direction of reefs, which grow in a north-north-west direction.

Sediment also has a considerable influence on coral reef communities; the cloudier the water, the less sunlight can penetrate and the lower the productivity of the zooxanthellae. In stormy weather sediment can erode the reef structure.

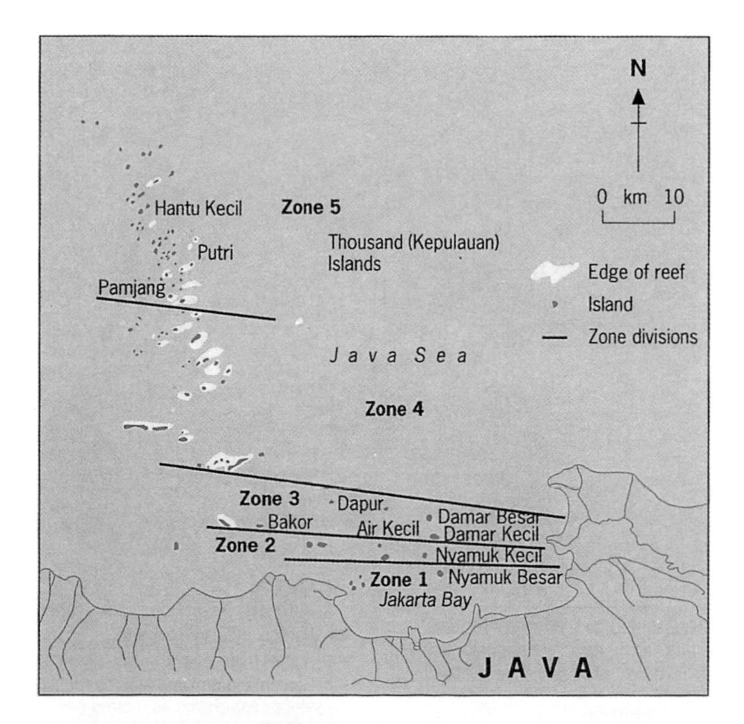

Figure 14.17 Zones within the Thousand (Kepulauan) Islands (*Source:* UNESCO, 1986)

Figure 14.18 Indonesia

El Nino

During 1997 another El Niño effect changed the temperature of the water off the coast of Indonesia. More nutrients were circulated because of the cooler waters, resulting in an imbalance in the coral reefs. Phytoplankton grew and decreased the amount of sunlight that reached the algae, bleaching some corals.

In addition, the drier climate dried out the rainforest floor, making it easy to set alight. Companies that wanted to clear the forest for oil palms used El Niño as an excuse to burn the forest. This lead to a quick release of acid deposits into the sea when the rains came. The increased acidity and turbidity affected the reefs and further destroyed the corals.

?

9 What inputs could the monsoon currents bring to coral reef systems (Fig. 14.19)?

10a Using Figure 14.20, describe the difference between the winds occurring in zones 4 and 5 and those occurring in zones 1 to 3.

b How might this affect the growth of reefs?

Table 14.3 Sources and locations of coral damage throughout Indonesia (*Source:* UNESCO, 1986)

Problem	Location	Remarks
Blasting of coral	East Indonesia (Spermonde Maluku), Kepulauan Seribu	Reefs do not recover from this
Poisoning of reef fish	Kepulauan Seribu, Bali, East Indonesia (Spermonde)	Spear guns with poison tips are used to catch fish
Fishing		
Babu trap	Found at many sites throughout Indonesia	Overfishing can affect nutrient cycling and the whole reef community
Muroami trap	Kepulauan Seribu, Madura	One type of trap is disguised by a covering of coral
Spearing	Kepulauan Seribu, Bali	
Tourism		
Sub-aqua diving	Kepulauan Seribu, other sites	Underwater corals are used as resting places and coral is broken off
Collection of coral and shells and aquarium fish	Pangdandaran, Lampung, Bali, Bandegan, Pasir putih	Used for souvenirs
Boat anchor damage	Bali, Kepuilauan Seribu	
Reef walking	Kepulauan Seribu, Bali	Physical damage to corals
Coral mining	Ball, Carlta, Seram	
Construction industries	Kepulauan Seribu, Karimum Jawa, East Indonesia, Maluku	
Giant clam mining	Kepulauan Seribu, Karimum Jawa	
Pollution		
Sewage	Jakarta, Bintang	
Thermal effluent	Surabaya, Bintang	Waste waters reported to give sea-water temperature of 35–40°C
Oil exploration	Kepulauan Seribu	
Sedimentation from agricultural runoff	Carita, Beandengan	Affects productivity
Dredging		
Dredging for minerals, e.g. tin	Bilitung, Bangka, Lingga, Sinkep	Billiton now using one of their largest dredgers in this area: capacity 12 000 tons
Dredging channels	Bintang	The area of dredging and reclamation is 3 million m^3

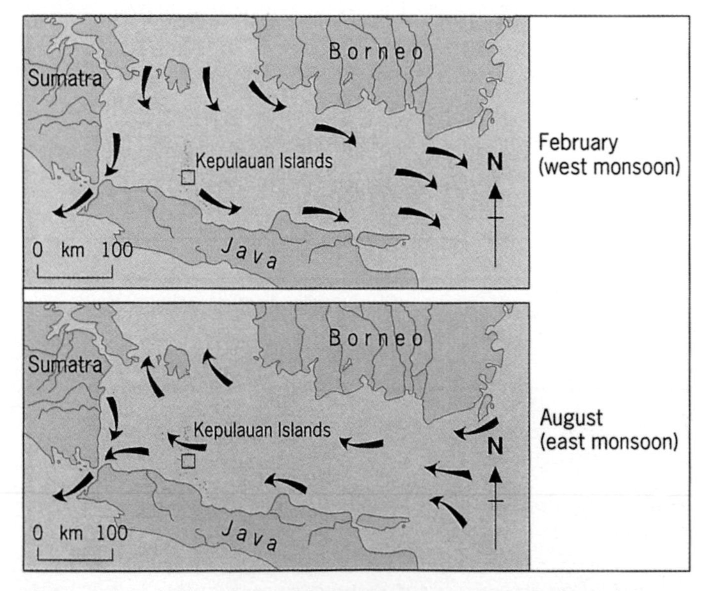

Figure 14.19 The relationship between monsoon winds and currents in the Java Sea, Indonesia (*Source:* UNESCO, 1986)

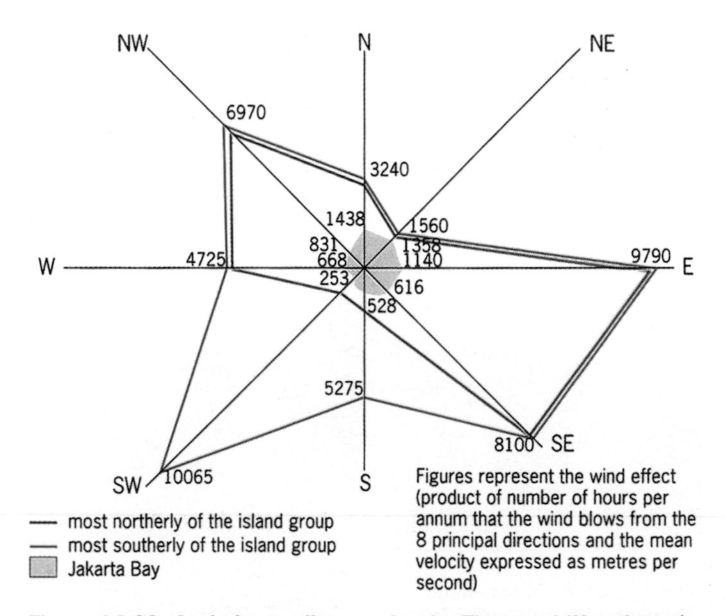

Figure 14.20 A wind rose diagram for the Thousand (Kepulauan) Islands (*Source:* UNESCO, 1986)

?

11a Using figures from Table 14.4, draw a scattergraph to show the relationship between salinity and distance from the shore.
b Describe the relationship that your graph shows.
c Explain your findings using Figure 14.25.
d What effect might this have on coral reef communities?

12a Using Table 14.5, describe how the coral varies among zones.
b Using Figure 14.21, describe how the fish populations relate to coral distribution.

13a Explain how the loss of a coral reef (Fig. 14.22) causes an island to disappear.
b From Table 14.6 identify the islands that are at risk in Jakarta Bay.
c From Figure 14.23 describe the trend for coral use in the area.
d From Table 14.4 draw a scattergraph to show the amount of flotsam (rubbish) in relation to distance from the shore.
e Describe the relationship between these two variables.
f To what extent can you use the information about flotsam to explain what is happening on the reefs?

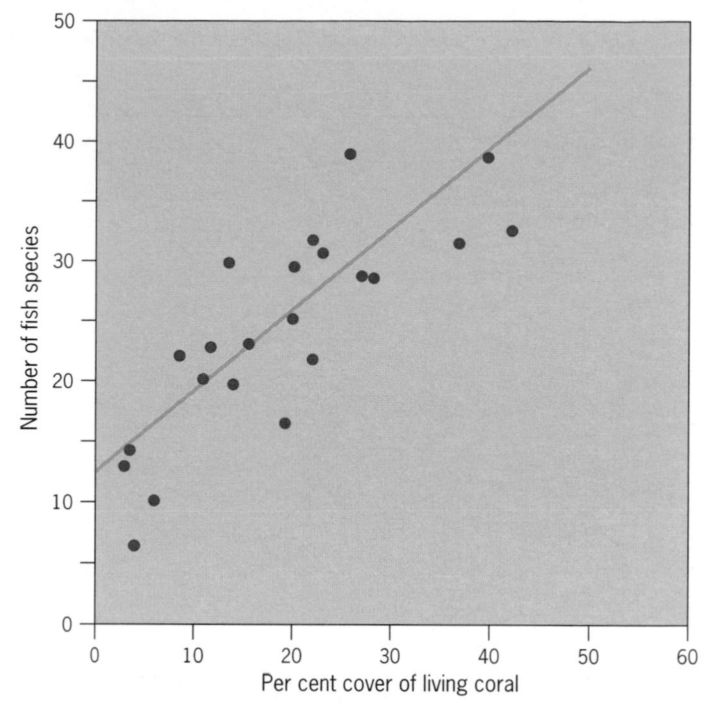

Figure 14.21 The relationship between the number of fish on reef slopes and percentage coral cover in reefs of the Thousand (Kepulauan) Islands (*Source:* UNESCO, 1986)

Table 14.4 Characteristics of reefs around the Thousand (Kepulauan) Islands (*Source:* UNESCO, 1986)

Zone	Island	Distance from Jakarta (km)	Distance from shore: Temperature (°C) 0m	Temperature (°C) 3m	Salinity (%) 0m	Salinity (%) 3m	Water turbidity (Secchi disc)	Percentage Coral cover	Number of inhabitants	Rubbish (flotsam) per metre on strandline (plastic bags, footwear, bottles). Number of items (mean) with standard deviation
1	Untung Jua	5	26.9	28.8	29.6	28.6	6	30	100	9.9±6.8
1	Nyamuk Besar	7	26.1	26.0	33.5	33.6	8	1	0	8.5±3.1
2	Air Besar	8	26.2	30.1	32.5	30.5	9	20	10	8.7±2.6
2	Nyamuk Kecill	10	28.2	28.1	31.9	31.9	9	15	10	5.5±3.0
2	Ubi Besar	5	26.6	31.1	31.4	29.6	1	1	0	10.2
3	Bokor	8	29.0	28.8	31.6	31.7	5	20	0	10.4±3.0
3	Damar Besar	15	25.4	26.1	33.2	33.2	9	40	20	4.1±1.9
3	Damar Kecil	11	26.0	25.9	33.2	33.2	12	25	10	5.5±3.0
3	Dapur	11	28.2	28.2	32.1	32.0	11	25	0	1.7±0.7
5	Pamjang	41	27.1	27.1	33.0	32.9	13	30	5	0.3±0.2
5	Putri	46	29.0	29.0	32.0	32.0	14	30–80	20+tourists	0.1±0.1
5	Hantukeal	52	29.0	29.0	32.0	32.2	13	15	0	0.1±0.1

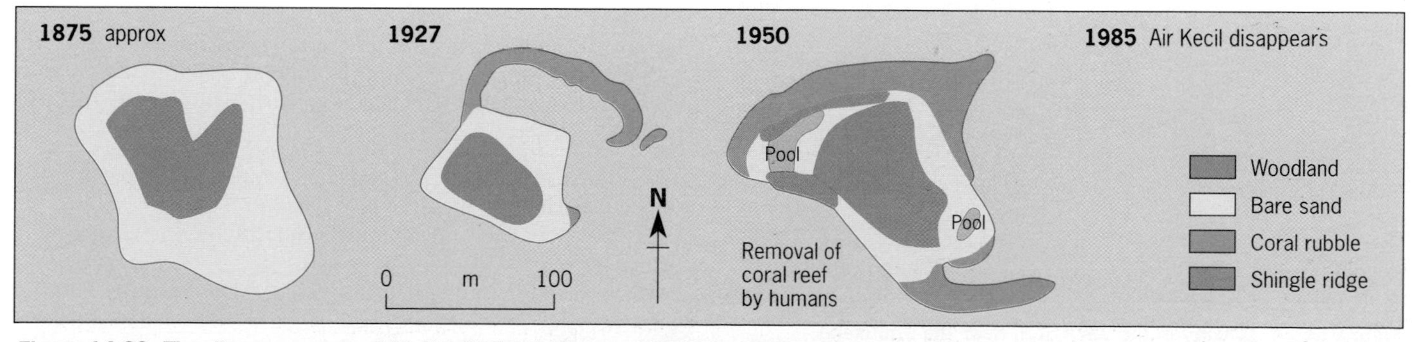

Figure 14.22 The disappearance of Air Kecil Island (*Source:* UNESCO, 1986)

Table 14.5 Details of coral communities in the five zones within the Thousand (Kepulauan) Islands (*Source:* UNESCO, 1986)

Zone	Percentage cover	Percentage cover living	Species numbers	Number of colonies	Average size (cm)
1	61	15	30	45	10.0
2	88	45	67	103	13.2
3	196	148	102	235	18.9
4	272	218	171	341	19.2
5	279	259	128	283	27.6

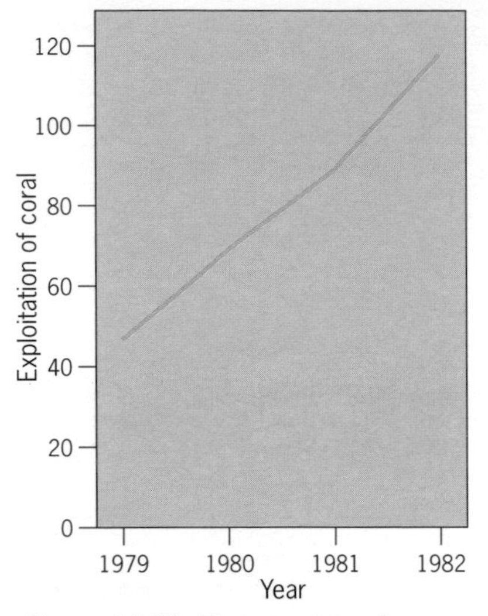

Figure 14.23 The use of coral per person per year in the Thousand (Kepulauan) Islands from 1979 to 1982 (*Source:* UNESCO, 1986)

Table 14.6 Status of coral reefs and islands in the Bay of Jakarta (*Source:* UNESCO, 1986)

Name of island	Almost all/ totally destroyed		Majority destroyed		Partly destroyed		Good condition	
	Reef	Island	Reef	Island	Reef	Island	Reef	Island
Nyarnuk Besar	•			•				
Nyamuk Kedl	•			•				
Damar Besar			•			•		
Damar Kecil					•			•
Bidadari			•			•		
Cipir			•			•		
Kapal			•			•		
Kelor	•			•				
Air Besar			•			•		
Air Kecil	•	•						
Ubi Besar	•	•						
Utung Jawa					•			•
Rambut							•	•
Dapur			•	•				
Bokor							•	•

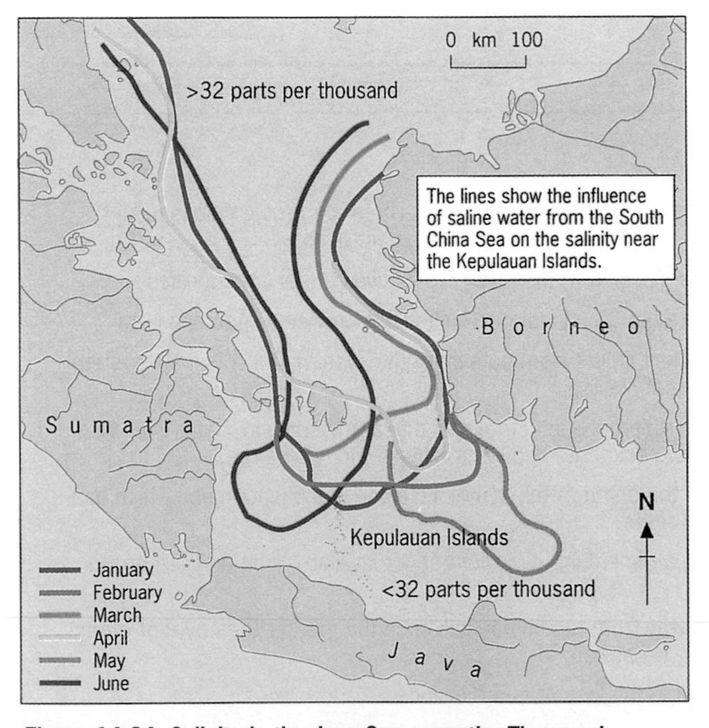

Figure 14.24 Salinity in the Java Sea, near the Thousand (Kepulauan) Islands (*Source:* UNESCO, 1986)

The marine park

A marine park has been established in the areas around the Thousand Islands (Fig. 14.25).

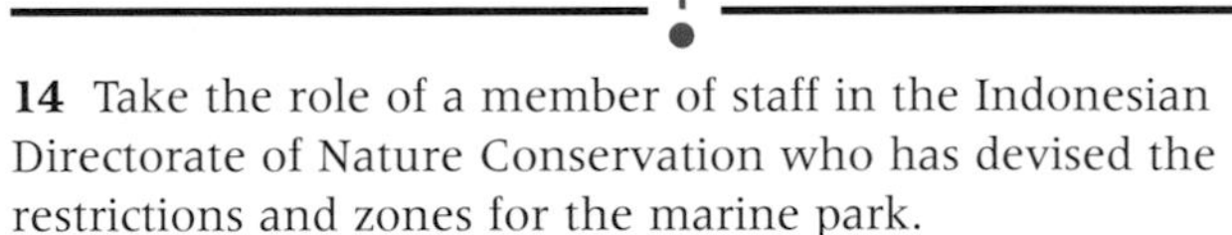

14 Take the role of a member of staff in the Indonesian Directorate of Nature Conservation who has devised the restrictions and zones for the marine park.

a Explain the restrictions on fishing to a local fisherman who knows that there are more fish as you move further north.

b Explain the restrictions imposed on extracting coral and sediments to a person who has been contracted to supply such materials for the construction of the runway of a nearby airport.

15a Explain to a European tourist operator the opportunities for **ecotourism** in the area (see Chapter 3).

b Explain how the tour operator could follow the key principles of ecotourism (Table 3.8).

c What would the advantages to the local community be (see Table 3.9)?

Wilderness zone
Set up so that users can appreciate the park in a natural setting, for protecting reefs and their habitats and stocks (for replenishing surrounding fishing grounds).

Intensive use zone
Set up for tourist and recreational development; however, reefs are protected from the adverse effects of such activities.

Coral sanctuary
Set up to protect representative samples of reef life for replenishing damaged or depleted communities.

Buffer zone
Set up for the continuation of traditional or established uses and activities other than those which are damaging to the environment, incompatible with park regulations or other laws. For example, regulations prohibit harvesting turtles or turtle eggs, mining coral and fishing with poisons and spearguns.

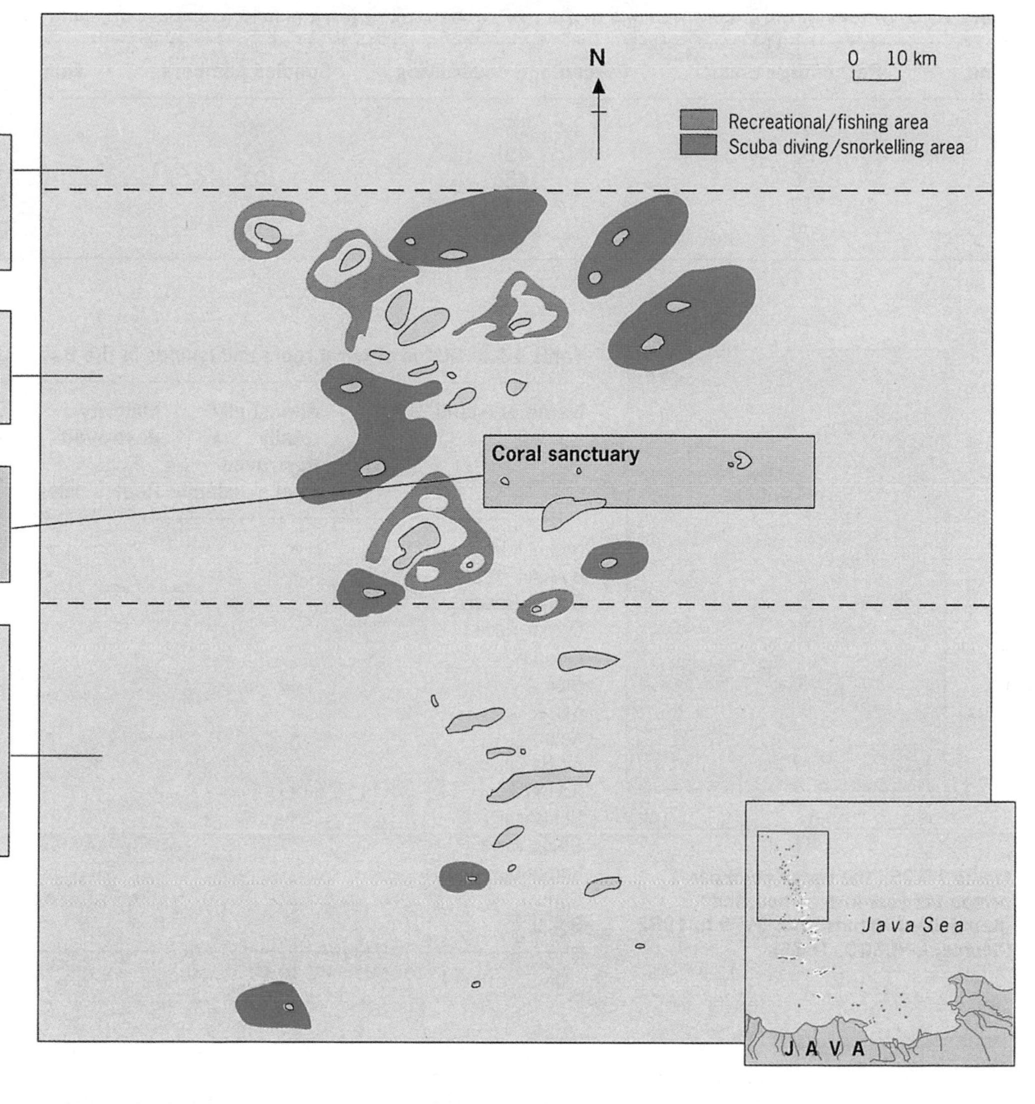

Figure 14.25 Zones of use in the Pilau Seribu Marine National Park (*Source:* UNESCO, 1986)

Summary

- Reefs are sources of food, minerals and tourism. There are problems associated with the sustainable management of some of these activities.
- Coral reefs are rare but are one of the most productive natural communities.
- High productivity is linked to the symbiotic relationship between algae and coral.
- Coral reefs rely on nutrient inputs from outside the system, but within the system there is very short nutrient cycling.
- There is a close relationship between coral reefs, sea grasses and mangroves with movements of nutrients, sediments and living things among all three.
- Coral bleaching due to the death of the symbiotic algae is a phenomenon which has been observed in coral reefs.
- Corals are sensitive to sediment inputs which reduce available light and can erode reefs.
- The further reefs are away from the human influence the greater the cover of coral species and other living creatures.
- Loss of reefs can result in the loss of entire islands.

?

16 Essay: Choose two strategies to protect coral reef ecosystems and evaluate their potential impact on people and environments.

15 Land-use conflict in the Coto Doñana

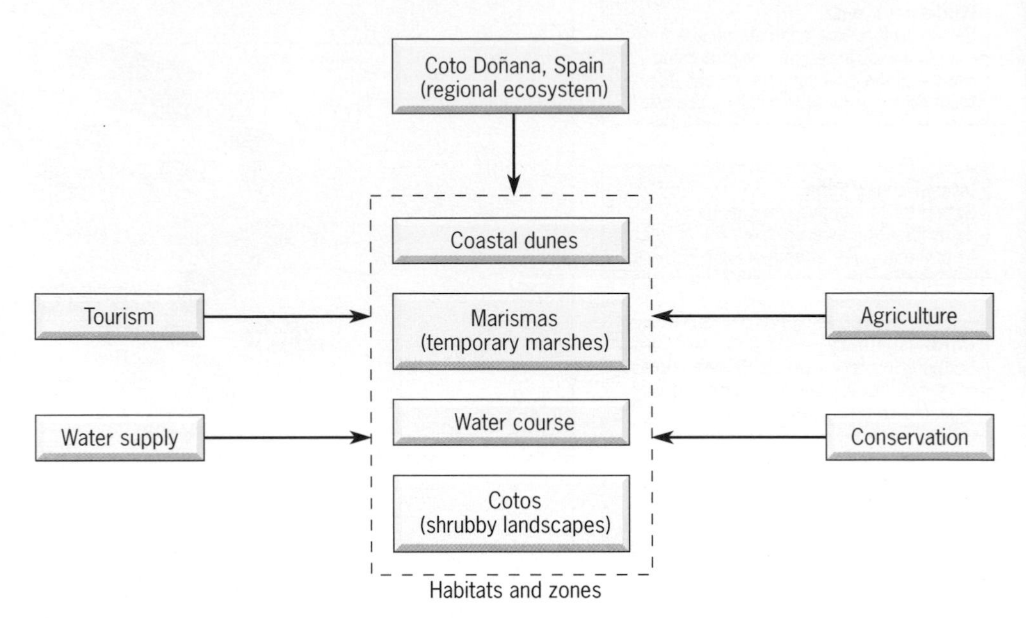

15.1 Introduction

Assume the role of a research student who is investigating the background to complaints made about conflicting land uses in the Coto Doñana area (Fig. 15.1). Produce a report for the local mayor which includes the tasks in Questions 1–4 on the left.

You have received a letter (Fig. 15.2) listing the information that is available to you.

?

1 Identify competing land uses in the Coto Doñana area and the reasons for the complaints.

2 For each land use, evaluate the advantages and disadvantages of increasing the areas of each type of land use.

3 Complete a set of recommendations for the **sustainable management** of the Coto Doñana under the following headings:
a Tourism
b Water supply
c Agriculture
d Conservation.

4 Outline, with reasons, the interest groups who would support or oppose your proposals.

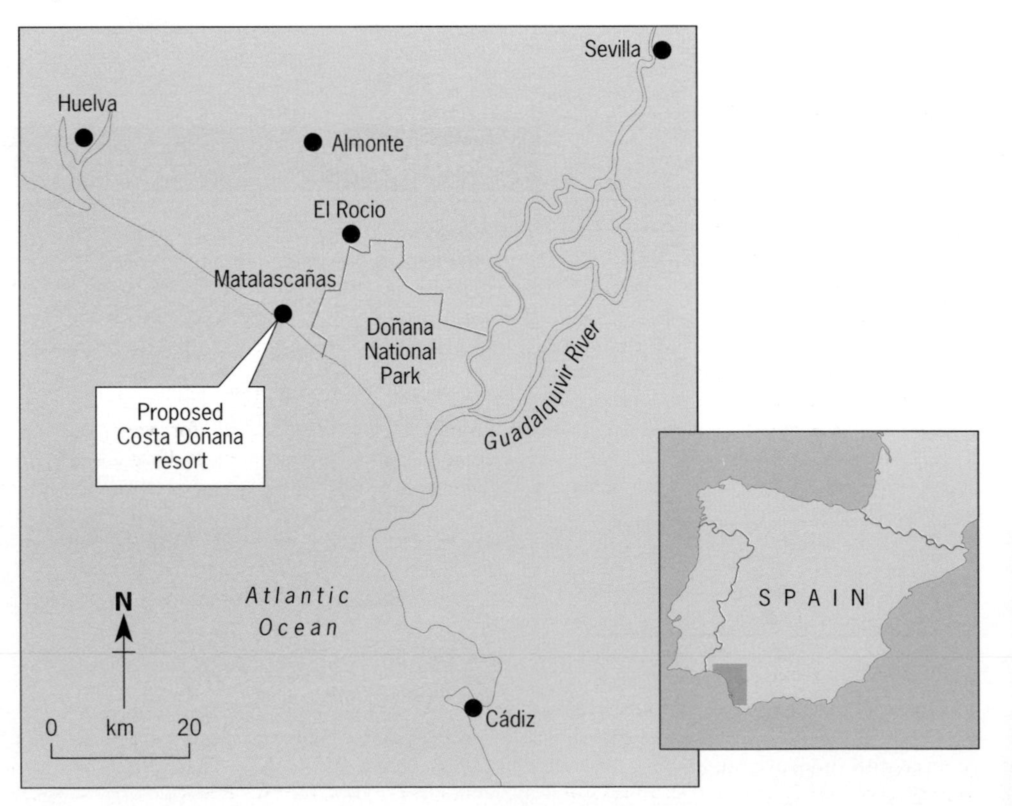

Figure 15.1 The Coto Doñana

Dear Sir or Madam,

I would like you to investigate the reasons behind the two complaints we have received about land use in the Coto Doñana (Figs. 15.3 and 15.4).

The area is of great conservation importance. From at leaast the 13th century, the Kings of Spain used the Coto Doñana for hunting animals such as deer, lynx and wild boar. This effectively ensured its survival until the twentieth century.

In 1989, the Spanish Government recognised its ecological importance by designating part of it as a Biosphere Reserve, which theoretically ensures long-term protection. In 1982, it was listed as a Ramsar site and that charges the Spanish Government with the job of conserving the wetlands and waterfowl.

Large areas of marsh have been turned into agricultural land, and irrigation schemes prevent water from flowing into the marismas. Vegetables and fruits are grown for export to other European countries. At least 20 000 ha of marsh land has been converted for growing rice. In 1986, 30 000 birds died, almost certainly due to poisoning from illegal pesticides.

Since the beginning of the 1960s, the small fishing village of Matalascañas has been developed into a seaside resort, catering for 40 000 people during peak times. It has been estimated that every visitor uses at least 275 litres of water every day. In addition, the resort has two square kilometres of gardens and lawns which are watered regularly. It is estimated that four billion litres of water are used every year by Matalascañas. The water comes from a series of wells sunk into aquifers which would normally supply the Coto Doñana.

In the 1980s a proposal was made to develop 78 ha of coastal dunes to the north of Matalascañas, as a tourist resort called the Costa Doñana. The resort would eventually have a population of 32 000. It was also proposed that Europe's largest golf complex would be built here. This would require large amounts of water to keep the golf courses green. The demand for water could be 10 billion litres a year.

I have enclosed a number of documents which will help you to investigate the conflict in this area:

- Complaint A (Fig. 15.3)
- Complaint B (Fig. 15.4)
- Greenpeace newspaper extract (Fig. 15.5)
- An annotated map showing the main ecological zones of the Coto Doñana (Fig. 15.6)
- A profile of Coto Doñana habitats (Fig. 15.7)
- A diagram of the human threats to the ecosystems and wildlife of the Coto Doñana (Fig. 15.8)
- An advertisement for the Coto Doñana (Fig. 15.9)
- Photographs of the area and its wildlife (Figs 15.10–15.17)
- A newspaper article about the conflict (Fig. 15.18).

I very much look forward to reading your report.

Yours faithfully,

Gabriela Marquez
Mayor

Figure 15.2 Letter to the student

Dear Sra Marquez,

Our fire alarm went off last night at 2 a.m. and we dashed out to start up our fire engines only to discover that they had been set on fire. This was a clear case of arson and we do not have any means of protecting the wildlife reserve of Doñana should a fire break out. We believe that this was the work of developeres who want to gain a fast buck from tourism. We have received telephone calls warning us that when the wetlands dry up in the summer, the arsonists will return.

We are already battling against pollution and water drainage on our wetlands areas. We do not want the extra hassle from these vindictive people.

Sincerely,

Juan Pablo
Warden of Coto Doñana National Park

Figure 15.3 Complaint A

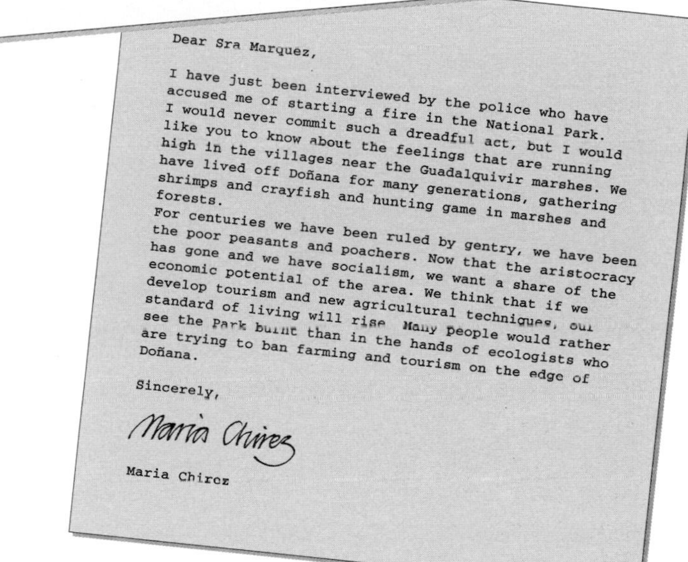

Dear Sra Marquez,

I have just been interviewed by the police who have accused me of starting a fire in the National Park. I would never commit such a dreadful act, but I would like you to know about the feelings that are running high in the villages near the Guadalquivir marshes. We have lived off Doñana for many generations, gathering shrimps and crayfish and hunting game in marshes and forests.

For centuries we have been ruled by gentry, we have been the poor peasants and poachers. Now that the aristocracy has gone and we have socialism, we want a share of the economic potential of the area. We think that if we develop tourism and new agricultural techniques, our standard of living will rise. Many people would rather see the Park burnt than in the hands of ecologists who are trying to ban farming and tourism on the edge of Doñana.

Sincerely,

Maria Chirez

Maria Chirez

Figure 15.4 Complaint B

Figure 15.5 Greenpeace newspaper cutting, 1998

Greenpeace sends ship to toxic spill area in southern Spain

29 April 1998, Amsterdam, Seville

Greenpeace today started an environmental impact assessment of the toxic spill in southern Spain. The flagship of the environmental activists, the *MV Greenpeace*, has arrived at the mouth of the Guadalquivir River near the Coto Doñana National Park and has sent inflatables upstream to assess the damage to the river ecosystem.

Greenpeace is working with local ecologists in fighting the effects of the toxic flood which left a trail of dead fish and poisoned soil behind it.

The spill, estimated to contain five million cubic metres of liquid waste contaminated with residues of zinc, lead and other metals, occurred Saturday at the Canadian-Swedish owned Aznalcollar mine after a residue pool dam ruptured.

Although the huge tide of poisonous waste was diverted away from one of the most prized National Parks in Europe, the ecological impact may still be disastrous. 'The toxic sludge could contaminate the groundwater of the region for years to come,' said Juantxo Lopez de Uralde for Greenpeace Spain.

Greenpeace is demanding a full investigation of the mining company operating the Los Frailes mine responsible for the spill. The mine produces zinc, lead, copper and silver.

The regional farmers' association estimates that at least 6 000 hectares of farmland has been affected, resulting in the immediate crop losses of two billion pesetas (13 million US dollars). The long-term economic damage to the farmers is estimated to be around a further ten billion pesetas (67 million US dollars) within the next couple of years.

The long-term effects on the region and the national park, home to six million migratory birds as well as the habitat for rare species, such as lynx, otters and imperial eagles, could be disastrous when the heavy metals find their way into groundwater reserves and the food chain.

One of Greenpeace's key demands for the 1998 International Year of the Ocean is to ban dumping and discharging of toxic waste into the marine environment. 'This disaster proves that the only way to prevent major ecological destruction from huge amounts of toxins is to stop producing them,' said Lopez de Uralde.

Coastal dunes
Doñana conserves one of the few systems of moving sand dunes that exist in the Iberian peninsula. They extend along a line parallel to the coast with a length of about 20 km and a maximum width of 5 km. The moving sand dunes form a series of active fronts named sand dune trains, gradually getting higher towards the interior, and leaving behind depressions or 'corrales' covered with a scrub-like vegetation including pine trees. The corrales represent an unstable and fragile ecosystem which requires strict protective measures to avoid changes in the water dynamics. Along the north-east side of the dune system the dune water table discharges into a series of ponds called 'lagunas'.

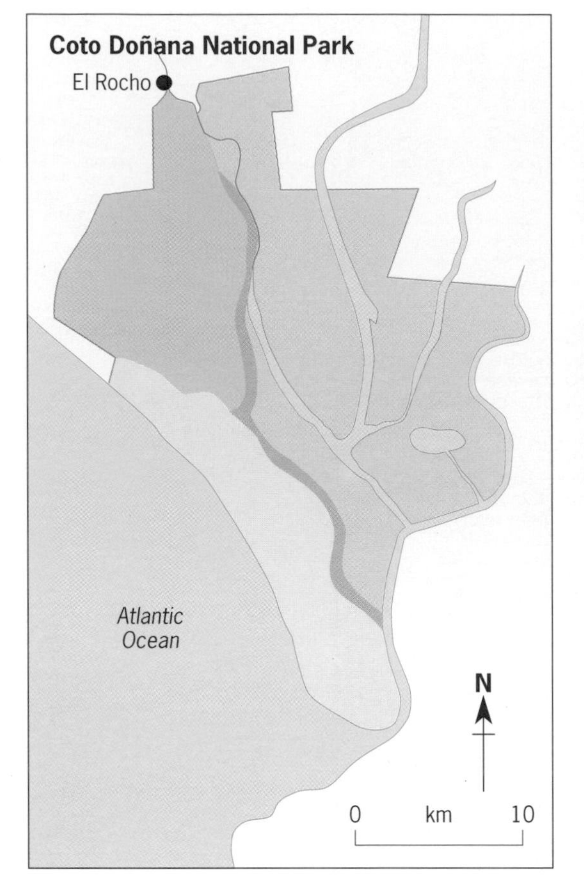

Marismas
The 'marismas' are fresh or slightly brackish temporary marshes lying on clay. Their average altitude is 3 metres above sea level. Water, salinity and topography act together to create a mosaic of varied biotopes. This deversity contributes to the ecological richness of the marismas. The water originates directly from rainfall and to a lesser extent from some small streams which have their watershed on the steady sands. Some of the streams that used to feed the marismas have been diverted to provide water for agriculture. The water regime is a succession of cyclic flooding and desiccation depending upon the precipitation–evapotranspiration balance as well as upon input and output by runoff. The seasonal flooding is one of the major factors contributing to the Coto Doñana's international importance, as up to 500,000 birds can overwinter there.

Cotos
Cotos are shrubby landscapes established on fairly flat acid sands between 5 and 20 metres above sea level. The cotos consist of scrub and woodland, including trees such as cork oak, juniper, stone pine and the introduced eucalyptus. The plant communities and ecological equilibrium of the Cotos are strongly dependent upon the groundwater level. A lowering of the water table could be detrimental to the whole ecosystem.

Ecotone
The ecotone is formed by the the gradient between the marshland and the drier sand deposits. It is a narrow (less than 1 km) sandy strip where the groundwater is above or close to the surface. In winter, water seeps over this ecotone and supplies brooks flowing to the marismas. In summer, water is retained in small pools. The availability of water throughout the year makes it one of the biologically richest parts of the Doñana.

Figure 15.6 Ecological zones of the Coto Doñana

Ocean → Guadalquivir River

15–30 m above sea level

Diversity of habitat; a warm damp climate; and a location where the fauna of Africa and Europe overlap to make the Coto Doñana biologically diverse.

Beach	Dunes	Pine woods	Pond	Heathland	Cork-oak savanas	Marsh margin	Marsh
	Marram grass	Stone pine Brambles Juniper	Grasses Rushes Reeds	Tree heath Gorse Broom	Cork-oak Tree heath Grasses	Bracken Brambles Grasses	Reeds Rushes
Grey plover Kentish plover Caspian tern Oystercatcher Sanderling	Lizards Snakes Crow	Short-toed eagle Wood lark Red-necked nightjar Spanish imperial eagle	Great crested grebe Purple heron Mallard Marbled teal Coot	Red-legged partridge Stone curlew Thekla lark Magpie Spotted cuckoo Great gray shrike	Kite Kestrel Barn owl Green woodpecker Melodious warbler Spoonbill	Warblers Yellow wagtail Pratincole Short-toed lark Rabbit	Coot Moorhen Purple gallinule Black-winged stilt Marsh harrier

Figure 15.7 Plants, animals and habitats of the Coto Doñana (after Mountfort, 1957)

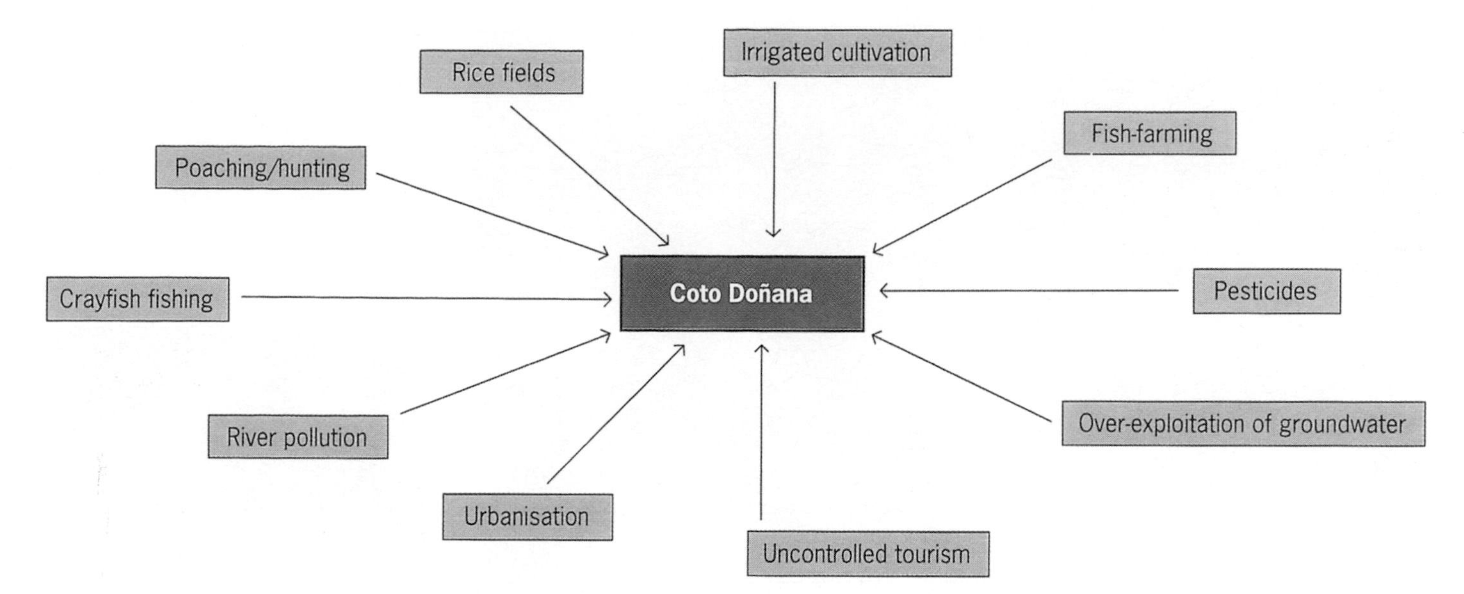

Figure 15.8 Human threats to the ecosystems and wildlife of the Coto Doñana

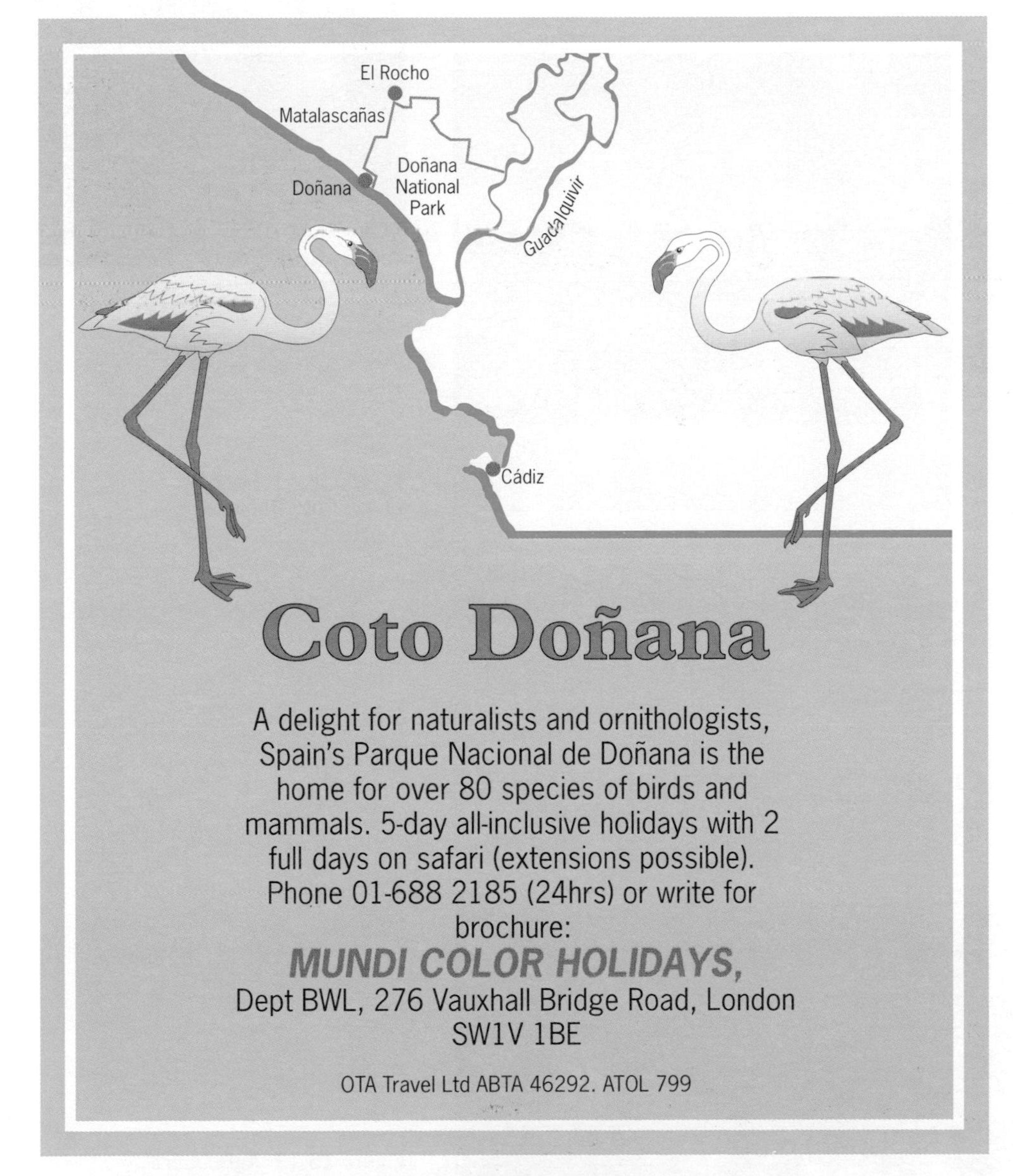

Figure 15.9 Ecotourism advertisement for the Coto Doñana

Figure 15.10 Flamingoes in the *marismas*

Figure 15.11 Bee eater

Figure 15.12 Interior of the Coto Doñana

Figure 15.13 Agriculture adjacent to the Coto Doñana

Figure 15.14 Large hotel at Matalascañas

Figure 15.15 Ecosystem in the Coto Doñana

Figure 15.16 Building at El Rocio

Figure 15.17 View of part of Matalascañas

Resort plan puts birds in danger

European conservationists are angered by a tourist development project in southern Spain which threatens to destroy a bird sanctuary. Edward Owen writes from Madrid.

Europe's biggest nature reserve and one of the remaining unspoiled stretches of Spanish coastline is threatened with tourist development, and many British migratory birds could lose a feeding ground as a result.

Their habitat is in danger of destruction by property developers who want to build a resort near the reserve. The park's wildlife also includes some rare species such as the imperial eagle and the Iberian lynx.

But there is hope following a report by 11 international experts on a development strategy for the park. European conservationists have welcomed the report, which was commissioned by the regional Andalusian government in Seville, and recently launched by Manuel Chaves, president of Andalusia.

The report, ordered after protests from ecologists at the Coto Doñana project, took a year to prepare and compares the scheme to the old fable of the goose and the golden eggs, because it would devastate its main *raison d'être*, the bird sanctuary. The report proposes total protection for the Asperillo dunes, where the 32,000-bed resort is planned with golf courses and sports facilities, and advises against urban development.

Mr Rose, senior international officer for the RSPB, said 'My immediate reaction is that the report is very, very positive'. His society helped the Spanish Ornithological Society spearhead a campaign to save the 400 square miles of wetlands, woods and dunes west of the Guadalquivir river estuary. 'My messsage now to Chaves is to put his money where his mouth is. He said he would abide by what the experts said.'

Mr Rose said the Coto Doñana project, reportedly backed by local businessmen connected with the Expo '92 world fair in Seville, completely misunderstands the value of Doñana's wildlife, scenery and villages. The area has already been blighted by Matalascañas, a ramshackle coastal development from the Franco era.

Javier Castroviejo, a biologist and former park director who is president of the Friends of Doñana Association, said: 'The report is a good declaration of principles, but has no legal effect. It states what we and many others including the European Community have been saying: if you take the water from the park to run the resort, the wetlands disappear.'

The report calls for a £237 million investment programme in the region by Spain and the EC with the emphasis on ecological tourism with accommodation in village inns and farmhouses, and better organised visits to a better managed park.

Figure 15.18 Newspaper article outlining the Coto Doñana conflict (*Source: The Times*, 10 April 1992) © Times Newspapers Ltd

Appendices

Al Assessing reactions to a forest trip

Table 1 Possible perceived benefits of a forest trip (*Source:* T. Lee,1989)

	Not very important 1	2	3	Very important 4	5
Peace and quiet					
Privacy					
Complete break from worries					
Chance to observe wildlife					
Heavy exercise (walking or other)					
Social/family outing					
Good for walking dogs					
Escape from city life					
Beautiful scenery					
Open to everyone					
Good for children to play					

Table 2 Possible affective reactions to a forest trip (*Source:* T. Lee,1989)

	Not very important 1	2	3	Very important 4	5
If alone, worried about being alone					
Afraid of trespassing					
Vulnerable					
Secure					
Uneasy					
Happy					
Afraid of getting lost					
Close to nature					
Free to explore					
Uplifted/revived					
In touch with the past					
Relaxed					
Bored					
Hemmed in					

A2 *t*-test calculation

In Chapter 5, question 17, the difference between the streams with and without dippers can be compared statistically and the statistical significance of the various factors assessed.

Example

The following example shows the t-test for the percentage catchment of the streams which is afforested.

$\underline{SD}$ = standard deviation $\quad$ $\underline{SE}$ = standard error

x = mean $\quad$ n = number in population

Standard error of the mean of population 1:

$$SE_1 = \frac{SD_1}{n_1} = \frac{26.9}{21} = 1.28$$

Standard error of the mean of population 2:

$$SE_2 = \frac{SD_2}{n_2} = \frac{26.9}{53} = 0.51$$

Standard error of the difference between means of populations 1 and 2:

SE difference =

$$\sqrt{SE\bar{x}_1^2 + SE\bar{x}_2^2} = \sqrt{1.64 + 0.26} = \sqrt{1.9 + 1.38}$$

$$t = \frac{\text{difference between means of population 1 and 2}}{\text{standard error of the difference}}$$

$$= \frac{\bar{x}_1 - \bar{x}_2}{SE_{diff}} = \frac{(8.4 - 36.3)}{1.38} = 20.28$$

To make sure that the result of the t-test is not due purely to chance, we use a t-table to test its statistical significance. To discover if the result is significant at the 0.999 confidence limit, we enter the table under the column 0.999.

degrees of freedom, $v = n_1 + n_2 - 2$
$= 21 + 53 - 2$
$= 72$

This value is over 60. From the table we can see that, in order for the difference for sample means to be significant, the critical value must be more than 3.29. Since the result is 20.28, this shows that the probability of the differences in dipper populations being due to chance is very small indeed.

Table 11.7 Critical values on student's *t* - distribution

	Confidence limits				
	0.90	**0.95**	**0.98**	**0.99**	**0.999**
	Two-tailed significance levels (one-tailed levels in parentheses)				
Degrees of freedom *v*	0.10 (0.05)	0.05 (0.025)	0.02 (0.01)	0.01 (0.005)	0.001 (0.0005)
1	6.31	12.71	31.81	63.66	636.6
2	2.92	4.30	6.97	9.93	31.60
3	2.35	3.18	4.54	5.84	12.92
4	2.13	2.78	3.75	4.60	8.61
5	2.02	2.57	3.37	4.03	6.86
6	1.94	2.45	3.14	3.71	5.96
7	1.90	2.37	3.00	3.50	5.41
8	1.86	2.31	2.90	3.36	5.04
9	1.83	2.26	2.82	3.25	4.78
10	1.81	2.23	2.76	3.17	4.59
11	1.80	2.20	2.72	3.11	4.44
12	1.78	2.18	2.68	3.06	4.32
13	1.77	2.16	2.65	3.01	4.23
14	1.76	2.15	2.62	2.98	4.14
15	1.75	2.13	2.60	2.95	4.07
16	1.75	2.12	2.58	2.92	4.02
17	1.74	2.11	2.57	2.90	3.97
18	1.73	2.10	2.55	2.88	3.92
19	1.73	2.09	2.54	2.86	3.88
20	1.73	2.09	2.53	2.85	3.85
21	1.72	2.08	2.52	2.83	3.82
22	1.72	2.07	2.51	2.82	3.79
23	1.71	2.07	2.50	2.81	3.77
24	1.71	2.06	2.49	2.80	3.57
25	1.71	2.06	2.49	2.79	3.73
26	1.71	2.06	2.48	2.78	3.71
27	1.70	2.05	2.47	2.77	3.69
28	1.70	2.05	2.47	2.76	3.67
29	1.70	2.05	2.46	2.76	3.66
30	1.70	2.04	2.46	2.75	3.65
40	1.68	2.02	2.42	2.70	3.55
60	1.67	2.00	2.39	2.66	3.46
over 60		approximates to the normal distribution			
z	1.64	1.96	2.33	2.58	3.29

Table 11.8 Significance level (one-tailed test) (*Source:* Siegel, 1956; after Olds, 1938 and 1949)

n	**0.05**	**0.01**
4	1.000	
5	0.900	1.000
6	0.829	0.943
7	0.714	0.893
8	0.643	0.833
9	0.600	0.783
10	0.564	0.746
12	0.506	0.712
14	0.456	0.645
16	0.425	0.601
18	0.399	0.564
20	0.377	0.534
22	0.359	0.508
24	0.343	0.485
26	0.329	0.465
28	0.317	0.448
30	0.306	0.432

A3 Spearman's rank correlation

The Spearman rank correlation coefficient uses data measured on an ordinal or rank scale, and is particularly useful in surveys of attitudes or decision-making, where respondents are often asked to rank their preferences.

n = number of pairs
d = difference in rank of each pair of values

Spearman rank correlation coefficient $(r_s) = 1 - \left[\frac{6\Sigma d^2}{n_3 - n}\right]$

When you have calculated r_s, you need to find its statistical significance. This enables you to discover whether or not r_s is the result of a chance association. To do this, you need to consult a significance table.

The critical values of r_s for $n = 4$ to $n = 30$ are at the 0.05 and 0.01 levels of significance. The larger the value of r_s, the more significant the result. For numbers of pairs greater than $n = 30$, the value of r_s changes very little.

To find out the value of environmental costs and benefits, contingent valuation can be used. This method involves setting up a questionnaire which asks people, through structured questions, their willingness to pay and/or accept compensation for changes to environmental resources.

A4 Ideas for fieldwork

1 Evaluate the management of competing land uses along an estuary.
2 Compare the wildlife, timber, recreation and land use value of coniferous and deciduous woodlands.
3 Evaluate multi-use management in temperate deciduous woodlands.
4 Assess the uptake and impact of Woodland Grant Schemes.
5 Assess the effect of grazing on woodlands.
6 Evaluate sand dune ecosystems and their management.
7 Assess the impact of vegetation management on grass verges/footpaths/urban parks/reserves.
8 Compare climate and urban ecology.
9 Compare the management of landscapes for different purposes and their impact on wildlife.
10 Evaluate overgrazing in upland areas and its impact on vegetation.
11 Assess the impact of trampling on ecosystems.
12 Evaluate the effect of EU legislation on the wildlife of a farm.
13 Assess the management of hedgerows and their value to wildlife.
14 Assess the water quality of streams running through agricultural areas. (NB: Safety)
15 Assess how much people will pay to protect wildlife sites using contingent valuation and cost-benefit analysis.
16 Compare the wildlife value from the inner city to the suburbs.
17 Assess strandline rubbish, its origins and significance
18 Evaluate ecotourism and its costs and benefits.
19 Assess the influence of human disturbance on vegetation, e.g. tipping, excavating, etc. (NB: Safety)
20 Compare a silage field with a traditionally managed meadow.

Glossary

Abiotic Non-living. The abiotic parts of ecosystems are components such as rocks, water and gases.

Acid rain A better term is *acid deposition*. This is (a) the wet deposition of acid substances as rain and other forms of precipitation, or (b) dry deposition of acid-forming substances such as sulphur dioxide.

Adaptation Any characteristic that improves the chances of an organism or species of passing on its genes to the next generation.

Afforestation The large-scale planting of trees in an area which was not forested to begin with.

Albedo The percentage of solar radiation reaching the earth's surface which is reflected back into the atmosphere. Light surfaces reflect more than dark areas and therefore have a higher albedo.

Biodiversity The variety of species within biological communities.

Biological oxygen demand Used as a measure of organic pollution in water. It represents the amount of biochemically degradable substances in the water or effluent sample. A test sample is stored in darkness for five days at 20°C and the amount of oxygen taken up by the micro-organisms present is measured in grams per cubic metre.

Biomass The total mass of plant and animal life occurring in an ecosystem.

Biome World-scale community of plants and animals, characterised by a particular vegetation type and covering a large geographical area, e.g. tropical rainforest, tundra.

Biosphere The part of the earth and its atmosphere capable of supporting life. The global ecosystem.

Biosphere Reserve UNESCO's Man and the Biosphere Programme which began in 1970 aimed to conserve representative natural areas throughout the world by establishing a network of biosphere reserves. Each reserve must have long-term legal protection and be large enough to be an effective conservation unit and accommodate different uses without conflict.

Biotic Living. The biotic parts of ecosystems are plants, animals and micro-organisms.

Boreal Where trees which are mainly coniferous replace deciduous vegetation when the growing season falls below six months and the frost-free period below four months.

Broadleaved With flat, broad leaves, e.g. oak, beech.

Carnivore Animal or plant which eats animals.

Chernozem Black or dark brown soil, rich in humus, which occurs in climates where the mean annual evaporation is more than the rainfall. The A-horizon is deep, and the B-horizon is rich in calcium compounds, making the soil alkaline. Chernozems are associated with temperate grasslands.

Climatic climax community The community which forms naturally under given climatic conditions. It is the final seral stage, is self-duplicating and not subject to further change unless there is a change in the climate.

Co-evolution The evolution of unrelated organisms, such as a plant and animal that have developed physiological relationships which allow both species to benefit.

Colonisation The arrival and establishment of plants or animals in an area.

Competition The interaction among plants or animals seeking resources for survival. **Interspecific competition** occurs among different species. **Intra-specific competition** occurs among individuals within a species

Coniferous Cone-bearing.

Consumers Living things which eat other living things.

Convergence The process whereby unrelated species or communities have evolved to resemble each other because they experience similar environmental variables.

Coppice An area of woodland, usually hazel, but also oak, alder or sweet chestnut, which has been managed by cutting the trees to their stumps and allowing regeneration from these stumps. There is usually regrowth at different stages within a wood, because blocks are cut in rotation.

Deciduous Experiencing leaf-fall once a year.

Decomposer An organism such as a bacterium or fungus which breaks down organic matter and releases nutrients which can be taken up by other organisms.

Deflected climax community A climatic community that is changed by a human factor such as grazing, after which it will not become a climatic climax community again.

Deforestation Cutting down of large areas of forest.

Detritivore A living thing which eats dead organic material.

Disturbance Damage to the ecosystem, either natural (such as a volcanic eruption) or created by people.

Dynamic equilibrium Balance between the inputs and outputs to a system, steady state.

Ecology The study of the relationships between living things and their environment.

Ecosystem A self-regulating biological community in which living things interact with the environment. Ecosystems can be small or large, e.g. a tree, a tropical rainforest, depending on the interests of the person who defines the ecosystem.

Ecotone A habitat which is in a transition area between others and contains characteristics of each bordering environment.

Ecotourism A form of tourism which aims to conserve fragile ecosystems and market their appeal while providing income for local people.

Environmentally Sensitive Area An area of landscape and wildlife value within the EU. Farmers are given annual payments to change farming practices to protect and enhance existing landscapes and wildlife habitats.

Eutrophication The addition of nutrients to an aquatic (water) environment, particularly nitrogen and phosphorus.

Evapotranspiration Transfer of water to the atmosphere by evaporation and plant transpiration.

Fauna The total population of animals that lives in an area.

Feedback A proportion of a system's outputs which feed back into its inputs. See Positive feedback and Negative feedback.

Ferrisol Also called a red-yellow latosol. A zonal soil developed under free-draining conditions in tropical rainforests. Ferrisols have a deeply weathered profile with limited minerals and nutrients. This is due to weathering and leaching which removes all bases, leaving free iron and aluminium oxides, plus small amounts of titanium, chromium, nickel oxides plus fine quartz and kaolinite. They are acid to very acid (pH 4-5).

Flora The total grouping of plants that lives in an area.

Food web/chain A hierarchy of organisms in a community, each member of which feeds on another in the chain and is in turn eaten. Where organisms eat more than one type of food, this links food chains together to form a network called a food web.

Forestry Commission A UK government department established in 1919 to increase home-grown timber production. In 1992 the Commission split into the Forestry Authority and Forest Enterprise. The Forestry Authority administers grants, monitors the industry and provides research. Forest Enterprise is responsible for multiple-purpose management of the Commission's own forests and woodlands.

Gene pool The sum of all genes and hereditary information that a population of a species has.

Greenhouse effect A gradual increase in the air temperature of the lower atmosphere. About 30 gases produced by human activity have been identified as contributing to the greenhouse effect. The main ones are carbon dioxide, methane, chlorofluorocarbons and nitrous oxide. It is called the greenhouse effect because the gases act like glass in a greenhouse which helps to trap the sun's energy.

Habitat The natural home of a plant or animal species or community.

Hardwood Wood which has a high density, such as oak and beech (720kg per cubic metre).

Herbivore A plant-eating animal.

Horizon A layer within a soil. Each horizon has a specific characteristic: the H-horizon is the humus; the A-horizon is the upper layer, mixed mineral and organic matter; the B-horizon is the subsoil; the C-horizon lies just above the parent material, e.g. bedrock.

Humus Thoroughly decomposed organic matter which is present in the soil. It is colloidal and dark brown in colour.

Hydrosphere The layer of water on the earth's surface, from within the bedrock to the atmosphere.

Inputs Energy or matter that enters a system.

Intensification Maximising the yield per unit of farmland. In the economically developing world, intensification is labour-intensive. In the economically developed world intensive commercial farming is capital-intensive, involving machinery, the addition of fertilisers and pesticides.

Invertebrates Animals without a backbone. This includes all animals except fish, amphibians, reptiles, birds and mammals.

Less Favoured Areas Areas in the EU which are given extra funding because they are difficult environments to live and work in.

Lithosphere The rigid outer layer of the earth's crust.

Management Control of resources for a particular purpose, such as recreation, conservation and landscape value. Where the resources are managed for more than one purpose, this is known as multi-use management.

Negative feedback The feedback mechanism in a system which keeps the system stable or in dynamic equilibrium. It decreases the amount of change by reducing some of the inputs.

Niche The specific role of a particular species in its habitat.

Outputs Energy or matter that moves out of a system.

Photosynthesis The process by which plants combine carbon dioxide and water to form organic compounds such as sugars, using energy obtained from sunlight. Oxygen is given out in the process.

Physiological drought Lack of water for plants which occurs without climatic drought, for example where ground is frozen or on sandy soils.

Phytoplankton Microscopic aquatic plants which float in water. They are extremely productive and form the basis of all marine food chains.

Plagioclimax community A stable plant community formed when a climatic climax community has been changed by people.

Plate tectonics The theory that the earth's crust is made up of a series of plates which move as a result of convection currents within the earth.

Podsol A soil which occurs under coniferous forests, heath and moorland. The soil is acid and has low levels of nutrients. Minerals are leached downwards and can collect to form an impermeable iron pan.

Positive feedback The feedback mechanism in a system which causes the system to become unstable or break down. It increases the amount of change, by raising some of the inputs.

Primary productivity The rate at which green plants transfer energy from sunlight into organic matter. Gross primary productivity (GPP) is measured in tonnes per ha per year and is a measure of all the photosynthesis that occurs in a system. Net primary productivity (NPP) is the energy fixed in photosynthesis (the GPP) minus the energy lost by respiration in the plants.

Primary woodland/forest A woodland community which is a climatic climax, unaltered by people.

Producer A green plant.

Radiation (short-wave and long-wave) Radiation is the transfer of the sun's energy to the earth. Radiation which enters the atmosphere is short-wave (0.15 to 3.0 in), while long-wave radiation is the heat energy which returns to the atmosphere from the earth's surface.

Relict communities Communities which have survived from past climatic eras.

Renewable energy Energy sources, such as wind or tidal energy, that are not likely to run out.

Resource partitioning The process which occurs through interspecific competition and leads to development of different niches within an ecosystem.

Secondary vegetation A plant community formed when a climatic climax community is changed so that it begins succession again.

Seral stage A stage in the succession of a community from colonisation to climax. Any plant community which is not the climax community but forms part of a succession is called a seral community.

Set-aside grant A payment given by the EU to farmers to take arable land out of production and leave it fallow. The aim is to reduce EU over-production.

Softwood Wood of a relatively low density compared to other types - for example, Norway spruce (380kg per cubic metre), Scots pine (540 kg per cubic metre).

Specialisation The move from farming a variety of crops and animals to the production of only one type of crop or animal

Species Organisms of which there are many individuals but only one type. Each species belongs to a genus (plural: genera). This genus is part of family.

SSSI (Sites of Special Scientific Interest) The best examples of British natural heritage, wildlife, geology and landforms. SSSI land is notified as being of special national conservation interest under the Wildlife and Countryside Act 1981 or the National Parks and Access to the Countryside Act 1949. These areas are designated by government-run departments in Scotland (Scottish Natural Heritage), England (English Nature) and Wales (Countryside Council for Wales). In Northern Ireland they are called ASSIs and are administered by the Department of the Environment. SSSI are protected by a series of management agreements with the owners of the land.

Stores Places within a system where materials or energy are held for a time.

Stratosphere The level of the earth's atmosphere above the troposphere, approximately 8 to 50 km in depth, in which temperature is fairly constant and where the ozone layer is found.

Stratum specificity The way in which a particular species adapts to occupy a niche in a particular layer (stratum) within an ecosystem.

Stress A physical or climatic factor which makes colonisation and habitation difficult. These stresses mean that some plants and animals need special adaptations to live in such environments.

Sub-climax community A community formed when different physical conditions (such as altitude, drainage, rock and soil types) prevent a climatic climax community from forming, for example, lowland peat bogs.

Succession The progression from the first colonisation of an area to the climax population, for example, lichens and mosses to soils and grasses to trees and shrubs. Succession which begins on bare ground is called a primary succession. Succession which begins after the destruction of existing vegetation is called a secondary succession.

Sustainability The management of resources where the ability of the system to replace itself is greater than the level of exploitation.

Symbiosis An association of dissimilar organisms from which both participating species gain some benefit.

Transpiration The process by which plants lose water into the atmosphere. Over 90 per cent of transpiration takes place from the leaves of plants; the remainder takes place from the stem and bark.

Trophic level A feeding level within a food chain. The first trophic level is made up of primary producers.

Zonation The variation of communities with space. A physical factor can cause a notable change in the distribution (zoning) of a plant community. For example, changes in flooding and salt concentration along a sea shore produce changes in plant zonation.

Zooplankton Microscopic aquatic animals which float in water. Zooplankton feed on phytoplankton and are a major source of food for fish.

References

Abbott, J (1990) *RSPB GCSE Woodlands Guide*, Royal Society for the Protection of Birds.
Allport, Ausden et al (1989). *The birds of the Gola Forest and their conservation*, Birdlife International.
Bain, C & Eversham (1991). *Thorne and Hatfield Moors papers*, vol. 2, Thorne and Hatfield Moors Conservation Forum.
Beddington, J R & May, R M (1982). 'The harvesting of interacting species in a natural ecosystem'. In *Scientific American*, NY.
British Trust for Ornithology (1977). *Atlas of Breeding Birds in Britain and Ireland*.
Brown, B (1987). Reprinted from *Marine Pollution Bulletin*, vol 18, no 1, 'Worldwide death of corals – natural cyclical event or man-made pollution?' Brown B, pp 9-13. Copyright (1987), with kind permission from Elsevier Science Ltd, The Boulevard, Langford Lane, Kidlington OX5 lGB, UK.
Bryant, D M (1987). 'Wading birds and wildfowl'. In *Proceedings of the Royal Society of Edinburgh, Section B: Biological Sciences*, vol. 93, parts 3 & 4; Royal Society of Edinburgh.
Carter, R (1989). *Coastal Environments*, Academic Press, London.
Chapman, W B (1973). *Journal of Natural Ecosystems*. Reprinted with permission from *Nature*. Copyright (1973) Macmillan Magazines Ltd.
Collins, N M & Morris, M G (1985). 'Threatened swallow tail butterflies of the world.' In *IUCN Red Data Book*, International Union for the Conservation of Nature, Switzerland.
Countryside Commission (1983). *Mid-Wales uplands study*, Ref: CCP 177.
Cousins, S H (1982). 'Species size distributions of birds and snails in an urban area.' In *Urban Ecology: the Second European Ecological Symposium*, Blackwell Scientific Publications Ltd.
Cox, C & Moore, P D (1980). *Biogeography: An ecological and evolutionary approach*, Blackwell Scientific Publications Ltd.
Dee Estuary Conservation Group (1976). *The Dee Estuary*, Royal Society for the Protection of Birds.
Duvigneaud, P & Denaeyer-De Smet, S (1970). 'Nutrient cycling.'As reprinted in *Soils, Vegetation and Ecosystems* (1988), Oliver & Boyd.
Edwards, P S (1977).'Studies of mineral cycling in a mountain rainforest in New Guinea: the production and disappearance of litter.' In *Journal of Ecology*, vol. 65, pp 971-972, Blackwell Scientific Publications Ltd.
Ena et al (1983). 'The great bustard.' In *Bustard Studies*, vol. 2, pp 35-53, Bustard Study Group, Birdlife International, Cambridge.
English Nature (1991). *Nature conservation and estuaries in Great Britain*.
English Nature (1988). *The Flow Country – the peatlands of Caithness and Sutherland*.
Forestry Commission (1993). *Forests and water guidelines* (3rd edn).
Forestry Commission (1994). *Forestry facts and figures* 1992-93.
Frontiers (1978). 'Nitrogen cycling in a short grass prairieland.' In *Frontiers*, Academy of Natural Sciences, Philadelphia.
Gersmehl, P J (1976).'An alternative biogeography.' In *Annals of the AAG*, vol. 66, pp 223-41.
Gordon, Kater & Schwaar (1979). *Vegetation and land use in Sierra Leone: a reconnaissance survey*, UNDP/Food and Agriculture Organisation Technical Report 2.
Grime, J P (1974). 'Vegetation classification by reference to strategies.' Reprinted with permission from Nature (vol. 250, pp 26-31). Copyright (1974) Macmillan Magazines Ltd.
Grime, J P (1977), *American Naturalist*, vol. 111, pp 1169-94, University of Chicago Press.
Haggett, P (1983). *Geography: A Modern Synthesis*, HarperCollins, NY.
Harrison, C M (1980-81). Reprinted from *Biological Conservation*, Volume 19, 'Recovery of lowland grassland and heathland in southern England from disturbance by seasonal trampling', pp 119-130., Copyright (1980-81), with kind permission from Elsevier Science Ltd, The Boulevard, Langford Lane, Kidlington OX5 1GB, UK.
HMSO (1990). *Energy Information Bulletin*, March 1990. Reproduced with the permission of the Controller of Her Majesty's Stationery Office.
Horbert et al (1982). 'Ecological contributions to urban planning.' In *Urban Ecology: the Second European Ecological Symposium*, Blackwell Scientific Publications Ltd.
ICES (1989) *Bulletin statistique*, International Council for the Exploration of the Sea, Denmark.
IPCC (1987) *Guide to Irish peatlands*, Irish Peatland Conservation Council, Dublin.
Krebs, C (1985). *Ecology, The experimental analysis of distribution and abundance*, HarperCollins, NY.
Kunick, W (1982). 'Comparison of flora of some cities of the central European lowlands.' In *Urban Ecology: the Second European Ecological Symposium*, Blackwell Scientific Publications Ltd.
Longman & Jenik (1987). *The tropical forest and its environment*, Longman UK.
Lousley, J (1976). *Wild flowers of chalk and limestone*, HarperCollins, UK.
Lululay, A (1989). *Perceptions of forest depletion and wildlife conservation: a case study of Gaura and Nongowa chiefdoms*, University of Sierra Leone.
MAFF (August 1993). *Agriculture and England's environment*, Ministry of Agriculture, Fisheries and Food News Release © Crown Copyright.
Milne & Dunnet (1972). 'The fauna of the Ythan estuary.' In *The estuarine environment*, Elsevier Applied Science Publishers.
Minshull, G N (1978). *The new Europe*, Hodder & Stoughton Educational.
O'Riordan, T (1981). 'Teaching of geography in higher education.' In *Journal of Geography and Higher Education*, vol. 5, no. 1.
O'Riordan, T (1987). 'Agriculture and environmental protection.' In *Geography Review*, vol. 1, no. I p36ff; Philip Allan Publishers.
Pearsall, W H (1950). *Mountains and moorlands*, HarperCollins, UK.
Polunin, 0 & Huxley, A (1981). *Flowers of the Mediterranean*, Chatto and Windus.
Richards, P & Glyn Davies, A (1991). *The Rainforest in Mende Life*, University College London.
RSPB (1989). *RSPB Nature Conservation Review*, 1989, no. 3, Royal Society for the Protection of Birds.
RSPB (1991a). *RSPB Nature Conservation Review*, 1991, no. 5, Royal Society for the Protection of Birds.
RSPB (1991b). *Non-commercial woodlands in the Brecon Beacons: a survey for the National Park Authority*, Royal Society for the Protection of Birds, Wales.
RSPB (1993). Data from the Royal Society for the Protection of Birds 'Estuaries inventory project' (unpublished).
Seager et al (1992). 'Assessment and control of farm pollution.' In *IWEM journal*, February 1992, pp 48-54; Institution of Water and Environmental Management, London.
Shreeve, T (1983). *Discovering ecology*, Longman UK.
Small, D (1952). *Some ecological and vegetational studies in the Gola Forest reserve*, D Small, Sierra Leone.
Smart, N & Andrews, J (1985). *Birds and broadleaved woodland handbook*, Royal Society for the Protection of Birds.
Smit & Piersma (1989). *Flyways and reserve networks for waterbirds*, IWRB Special Publication No. 9, IWRB, Slimbridge.
Speck & Carter (1979). *Sketch map geographies Book 4*, Longman Education.
Sullivan (1977). *Plant communities of the Scottish Highlands*, HMSO Nature Conservancy Monograph, No. 1.
Taylor, J P & Dixon, J B (1990). *Agriculture and the environment: towards integration*, Royal Society for the Protection of Birds.
Tait (1972). *Elements of marine ecology*, Butterworth-Heinemann Ltd.
Thomas et al (1981). 'The demography and flora of the Ouse Washes.' In *Biological Conservation*, vol. 21 pp 197-229, Elsevier Applied Science Publishers.
Tubbs, C R (1991). *British Wildlife*, vol. 2, no. 5 British Wildlife Publishing.
Tyler, S (1985-7). Data from RSPB Wales, with thanks to Dr Stephanie Tyler.
UNESCO (1983). *Coral reefs, seagrass beds and mangroves: their interaction in the coastal zones of the Caribbean*, UNESCO Reports in Marine Science 23, UNESCO, Paris.
UNESCO (1984). *Productivity and processes in island marine ecosystems*, UNESCO Reports in Marine Science 27, UNESCO, Paris.
UNESCO (1986). *Human induced damage to coral reefs*, UNESCO Reports in Marine Science 40, UNESCO, Paris.
UNESCO (1986). *Caribbean coastal marine productivity*, UNESCO Reports in Marine Science 41, UNESCO, Paris.
Van Dyne, G N & Reymeyer (1980). *Grasslands, systems analysis and man*, Cambridge University Press.
Vedel & Lange (1960). *Trees, shrubs and woodland*, Methuen.
Webb, N (1989). *Heathlands*, HarperCollins, UK.
Whittaker, R H (1975). Reprinted with the permission of Macmillan College Publishing Company from *Communities and ecosystems* 2nd Edition by Robert H Whittaker. Copyright © 1970, 1975 by Robert H Whittaker by Macmillan College Publishing Company, Inc.
Wright, P (1993). 'Ecotourism: ethics or ecosell?' In *Journal of Travel Research*, Winter 1993.

Published by Collins Educational
An imprint of HarperCollins Publishers
77-85 Fulham Palace Road
London W6 8JB

www.CollinsEducation.co.uk
On-line support for schools and colleges
You might also like to visit www.fireandwater.co.uk

ISBN 0 00 326652 4

Judith Woodfield, Geoff Abbott, Barrie Cooper, Richard Farmer, Fay Pascoe and Chris Skinner assert their moral right to be identified as the authors of this work.

Editing/Project management: Melanie McRae
Design: Sally Boothroyd
Picture research: Caroline Thompson
Artwork: Jerry Fowler and Contour Publishing

Printed and Bound by Scotprint, UK

Acknowledgements

Every effort has been made to contact the holders of copyright material, but if any have been inadvertently overlooked the publishers will be pleased to make the necessary arrangements at the first opportunity.

The publishers would like to thank the following for permission to reproduce photographs (T = Top, B = Bottom, C = Centre, L= Left, R = Right):

Des Bowden, Fig. 3.10;
N Tomalin/Bruce Coleman Ltd, Fig. 3.23;
Forest Life Picture Library, Fig. 5.13;
Greenpeace, Fig.12.21;
Robert Harding Picture Library, Fig. 11.12;
Kemira Fertilisers, Fig. 7.9;
NHPA/A Callow, Fig. 6.35; L Campbell, Fig. 7.13; M Garwood, Fig. 8.14, B Hawkes, Fig. 10.18, Sorensen & Olsen, Fig. 5.1, D Woodfall, Fig. 4.18;
www.osf.uk.com/Kathie Atkinson, Fig. 13.4TL, Doug Allan, Fig.13.4TR; Kjell Sandved, Fig.13.4BM;
Photo Flora, Fig. 6.36;
Planet Earth Pictures, Figs 14.5, 14.7, 14.12;
Robert Prosser, Fig. 3.33;
RSPB, Figs 1.6, 4.3, 4.4, 4.5, 4.6, 4.7, 4.8, 4.28, 4.29, 4.31, 4.32, 5.7, 5.8, 5.10, 5.14, 5.18, 5.23, 6.5, 6.6, 6.19, 6.20, 6.21, 6.22, 6.31, 6.32, 6.34, 7.1, 7.2, 7.7, 7.12, 9.5, 9.22, 10.3, 10.9, 10.10, 10.19, 10.21, 11.1, 11.2, 11.11, 12.16, 12.17, 14.6;
RSPB/G Abbott, Figs 6.1, 6.2, 6.3, 6.4, 6.9, 6.10, 6.11, 6.30, 8.13, 9.6, 9.9, 9.10, R Berry, Figs1.1, 1.13, 12.1, A Brown, Figs 2.5, 2.6, B Cooper, Figs 15.10, 15.11, 15.12, 15.13, 15.14, 15.15, 15.16, 15.17, E Coppola/A Petretti, Fig. 8.20, D Elcome, Fig. 10.12, D Heaver, Fig. 9.17, RSPB International, Figs 3.7, 3.8, 3.9, 3.12, 3.15, 3.16, 3.17, 3.18, 3.19, 3.21, 3.25, R Lovegrove, Figs 4.21, 4.22, M A Naveso, Figs 8.21, 8.24, 8.26, 8.27, 8.28, C Nicholson, 6.46, F Pascoe, 2.16, 2.19, 4.17, 6.8, 6.12, 6.14, 6.15, 14.15, S Pitts, Figs 7.10, 7.14, I Proctor, Figs 5.24, 5.25, SEO, Fig. 8.19, A Sanchez, Fig. 8.17;
Still Pictures/A Crump, Fig. 14.2, M Edwards, Figs 2.17, 3.3, N Dickinson, Fig. 3.1, M Gunther, Fig. 14.16;
M King/Swift Picture Library, Fig. 8.4;
C & S Thompson, Figs 6.31, 9.15, 10.11;
West Air Photography, Fig. 5.22;
Andy Williss, Fig. 14.1;
David Woodfall, Figs 7.11, 9.16, 10.15;
Judith Woodfield, Figs 6.24, 6.26, 6.27, 6.29;
Cover photograph: Gary Bell/Planet Earth Pictures

Additional contribution
The publishers would like to acknowledge assistance given by:
Dr Nigel Stringer, Countryside Council for Wales
Dr S Ormerod, UWIST
Liz Roblin, National Rivers Authority
David Jenkins, Coed Cymru